Surface and Thin Film Analysis

Edited by H. Bubert and H. Jenett

Surface and Thin Film Analysis

Principles, Instrumentation, Applications

Edited by H. Bubert and H. Jenett

Dr. Henning Bubert
Dr. Holger Jenett
Institute of Spectrochemistry and
Applied Spectroscopy (ISAS)
Bunsen-Kirchhoff-Straße 11
44139 Dortmund
Germany

Library of Congress Card No.: applied for
A catalogue record for this book is available from the **British Library.**

Die Deutsche Bibliothek – CIP Cataloguing-in-Publication-Data
A catalogue record for this publication is available from Die Deutsche Bibliothek

Printed in the Federal Republic of Germany.
Printed on acid-free paper.

Typesetting ProSatz Unger, Weinheim
Printing betz-druck gmbh, Darmstadt
Bookbinding J. Schäffer GmbH & Co. KG, Grünstadt

ISBN 3-527-30458-4

Contents

Preface

The surface of a solid interacts with its environment. It may be changed by the surrounding medium either unintentionally (for example by corrosion) or intentionally due to technological demands. Intentional changes are made in order to refine or protect surfaces, i. e., to generate new surface properties. Such surface changes can be made by ion implantation, deposition of thin films or epitaxially grown layers, among others. In all these cases, it is necessary to analyze the surface, the layer or system of layers, the grain boundaries, or other interfaces in order to control the process which finally meets the technological requirements for a purposefully changed surface. A wealth of analytical methods is available to the analyst, and the choice of the method appropriate for the solution of his problem requires a basic knowledge on the methods, techniques and procedures of surface and thin-film analysis.

Therefore, this book is to give the analyst – whether a newcomer wishing to acquaint themself with new methods or a materials analyst needing to inform themself on methods that are not available in their own laboratory – a clue about the principles, instrumentation, and applications of the methods, techniques, and procedures of surface and thin-film analysis. The first step into this direction was the chapter *Surface and Thin Film Analysis* of *Ullmann's Encyclopedia of Industrial Chemistry* (Vol. B6, Wiley-VCH, Weinheim 2002) in which practitioners give briefly outline the methods.

The present book is based on that chapter. It has essentially been extended by new sections dealing with *electron energy loss spectroscopy* (EELS), *low-energy electron diffraction* (LEED), *elastic recoil detection analysis* (ERDA), *nuclear reaction analysis* (NRA), *energy dispersive X-ray spectroscopy* (EDXS), *X-ray diffraction* (XRD), *surface analysis by laser ablation* (LA), and *ion-beam spectrochemical analysis* (IBSCA). Thus, the book now comprises the most important methods and should help the analyst to make decisions. Except for *atomic force microscopy* (AFM) and *scanning tunneling microscopy* (STM), microscopic methods, as essential as they are for the characterization of surfaces, are only briefly discussed when combined with a spectroscopic method. Methods of only limited importance for the solution of very special problems, or without availability of commercial equipment, are not considered or only briefly mentioned in the sections entitled *Other Detecting Techniques* without updating or giving examples of their applications.

Furthermore, the objective was not to issue a voluminous book but a clearly arranged one outlining the methods of surface and thin film analysis. For a deeper understanding of any of these topics, the reader is referred to the special literature given in the references.

The editors are gratefully indebted to all contributors who were ready to redirect time from their research, educational, and private activities in order to contribute to this book. They also wish to thank Mrs Silke Kittel for her tireless help in developing our editorial ideas.

Autumn 2001

Henning Bubert
Holger Jenett

List of Authors

Prof. Dr. Heinrich F. Arlinghaus
Physikalisches Institut
Westfälische Wilhelms-Universität
Wilhelm-Klemm-Straße 10
48149 Münster
Germany
hearlin@uni-muenster.de

Prof. Dr. Peter Bauer
Institut für Experimentalphysik
Johannes Kepler Universität
4040 Linz
Austria
bauer@exphys.uni-linz.ac.at

Prof. Dr. Oswald Benka
Institut für Experimentalphysik
Johannes Kepler Universität
4040 Linz
Austria
bauer@exphys.uni-linz.ac.at

Prof. Dr. Michail Bolshov
Institute of Spectroscopy
Russian Academy of Sciences
142092 Troitzk, Moscow reg.
Russia
bolshov@isan.troitsk.ru

Dr. Henning Bubert
Institut für Spektrochemie und
Angewandte Spektroskopie
Bunsen-Kirchhoff-Straße 11
44139 Dortmund
Germany
bubert@isas-dortmund.de

Dr. Laszlo Fabry
Wacker Siltronic AG
Johannes-Hess-Straße 24
84489 Burghausen
Germany
laszlo.fabry@wacker.com

Prof. Dr. Gernot Friedbacher
Institut für Analytische Chemie
Technische Universität
Getreidemarkt 9/151
1060 Wien
Austria
gfried@email.tuwien.ac.at

Dr. P. Neil Gibson
Institute for Health and Consumer
Protection
Joint Research Centre
21020 Ispra (VA)
Italy
neil.gibson@jrc.it

Dr. Bernd Gruska
SENTECH
Carl-Scheele-Straße 16
12489 Berlin-Adlershof
Germany
marketing@sentech.de

Dr. Georg Held
Department of Chemistry
University of Cambridge
Lensfield Road
Cambridge CB2 1EW
United Kingdom
gh10009@cam.ac.uk

Dr. Wieland Hill
Lambda Physik AG
Hans-Böckler-Straße 12
37079 Göttingen
Germany
wieland.hill@arcormail.de

Dr. Karsten Hinrichs
Institut für Spektrochemie und Angewandte Spektroskopie
Albert-Einstein-Straße 9
12489 Berlin-Adlershof
Germany
hinrichs@isas-berlin.de

Prof. Dr. Herbert Hutter
Institut für Analytische Chemie
Technische Universität
Getreidemarkt 9/151
1060 Wien
Austria
h.hutter@tuwien.ac.at

Dr. Holger Jenett
Fakultät Chemie
Universität Bielefeld
Universitätsstraße 25
33615 Bielefeld
Germany
holger.jenett@uni-bielefeld.de

Dr. Siegfried Pahlke
Wacker Siltronic AG
Johannes-Hess-Straße 24
84489 Burghausen
Germany
siegfried.pahlke@wacker.com

Prof. Dr. Leopold Palmetshofer
Institut für Halbleiter- und Festkörperphysik
Johannes Kepler Universität
4040 Linz
Austria
l.palmetshofer@jk.uni-linz.ac.at

Dr. Alfred Quentmeier
Institut für Spektrochemie und Angewandte Spektroskopie
Bunsen-Kirchhoff-Straße 11
44139 Dortmund
Germany
quentmeier@isas-dortmund.de

Prof. Dr. John C. Rivière
Harwell Laboratory
AEA Technology
Didcot
Oxfordshire, OX11 OQ J
United Kingdom

Dr. habil. Arthur Röseler
Institut für Spektrochemie und Angewandte Spektroskopie
Albert-Einstein-Straße 9
12489 Berlin-Adlershof
Germany
roeseler@isas-berlin.de

Dr. Volker Rupertus
SCHOTT GLAS
Hattenbergstraße 10
55120 Mainz
Germany
rpr@schott.de

Dr. Reinhard Schneider
Institut für Physik
Humboldt-Universität zu Berlin
Invalidenstraße 110
10115 Berlin
Germany
reinhard.schneider@physik.hu-berlin.de

1
Introduction

John C. Rivière and Henning Bubert

Wherever the properties of a solid surface are important, it is also important to have the means to measure those properties. The surfaces of solids play an overriding part in a remarkably large number of processes, phenomena, and materials of technological importance. These include catalysis; corrosion, passivation, and rusting; adhesion; tribology, friction, and wear; brittle fracture of metals and ceramics; microelectronics; composites; surface treatments of polymers and plastics; protective coatings; superconductors; and solid surface reactions of all types with gases, liquids, or other solids. The surfaces in question are not always external; processes occurring at inner surfaces such as interfaces and grain boundaries are often just as critical to the behavior of the material. In all the above examples, the nature of a process or of the behavior of a material can be understood completely only if information about both surface composition (i. e. the types of atoms present and their concentrations) and surface chemistry (i. e. the chemical states of the atoms) is available. Occasionally, knowledge of the arrangement of surface atoms (i. e. the surface structure) is also necessary.

First of all, what is meant by a solid surface? Ideally the surface should be defined as the plane at which the solid terminates, that is, the last atom layer before the adjacent phase (vacuum, vapor, liquid, or another solid) begins. Unfortunately such a definition is impractical because the effect of termination extends into the solid beyond the outermost atom layer. Indeed, the current definition is based on that knowledge, and the surface is thus regarded as consisting of that number of atom layers over which the effect of termination of the solid decays until bulk properties are reached. In practice, this decay distance is of the order of 5–20 nm.

By a fortunate coincidence, the depth into the solid from which information is provided by the techniques described here matches the above definition of a surface almost exactly. These techniques are, therefore, surface-specific, in other words, the information they provide comes only from that very shallow depth of a few atom layers. Other techniques can be surface sensitive, in that they would normally be regarded as techniques for bulk analysis, but have sufficient sensitivity for certain elements that can be analyzed only if they are present on the surface only.

Why should surfaces be so important? The answer is twofold. Firstly, the properties of surface atoms are usually different from those of the same atoms in the bulk and, secondly, because in any interaction of a solid with another phase the surface atoms are the first to be encountered. Even at the surface of a perfect single crystal the surface atoms behave differently from those in the bulk simply because they do not have the same number of nearest neighbors; their electronic distributions are altered and hence their reactivity. Their structural arrangement is often also different. When the surface of a polycrystalline or glassy multielemental solid is considered, such as that of an alloy or a chemical compound, the situation can be very complex. The processes of preparation or fabrication can produce a material the surface composition of which is quite different from that of the bulk, in terms of both constituent and impurity elements. Subsequent treatment (e.g. thermal and chemical) will almost certainly change the surface composition to something different again. The surface is highly unlikely to be smooth, and roughness at both micro and macro levels can be present, leading to the likelihood that many surface atoms will be situated at corners and edges and on protuberances (i.e. in positions of increased reactivity). Surfaces exposed to the atmosphere, which include many of those of technological interest, will acquire a contaminant layer 1–2 atom layers thick, containing principally carbon and oxygen but also other impurities present in the local environment. Atmospheric exposure might also cause oxidation. Because of all these possibilities the surface region must be considered as a separate entity, effectively a separate quasi-two-dimensional phase overlaying the normal bulk phase. Analysis of the properties of such a quasi phase necessitates the use of techniques in which the information provided originates only or largely within the phase – i.e., the surface-specific techniques described in this article.

Nearly all these techniques involve interrogation of the surface with a particle probe. The function of the probe is to excite surface atoms into states giving rise to emission of one or more of a variety of secondary particles such as electrons, photons, positive and secondary ions, and neutrals. Because the primary particles used in the probing beam can also be electrons or photons, or ions or neutrals, many separate techniques are possible, each based on a different primary–secondary particle combination. Most of these possibilities have now been established, but in fact not all the resulting techniques are of general application, some because of the restricted or specialized nature of the information obtained and others because of difficult experimental requirements. In this publication, therefore, most space is devoted to those surface analytical techniques that are widely applied and readily available commercially, whereas much briefer descriptions are given of the many others the use of which is less common but which – in appropriate circumstances, particularly in basic research – can provide vital information.

Because the various types of particle can appear in both primary excitation and secondary emission, most authors and reviewers have found it convenient to group the techniques in a matrix, in which the columns refer to the nature of the exciting particle and the rows to the nature of the emitted particle [1.1–1.9]. Such a matrix of techniques is given in Tab. 1.1., which uses the acronyms now accepted. The meanings of the acronyms, together with some of the alternatives that have appeared in the literature, are given in Listing 1.

Tab. 1.1. Surface-specific analytical techniques* using particle or photon excitation. The acronyms printed in bold are those used for methods discussed in more details in this publication.

Detection	***Excitation****					
	Electrons, e^-		***Ions, Neutrals, A^+, A^-, A^0***		***Photons, hν***	
e^-	**AES** AEAPS **EELS** **EFTEM** **LEED**	**SAM** RHEED	**IAES** INS MQS		**XPS**	**UPS**
A^+, A^-, A^0	ESD	ESDIAD	**SIMS** GDMS **RBS** **ERDA**	**SNMS** FABMS **LEIS** **NRA**		
hν	**EDXS** SXAPS IPES	 DAPS BIS	**GD-OES** **IBSCA**		**TXRF** **XRD** **LA** **RAIRS** SHG **ELL**	 LIBS **SERS** SFG

* For meanings of acronyms, see Listing 1.
** Some of the techniques in Tab. 1.1 have angle-resolved variants, with the prefix AR, e.g. ARUPS, or use Fourier transform methods, with the prefix FT, e.g. FT-RAIRS.

Tab. 1.2. Surface-specific analytical techniques* using non-particle excitation.

Detection	***Excitation***		
	Heat, kT	***High electrical field, F***	***Mechanical force***
A^+	TDS	APFIM POSAP	
A^-	TDS		
e^-		IETS **STM, STS**	
(Displacement)			**AFM**

* For meanings of acronyms, see Listing 1.

Listing 1. Meanings of the surface analysis acronyms, and their alternatives, that appear in Tabs 1.1. and 1.2.

1 Electron Excitation

AES	Auger electron spectroscopy
AEAPS	Auger electron appearance potential spectroscopy
BIS	Bremsstrahlung isochromat spectroscopy (or ILS: ionization loss spectroscopy)
DAPS	Disappearance potential spectroscopy
EDXS	Energy-dispersive X-ray spectroscopy
EELS	Electron energy loss spectroscopy
EFTEM	Energy-filtered Transmission Electron Microscopy
ESD	Electron-stimulated desorption (or EID: electron-induced desorption)
ESDIAD	Electron-stimulated desorption ion angular distribution
IPES	Inverse photoemission spectroscopy
LEED	Low-energy electron diffraction
RHEED	Reflection high-energy electron diffraction
SXAPS	Soft X-ray appearance potential spectroscopy (or APS: appearance potential spectroscopy)
SAM	Scanning Auger microscopy

2 Ion Excitation

ERDA	Elastic Recoil Detection Analysis
GDMS	Glow discharge mass spectrometry
GD-OES	Glow discharge optical emission spectroscopy
IAES	Ion (excited) Auger electron spectroscopy
IBSCA	Ion beam spectrochemical analysis (or SCANIIR: surface composition by analysis of neutral and ion impact radiation or BLE: bombardment-induced light emission)
INS	Ion neutralization spectroscopy
LEIS	Low energy ion scattering (or ISS: Ion scattering spectroscopy)
MQS	Metastable quenching spectroscopy
NRA	Nuclear Reaction Analysis
RBS	Rutherford back-scattering spectroscopy (or HEIS: high-energy ion scattering)
SIMS	Secondary-ion mass spectrometry (SSIMS: static secondary-ion mass spectrometry) (DSIMS: dynamic secondary-ion mass spectrometry)
SNMS	Secondary neutral mass spectrometry

3 Photon Excitation

ELL	Ellipsometry
LA	Surface Analysis by Laser Ablation
LIBS	Laser-induced breakdown spectroscopy (or LIPS: Laser-induced plasma spectroscopy)
RAIRS	Reflection-absorption infrared spectroscopy (or IRRAS: Infrared reflection-absorption spectroscopy) (or IRAS: Infrared absorption spectroscopy) (or ERIRS: External reflection infrared spectroscopy)
SERS	Surface-enhanced Raman scattering
SFG	Sum Frequency Generation
SHG	Optical Second harmonic generation
TXRF	Total reflection X-ray fluorescence analysis
UPS	Ultraviolet photoelectron spectroscopy
XPS	X-ray photoelectron spectroscopy (or ESCA: electron spectroscopy for chemical analysis)
XRD	X-ray diffraction

4 Neutral Excitation

FABMS	Fast-atom bombardment mass spectrometry

5 Thermal Excitation

TDS	Thermal desorption spectroscopy

6 High Field Excitation

APFIM	Atom probe field-ion microscopy
IETS	Inelastic electron tunneling spectroscopy
POSAP	Position-sensitive atom probe
STM	Scanning tunneling microscopy
STS	Scanning tunneling spectroscopy

7 Mechanical Force

AFM	Atomic force microscopy

A few techniques, one or two of which are important, cannot be classified according to the nature of the exciting particle, because they do not employ primary particles but depend instead on the application either of heat or a high electric field. These techniques are listed in Tab 1.2.

2
Electron Detection

2.1
Photoelectron Spectroscopy

Henning Bubert and John C. Rivière

X-ray photoelectron spectroscopy (XPS) is currently the most widely used surface-analytical technique, and is therefore described here in more detail than any of the other techniques. At its inception by Siegbahn and coworkers [2.1] it was called ESCA (*electron spectroscopy for chemical analysis*), but the name ESCA is now considered too general, because many surface-electron spectroscopies exist, and the name given to each one must be precise. The name ESCA is, nevertheless, still used in many places, particularly in industrial laboratories and their publications. Briefly, the reasons for the popularity of XPS are the exceptional combination of compositional and chemical information that it provides, its ease of operation, and the ready availability of commercial equipment.

2.1.1
Principles

The surface to be analyzed is irradiated with soft X-ray photons. When a photon of energy $h\nu$ interacts with an electron in a level X with the binding energy E_B (E_B is the energy E_K of the K-shell in Fig. 2.1), the entire photon energy is transferred to the electron, with the result that a photoelectron is ejected with the kinetic energy

$$E_{kin}\,(h\nu, \mathrm{X}) = h\nu - E_B - \Phi_S \tag{2.1}$$

where Φ_S is a small, almost constant, work function term.

Obviously $h\nu$ must be greater than E_B. The ejected electron can come from a core level or from the occupied portion of the valence band, but in XPS most attention is focused on electrons in core levels. Because no two elements share the same set of electronic binding energies, measurement of the photoelectron kinetic energies enables elemental analysis. In addition, Eq. (2.1) indicates that any changes in E_B are reflected in E_{kin}, which means that changes in the chemical environment of an atom can be followed by monitoring changes in the photoelectron energies, leading to the

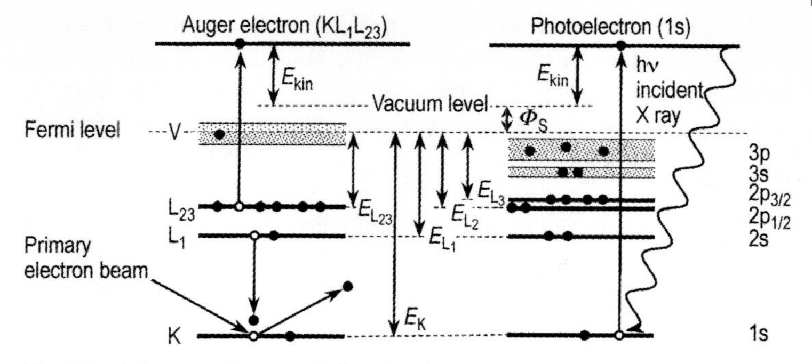

Fig. 2.1. Schematic diagram of electron emission processes in solids. Left side: Auger process, right side: photoelectron process. Electrons involved in the emission processes are indicated by open circles.

provision of chemical information. XPS can be used for analysis of all elements in the periodic table except hydrogen and helium.

Although XPS is concerned principally with photoelectrons and their kinetic energies, ejection of electrons by other processes also occurs. An ejected photoelectron leaves behind a core hole in the atom. The sequence of events following the creation of the core hole is shown schematically in Fig. 2.1 (right side). In the example, the hole has been created in the K-shell, giving rise to a photoelectron, the kinetic energy of which would be $(h\nu - E_K)$, and is filled by an electronic transition from the unresolved L_{23} shell. The energy $E_K - E_{L_{23}}$ associated with the transition can then either be dissipated as a characteristic X-ray photon or given up to an electron in the same or a higher shell, shown in this example also as the L_{23}. The second of these possibilities is called the Auger process after its discoverer [2.2], and the resulting ejected electron is called an Auger electron and has a kinetic energy given by:

$$E_{\text{kin}}\,(\text{KL}_1\text{L}_{23}) = E_K - E_{L_1} - E_{L_{23}} - E_{\text{inter}}\,(\text{L}_1\text{L}_{23}) + E_R - \Phi_S \tag{2.2}$$

where $E_{\text{inter}}\,(\text{L}_1\text{L}_{23})$ is the interaction energy between the holes in the L_1 and L_{23} shell and E_R is the sum of the intra-atomic and extra-atomic relaxation energies. X-ray photon emission (i. e. X-ray fluorescence) and Auger electron emission are obviously competing processes, but for the shallow core levels involved in XPS and AES the Auger process is far more likely.

Thus in all X-ray photoelectron spectra, features appear as a result of both photoemission and Auger emission. In XPS, the Auger features can be useful but are not central to the technique, whereas in AES (see Sect. 2.2), Eq. (2.2) forms the basis of the technique.

At this point the nomenclature used in XPS and AES should be explained. In XPS the spectroscopic notation is used, and in AES the X-ray notation. The two are equivalent, the different usage having arisen for historical reasons, but the differentiation is a convenient one. They are both based on the so-called j–j coupling scheme describing the orbital motion of an electron around an atomic nucleus, in which the total angular momentum of an electron is found by summing vectorially the individual electron spin and angular momenta. Thus if l is the electronic angular momentum quantum number and s the electronic spin momentum quantum number, the total angular momentum for each electron is given by $j = l + s$. Because l can take the values 0, 1, 2, 3, 4, ... and $s = \pm\frac{1}{2}$ clearly $j = \frac{1}{2}, \frac{3}{2}, \frac{5}{2}$ etc. The principal quantum number n can take values 1, 2, 3, 4, ... In spectroscopic notation, states with $l = 0$, 1, 2, 3, ... are designated s, p, d, f, ..., respectively, and the letter is preceded by the number n; the j values are then appended as suffixes. Therefore one obtains 1s, 2s, $2p_{1/2}$, $2p_{3/2}$.

In X-ray notation, states with $n = 1, 2, 3, 4$, ... are designated K, L, M, N, ..., respectively, and states with various combinations of $l = 0, 1, 2, 3$, ... and $j = \frac{1}{2}, \frac{3}{2}, \frac{5}{2}$ are appended as the suffixes 1, 2, 3, 4 .... In this way one arrives at K, L_1, L_2, L_3, M_1, M_2, M_3, etc. The equivalence of the two notations is set out in Tab. 2.1.

Tab. 2.1. Spectroscopic and X-ray notation.

Quantum numbers			***Spectroscopic state***	***X-ray state***
n	***l***	***j***		
1	0	1/2	1 s	K
2	0	1/2	2 s	L_1
2	1	1/2	2 $p_{1/2}$	L_2
2	1	3/2	2 $p_{3/2}$	L_3
3	0	1/2	3 s	M_1
3	1	1/2	3 $p_{1/2}$	M_2
3	1	3/2	3 $p_{3/2}$	M_3
3	2	3/2	3 $d_{3/2}$	M_4
3	2	5/2	3 $d_{5/2}$	M_5
etc.	etc.	etc.	etc.	etc.

In X-ray notation the Auger transition shown in Fig. 2.1 would therefore be labeled KL_2L_3. In this coupling scheme, six Auger transitions would be possible in the KLL series. Obviously, many other series are possible (e.g., KLM, LMM, MNN). These are discussed more fully in Sect. 2.2, dealing with AES.

The reasons why techniques such as XPS and AES, which involve measurement of the energies of ejected electrons, are so surface-specific should be examined. An electron with kinetic energy E moving through a solid matrix M has a probability of traveling a certain distance before losing all or part of its energy as a result of an inelastic collision. On the basis of that probability, the average distance traveled before such a collision is known as the inelastic mean free path (imfp) $\lambda_M(E)$. The imfp is a function only of M and of E. Figure 2.2 shows a compilation of measurements of λ made by Seah and Dench [2.3], in terms of atomic monolayers as a function of kinetic energy. Note that both λ and energy scales are logarithmic. The important con-

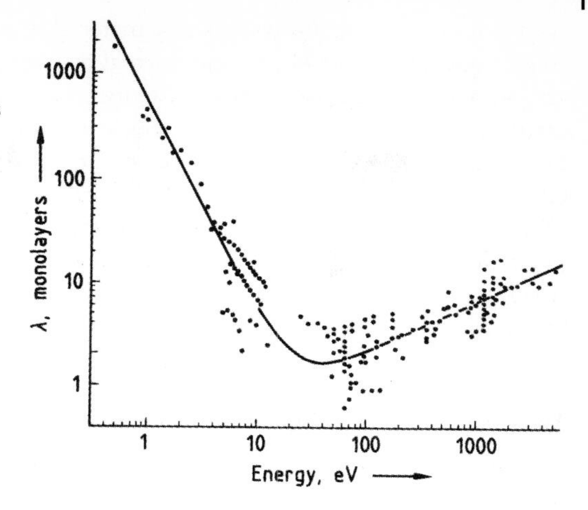

Fig. 2.2. Compilation by Seah and Dench [2.3] of measurements of inelastic mean free path as a function of electron kinetic energy. The solid line is a least-squares fit.

sequence of the dependence of λ on kinetic energy is that in the ranges of secondary electron kinetic energies used in XPS and AES, the values of λ are very small. In XPS, for example, typical energy ranges are 250–1500 eV, corresponding to a range of λ from about four to eight monolayers, whereas in AES, the energy range is typically 20 to 1000 eV, in which case λ would range from about two to six monolayers. What this means in practice is that if the photoelectron or the Auger electron is to escape into a vacuum and be detected, it must originate at or very near the surface of the solid. This is the reason why the electron spectroscopic techniques are surface-specific.

2.1.2
Instrumentation

2.1.2.1 Vacuum Requirements

Electron spectroscopic techniques require vacuums of the order of 10^{-8} Pa for their operation. This requirement arises from the extreme surface-specificity of these techniques, mentioned above. With sampling depths of only a few atomic layers, and elemental sensitivities down to 10^{-5} atom layers (i.e., one atom of a particular element in 10^5 other atoms in an atomic layer), the techniques are clearly very sensitive to surface contamination, most of which comes from the residual gases in the vacuum system. According to gas kinetic theory, to have enough time to make a surface-analytical measurement on a surface that has just been prepared or exposed, before contamination from the gas phase interferes, the base pressure should be 10^{-8} Pa or lower, that is, in the region of ultrahigh vacuum (UHV).

The requirement for the achievement of UHV conditions imposes restrictions on the types of material that can be used for the construction of surface-analytical systems, or inside the systems, because UHV can be achieved only by accelerating the rate of removal of gas molecules from internal surfaces by raising the temperature of the entire system (i.e. by baking). Typical baking conditions are 150–200 °C for several hours. Inside the system, any material is permissible that does not produce volatile components either during normal operation or during baking. Thus, for example, brass that contains the volatile metal zinc could not be used. The principal construction material is stainless steel, with mu-metal (76% Ni, 5% Cu, 2% Cr) used occasionally where magnetic screening is needed (e.g. around electron-energy analyzers). For the same reasons, metal seals, not elastomers, are used for the demountable joints between individual components – the sealing material is usually pure copper, although gold is sometimes used. Other materials that can be used between ambient atmosphere and UHV are borosilicate glass or quartz for windows, and alumina for electrical insulation for current or voltage connections.

2.1.2.2 X-ray Sources

The most important consideration in choosing an X-ray source for XPS is energy resolution. Eq. (2.1) gives the relationship between the kinetic energy of the photoelectron, the energy of the X-ray photon, and the binding energy of the core electron. Because the energy spread, or line-width, of an electron in a core level is very small, the line-width of the photoelectron energy depends on the line-width of the source, if no undue broadening is introduced instrumentally. In XPS the analyst devotes much effort to extracting chemical information by means of detailed study of individual elemental photoelectron spectra. Such a study needs an energy resolution better than 1.0 eV if subtle chemical effects are to be identified. Thus the line-width of the X-ray source should be significantly smaller than 1.0 eV if the resolution required is not to be limited by the source itself.

Other considerations are that the source material, which forms a target for high-energy electron bombardment leading to the production of X-rays, should be a good conductor – to enable rapid removal of heat – and should also be compatible with UHV.

Table 2.2 lists the energies and line-widths of the characteristic X-ray lines from a few possible candidate materials. In practice Mg Kα and Al Kα are the two used universally because of their line energy and width and their simple use as anode material.

For efficient production of X-rays by electron bombardment, exciting electron energies that are at least an order of magnitude higher than the line energies must be used, so that in Mg and Al sources accelerating potentials of 15 kV are employed. Mod-

Tab. 2.2. Energies and line-widths of some characteristic low-energy X-ray lines.

Line	*Energy [eV]*	*Width [eV]*
Y Mζ	132.3	0.47
Zr Mζ	151.4	0.77
Nb Mζ	171.4	1.21
Mg Kα	1253.6	0.70
Al Kα	1486.6	0.85
Si Kα	1739.5	1.00
Y Lα	1922.6	1.50
Zr Lα	2042.4	1.70

ern sources are designed with dual anodes, one anode face being coated with magnesium and the other with aluminum, and with two filaments, one for each face. Thus, a switch from one type of X-irradiation to the other can be made very quickly.

To protect the sample from stray electrons from the anode, from heating effects, and from possible contamination by the source enclosure, a thin (~2 μm) window of aluminum foil is interposed between the anode and the sample. For optimum X-ray photon flux on the surface (i. e. optimum sensitivity), the anode must be brought as close to the sample as possible, which means in practice a distance of ~2 cm. The entire X-ray source is therefore retractable via a bellows and a screw mechanism.

The X-radiation from magnesium and aluminum sources is quite complex. The principal Kα lines are, in fact, unresolved doublets and should correctly be labeled K$\alpha_{1,2}$. Besides the K$\alpha_{1,2}$ lines a series of further lines, so-called satellite lines, also exist of which the most important ones are K$\alpha_{3,4}$. The energy separations of the satellite lines for Mg and Al together with their intensities, related to K$\alpha_{1,2}$, are given in Tab. 2.3.

Tab. 2.3. Satellite lines* of magnesium and aluminum.

	Mg		*Al*	
X-ray line	*Separation from K$\alpha_{1,2}$ [eV]*	*Relative intensity [%] (K$\alpha_{1,2}$ = 100%)*	*Separation from K$\alpha_{1,2}$ [eV]*	*Relative intensity [%] (K$\alpha_{1,2}$ = 100%)*
Kα'	4.5	1.0	5.6	1.0
Kα_3	8.4	9.2	9.6	7.8
Kα_4	10.0	5.1	11.5	3.3
Kα_5	17.3	0.8	19.8	0.4
Kα_6	20.5	0.5	23.4	0.3
Kβ	48.0	2.0	70.0	2.0

* Data of Krause and Ferreira in: Briggs and Seah [2.4]

Removal of satellites, elimination of the bremsstrahlung background, and separation of the K$\alpha_{1,2}$ doublet can be achieved by monochromatization, shown schematically in Fig. 2.3. The X-ray source is positioned at one point on a spherical surface, called a Rowland sphere and a quartz crystal is placed at another point. X-rays from the source are diffracted from the quartz and, by placing the sample at the correct point on the Rowland sphere, the Kα_1 component can be selected to be focused on it. Quartz is a very convenient diffracting medium for Al Kα, because the spacing between the $10\bar{1}0$ planes is exactly half the wavelength of the X-radiation. Because the width of the Al Kα_1 line is <0.4 eV, the energy dispersion needed around the surface of the sphere implies that the Rowland sphere should have a diameter of at least 0.5 m. Although an XPS spectrum will be much "cleaner" when a monochromator is used, because satellites and background have been removed, the photon flux at the sample is much lower than that from an unmonochromatized source operating at the same power. Against this must be set the greatly improved signal-to-background level in a monochromatized spectrum.

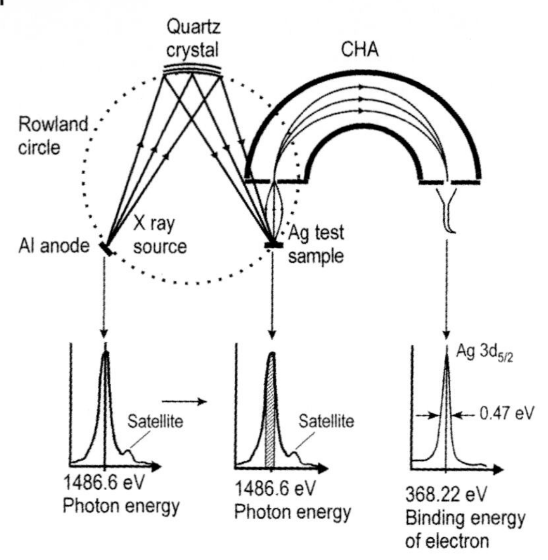

Fig. 2.3. Schematic diagram of X-ray monochromatization to remove satellites, eliminate bremsstrahlung background and separate the Al $K\alpha_{1,2}$ doublet. Courtesy of Kratos Analytical.

2.1.2.3 **Synchrotron Radiation**

The discrete line sources described above for XPS are perfectly adequate for most applications, but some types of analysis require that the source be tunable (i. e. that the exciting energy be variable). The reason is to enable the photoionization cross-section of the core levels of a particular element or group of elements to be varied, which is particularly useful when dealing with multielement semiconductors. Tunable radiation can be obtained from a synchrotron.

In a synchrotron, electrons are accelerated to near relativistic velocities and constrained magnetically into circular paths. When a charged particle is accelerated, it emits radiation, and when the near-relativistic electrons are forced into curved paths they emit photons over a continuous spectrum. The general shape of the spectrum is shown in Fig. 2.4. For a synchrotron with an energy of several gigaelectronvolts and a radius of some tens of meters, the energy of the emitted photons near the maximum is of the order of 1 keV (i. e., ideal for XPS). As can be seen from the universal curve, plenty of usable intensity exists down into the UV region. With suitable mono-

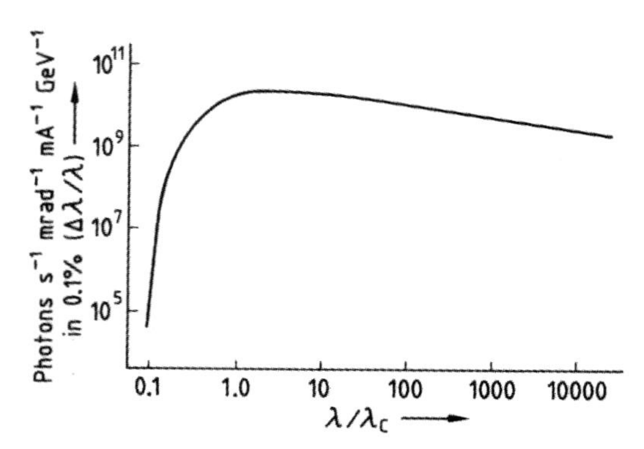

Fig. 2.4. Normalized spectrum of photon energies emitted from a synchrotron. λ_c = wavelength characteristic of the individual synchrotron.

chromators on the output to select a particular wavelength, the photon energy can be tuned continuously from approximately 20 to 500 eV. The available intensities are comparable with those from conventional line sources. The main advantage is the small beam diameter of the synchrotron radiation, which enables high spatial resolution.

2.1.2.4 **Electron-energy Analyzers**

In electron spectroscopic techniques – among which XPS is the most important – analysis of the energies of electrons ejected from a surface is central. Nowadays universally employed is the concentric hemispherical analyzer (CHA).

Energy analyzers cannot be discussed without discussion of energy resolution, which is defined in two ways. Absolute resolution is defined as ΔE, the full width at half-maximum (FWHM) of a chosen peak. Relative resolution is defined as the ratio R of ΔE to the kinetic energy E of the peak energy position (usually its centroid), that is, $R = \Delta E/E$. Thus absolute resolution is independent of peak position, but relative resolution can be specified only by reference to a particular kinetic energy.

In XPS, closely spaced peaks at any point in the energy range must be resolved, which requires the same absolute resolution at all energies.

The CHA is shown in schematic cross-section in Fig. 2.5 [2.5]. Two hemispheres of radii r_1 (inner) and r_2 (outer) are positioned concentrically. Potentials $-V_1$ and $-V_2$ are applied to the inner and outer hemispheres, respectively, with V_2 greater than V_1. The source S and the focus F are in the same plane as the center of curvature, and r_0 is the radius of the equipotential surface between the hemispheres. If electrons of energy $E = eV_0$ are injected at S along the equipotential surface, they will be focused at F if:

$$V_2 - V_1 = V_0\,(r_2/r_1 - r_1/r_2) \tag{2.3}$$

If electrons are injected not exactly along the equipotential surface, but with an angular spread $\Delta\alpha$ about the correct direction, then the energy resolution is given by:

$$\Delta E/E = (w_S + w_F)/4\,r_0 + (\Delta\alpha)^2 \tag{2.4}$$

where w_S and w_F are the respective widths of the entrance and exit slits. In most instruments, for convenience in construction purposes, $w_S = w_F = w$, whereupon the resolution becomes:

$$\Delta E/E = w/2\,r_0 + (\Delta\alpha)^2 \tag{2.5}$$

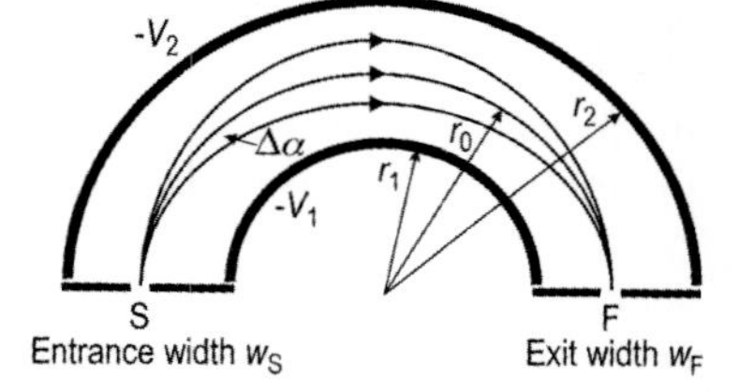

Fig. 2.5. Schematic diagram of a concentric hemispherical analyzer (CHA) [2.5].

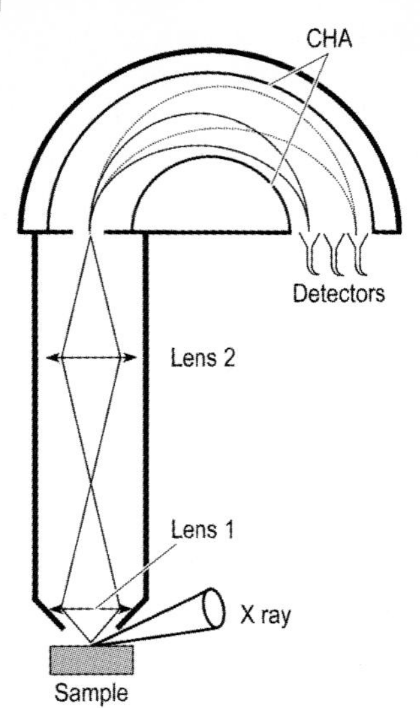

Fig. 2.6. Typical configuration of an XPS spectrometer.

In XPS the photoelectrons are retarded to a constant energy, called the pass energy, as they approach the entrance slit. If this were not done, Eq. (2.5) shows that to achieve an absolute resolution of 1 eV at the maximum kinetic energy of approximately 1500 eV (using Al Kα radiation), and with a slit width of 2 mm, would require an analyzer with an average radius of about 300 cm, which is impracticable. Pass energies are selected in the range 20–100 eV for XPS, which enables the analyzer to be built with a radius of 10–15 cm.

Modern XPS spectrometers employ a lens system on the input to the CHA, which has the effect of transferring an image of the analyzed area on the sample surface to the entrance slit of the analyzer. The detector system on the output of the CHA consists of several single channeltrons or a channel plate. Such a spectrometer is illustrated schematically in Fig. 2.6.

2.1.2.5 **Spatial Resolution**

A principal disadvantage of conventional XPS was lack of spatial resolution; the spectral information came from an analyzed area of several square millimeters and was, therefore, an average of the compositional and chemical analysis of that area. Many technological samples are, on the other hand, inhomogeneous on a scale much smaller than that of conventional XPS analysis, and obtaining chemical information on the same scale as the inhomogeneities would be very desirable.

One of the first steps to improve XPS spatial resolution was done by making use of the focusing properties of the CHA described above; this led to the development of the Escascope now developed into the present Escalab 250 (VG Scientific, East Grinstead, UK) that acts as an XPS microscope. Here, a Fourier-transformed image is generated at the entrance slit of the CHA by introducing a third lens into the transferring lens system. The CHA disperses this image energetically, so that a Fourier-transformed image exists at the exit slit, but only for the energy selected by the setting of the CHA. This image is then inverted into a real image, which can be detected in the plane of a channel plate and a spatial resolution of approximately 3 µm is obtained. A further development using the focusing properties of the transfer lens system is realized in the Axis instruments (Kratos Analytical, Manchester, UK). Here, the principle plane of the first lens is moved in direction to the specimen by introducing a magnetic lens below the specimen holder. The spatial resolution reached is reduced to approximately 15 µm without loss of intensity because of the simultaneously enhanced angle of acceptance, and is improved during the meantime. Such a magnetic lens was also introduced in the present Escalab from VG. A third development is realized in the Quantum 2000 (Physical Electronics, Eden Prairie, USA). Here, high-energy electrons of an electron gun are scanned over the surface of an Al anode, thereby generating an X-ray beam. This beam is then focused on to the specimen surface by an ellipsoidal monochromator and scans the specimen surface with the same rhythm as the electron beam is scanned. The X-ray beam is less than 10 µm in diameter and the scanned area is $1.4 \times 1.4\ mm^2$.

The thrust of development is toward ever better spatial resolution, which means the smallest possible spot size on the sample compatible with adequate signal-to-noise ratio, for acquisition within a reasonable length of time.

2.1.3
Spectral Information and Chemical Shift

Figure 2.7 shows a wide-scan or survey XPS spectrum, i.e. one recorded over a wide range of energies, in this instance 1000 eV. The radiation used was unmonochromatized Al Kα, at 1486.6 eV, and the surface is that of almost clean copper. Such a spectrum reveals the major, or primary, features to be found, but to investigate minor or more detailed features, spectra are acquired over much more restricted energy ranges, at better energy resolution; the latter are called narrow-scan spectra.

The primary features in Fig. 2.7 are peaks arising from excitation of core-level electrons according to Eq. (2.1). At the low-energy end, two intense peaks are found at

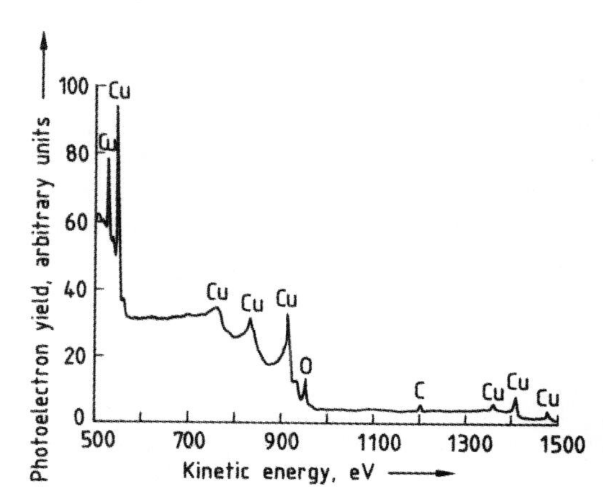

Fig. 2.7. Wide-scan spectrum from almost clean copper using Al Kα radiation.

approximately 553 and 533 eV, corresponding to photoelectrons from the $2p_{3/2}$ and $2p_{1/2}$ levels of copper, respectively, the separation of 20 eV being the spin–orbit splitting. At the high kinetic energy end, are three other copper photoelectron peaks at approximately 1363, 1409, and 1485 eV, which arise from the 3*s*, 3*p*, and 3*d* levels, respectively. The other primary features associated with copper are the three Auger peaks $L_3M_{4,5}M_{4,5}$, $L_3M_{2,3}M_{4,5}$, and $L_3M_{2,3}M_{2,3}$ appearing at 919, 838, and 768 eV, respectively. As pointed out in Sect. 2.1.1, the creation of a core hole by any means of excitation can lead to ejection of an Auger electron, so that in XPS Auger features make a significant contribution to the spectrum. If Auger and photoelectron peaks happen to overlap in any spectrum, they can always be separated by changing the excitation (e. g. from Al Kα to Mg Kα, or vice versa) because the Auger peaks are invariant in energy (Eq. 2.2) whereas the photoelectron peaks must shift with the energy of the exciting photons according to Eq. (2.1).

In addition to primary features from copper in Fig. 2.7 are small photoelectron peaks at 955 and 1204 eV kinetic energies, arising from the oxygen and carbon 1*s* levels, respectively, because of the presence of some contamination on the surface. Secondary features are X-ray satellite and ghost lines, surface and bulk plasmon energy loss features, shake-up lines, multiplet splitting, shake-off lines, and asymmetries because of asymmetric core levels [2.6].

Chemical shift is the name given to the observed shift in energy of a photoelectron peak from a particular element when the chemical state of that element changes. When an atom enters into combination with another atom or group of atoms, an alteration occurs in the valence electron density; this might be positive or negative according to whether charge is accepted or donated, causing a consequent alteration in the electrostatic potential affecting the core electrons. The binding energies of the core electrons therefore change, giving rise according to Eq. (2.1) to shifts in the corresponding photoelectron peaks. Tabulation of the chemical shifts experienced by any one element in a series of pure compounds of that element thus enables its chemical state to be identified during analysis of unknown samples. Many such tabulations have appeared; a major collection is found in Ref. [1.2]. The identification of chemical state in this way is the principal advantage of XPS over other surface-analytical techniques.

An example of a spectrum with a chemical shift is that of the tin 3*d* peaks in Fig. 2.8. A thin layer of oxide on the metallic tin surface enables photoelectrons from both the underlying metal and the oxide to appear together. Resolution of the doublet 3 $d_{5/2}$, 3 $d_{3/2}$ into the components from the metal (Sn^0) and from the oxide Sn^{n+} is shown in Fig. 2.8 B. The shift in this instance is 1.6–1.7 eV. Curve resolution is an operation that can be performed routinely by data processing systems associated with photoelectron spectrometers.

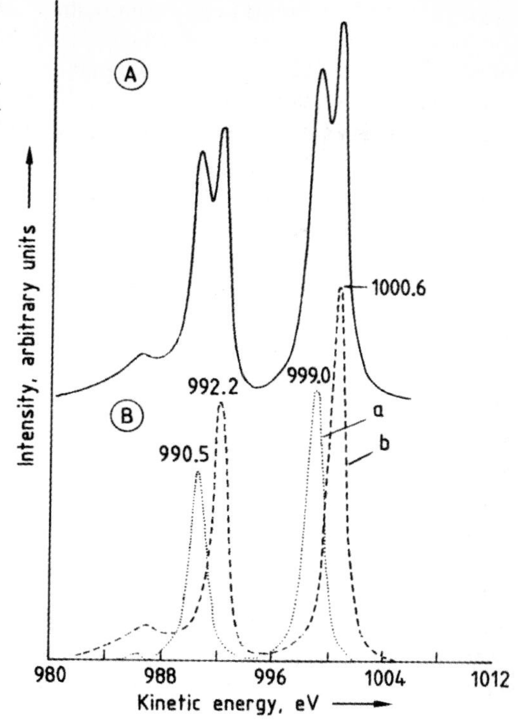

Fig. 2.8. Example of a chemical shift in the Sn 3*d* peak for a very thin layer of Sn oxide in Sn metal. (A) spectrum after linear background subtraction, (B) spectrum resolved into its respective components. (a) Sn^{n+}, (b) Sn^0.

2.1.4 Quantification, Depth Profiling and Imaging

2.1.4.1 Quantification

If an X-ray photon of energy $h\nu$ ionizes a core level X in an atom of element A in a solid, the photoelectron current I_A from X in A is:

$$I_A(X) = K\sigma_A(h\nu,X)\,\beta_A(h\nu,X)\,\bar{N}_A\,\lambda_M(E_A)\cos\theta \tag{2.6}$$

where $\sigma_A(h\nu, X)$ is the photoelectric cross-section for ionization of X (binding energy E_B) by photons of energy $h\nu$; $\beta_A(h\nu, X)$ the asymmetry parameter for emission from X by excitation with photons of energy $h\nu$; $\bar{N}_A$ the atomic density of A averaged over depth of analysis; $\lambda_M(E_A)$ the inelastic mean free path in matrix M containing A, at kinetic energy E_A, where $E_A = h\nu - E_B$; and θ is the angle of the emission to the surface normal. K is a constant of proportionality that contains fixed operating conditions such as incident X-ray flux, transmission of analyzer at kinetic energy E_A, and efficiency of detector at kinetic energy E_A, kept fixed during any one analysis.

Values of the total cross-section σ_A for Al Kα, radiation, relative to the carbon 1*s* level, have been calculated by Scofield [2.7], and of the asymmetry parameter β_A by Reilman et al. [2.8]. Seah and Dench [2.3] have compiled many measurements of the inelastic mean free path, and for elements the best-fit relationship they found was:

$$\lambda_M(E) = 538\,E^{-2} + 0.41\,(d_m E)^{1/2} \text{ monolayers} \tag{2.7}$$

where d_m is the monolayer thickness in nanometers.

In principle, therefore, the surface concentration of an element can be calculated from the intensity of a particular photoelectron emission, according to Eq. (2.6). In practice, the method of relative sensitivity factors is in common use. If spectra were recorded from reference samples of pure elements A and B on the same spectrometer and the corresponding line intensities are I_A^∞ and I_B^∞, respectively, Eq. (2.6) can be written as

$$\frac{I_A/I_A^\infty}{I_B/I_B^\infty} = \left(\frac{\lambda_M(E_A)\lambda_B(E_B)}{\lambda_M(E_B)\lambda_A(E_A)}\right)\left(\frac{R_B^\infty}{R_A^\infty}\right)\left(\frac{N_A\,N_B^\infty}{N_A^\infty\,N_B}\right) \tag{2.8}$$

where R_A^∞ and R_B^∞ are factors for the reference samples, depending on the roughness of the surfaces, and N_A^∞ and N_B^∞ are the atomic densities of the reference samples. From the above equation the ratio N_A/N_B can be obtained from the measured intensity ratio I_A/I_B. The accuracy of quantitative results is, however, only as good as the accuracy of the in-going parameter and the intensity values I_A and I_B. These intensities are simply the areas under the line peaks after background subtraction. Three different procedures are commonly in use – straight line subtraction, Shirley's method and Tougaard's method. With the last mentioned method, the measured spectrum is deconvoluted into the primary (true) excited spectrum by taking into account that the emitted electrons suffer an energy loss because of scattering during travelling in the solid.

A further critical point are the intensities I^∞ correlated to spectra of the pure elements. Calculated and experimentally determined values can diverge considerably, and the best data sets for I^∞ measured on pure reference samples still show a scatter of up to 10%. The use of an internal standard or a simultaneously measured external standard seems to be the most successful way to reducing the inaccuracy below 10%. (For a more detailed discussion of background subtraction and quantification see, e.g., Seah [2.9].)

2.1.4.2 **Depth Profiling**

Very often, in addition to analytical information about the original surface, information is required about the distribution of chemical composition to depths considerably greater than the inelastic mean free path. Non-destructive methods such as variation of emission angle or variation of exciting photon energy are possible, but they are limited in practice to evaluation of chemical profile within depths of only ca. 5 nm. To obtain information from greater depths requires destructive methods, of which the one used universally is removal of the surface by ion bombardment, also called sputtering. The combination of removal of the surface by sputtering and of analysis is termed depth profiling. Instrumentally, this is performed by means of ion guns which are described elsewhere (see Sect. 3.1, SIMS). Depth profiling in XPS is usually performed using noble gas, to minimize chemical effects. In compounds, preferential sputtering, e.g. O from TiO_2 surfaces and Be from Cu–Be alloys, can occur. This is readily comprehensible with regard to different surface binding energies and different momenta transferred to atoms with different masses. Preferential sputtering frequently causes a chemically modified surface even if the primary ion from the ion gun is not chemically reactive after its (immediate) neutralization in the solid.

The depth resolution achievable during profiling depends on many variables, and the reader is referred to a comprehensive discussion [2.9].

In XPS, chemical information is comparatively slowly acquired in a stepwise fashion along with the depth, with alternate cycles of sputtering and analysis. Examples of profiles through oxide films on pure iron and on Fe–12Cr–lMo alloy are shown in Fig. 2.9, in which the respective contributions from the metallic and oxide components of the iron and chromium spectra have been quantified [2.10]. In these examples the oxide films were only ~5 nm thick on iron and ~3 nm thick on the alloy.

Current practice in depth profiling is to use positively charged argon ions at energies between 0.5 and 10 keV, focused into a beam of 2–5-μm diameter, which is then ras-

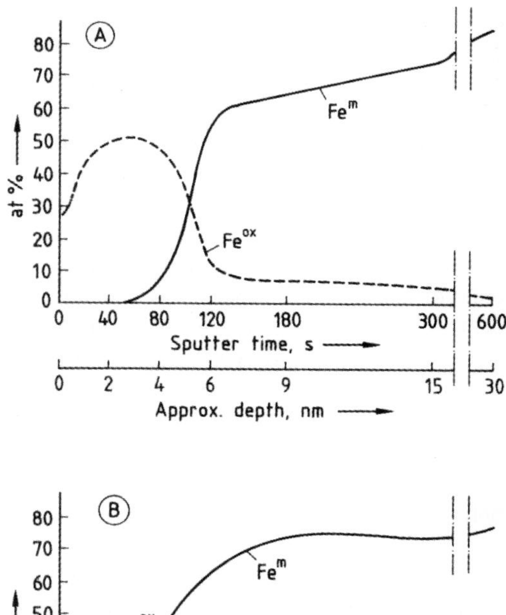

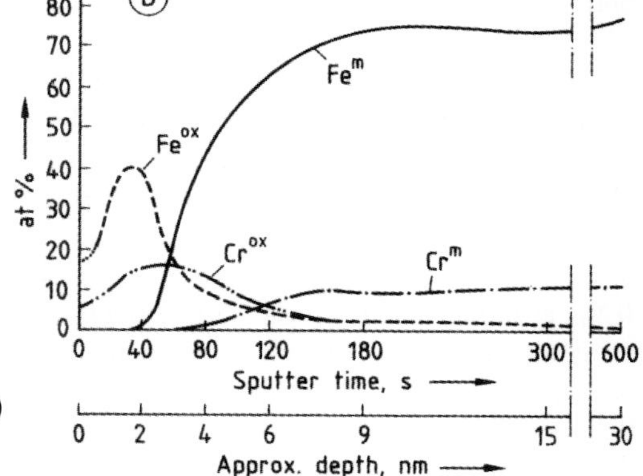

Fig. 2.9. Depth profiles of thin (3–5-nm) oxide films [2.10] on (A) pure iron, (B) Fe–12Cr–1Mo alloy.

tered over the area to be profiled. The area bombarded by the, typically, finely focused and rastered primary ion beam should be significantly greater than the analyzed area so that only the flat bottom of the sputtered crater is analyzed. Ion current densities are variable between 5 and 50 $\mu A\ cm^{-2}$. Uniformity of sputtering and, therefore, the depth resolution of the profile, can be improved by rotating the sample during bombardment.

During the sputtering process, many artifacts are introduced that can affect quantification; some of these are derived from the nature of the sample itself (e. g. roughness, crystalline structure, phase distribution, electrical conductivity), and others are radiation-induced (e. g. atomic mixing, preferential sputtering, compound reduction, diffusion and segregation, sputter-induced topography). This is an extensive subject in its own right; a discussion of all aspects is given elsewhere [2.11].

State-of-the-art for data evaluation of complex depth profile is the use of factor analysis. The acquired data can be compiled in a two-dimensional data matrix in a manner that the n intensity values $N(E)$ or, in the derivative mode $\mathrm{d}N(E)/\mathrm{d}E$, respectively, of a spectrum recorded in the ith of a total of m sputter cycles are written in the ith column of the data matrix $\mathbf{D}$. For the purpose of factor analysis, it now becomes necessary that the $(n \times m)$-dimensional data matrix $\mathbf{D}$ can be expressed as a product of two matrices, i. e. the $(n \times k)$-dimensional spectrum matrix $\mathbf{R}$ and the $(k \times m)$-dimensional concentration matrix $\mathbf{C}$, in which $\mathbf{R}$ in k columns contains the spectra of k components, and $\mathbf{C}$ in k rows contains the concentrations of the respective m sputter cycles, i. e.:

$$\mathbf{D} = \mathbf{R} \cdot \mathbf{C} + \mathbf{E} \tag{2.9}$$

$\mathbf{E}$ is an error matrix taking errors of measurement (e. g. random noise) into consideration. The term component describes such chemical or physical states the spectra of which cannot be generated by a linear combination of the other components. Thus, components can be elements, chemical compounds – stoichiometric or non-stoichiometric – or even states induced by physical processes, provided that the spectra differ significantly, e. g. in line shapes or line shifts.

Since the introduction of factor analysis for evaluation of depth profile data by Gaarenstroom [2.12], many papers have been published [e. g. 2.13–2.20]. With the help of factor analysis, three results can directly be obtained:

(1) the number k of the relevant components necessary to create the data matrix
(2) the so-called abstract spectrum matrix $\mathbf{R}^*$ and the abstract concentration matrix $\mathbf{C}^*$; and
(3) a new data matrix $\mathbf{D}^*$, which can be generated from $\mathbf{R}^*$ and $\mathbf{C}^*$ by leaving out the so-called noise components.

If the spectra of the k relevant components are known, the abstract matrices can be transformed into interpretable spectra and concentration matrices, respectively, and as a final result, one obtains the required component depth profile. If one or more of the spectra of the k relevant components are not known, a quantified component depth profile cannot be constructed.

New developments which have still to be checked for their usability in data evaluation of depth profiles are artificial neural networks [2.16, 2.21–2.25], fuzzy clustering [2.26, 2.27] and genetic algorithms [2.28].

2.1.4.3 **Imaging**

As already mentioned in Sect. 2.1.2.5 the Escalab acts as an XPS microscope and the result is called an elemental map or image. An example of chemical state imaging is given in Fig. 2.10 [2.29]. The material is a contaminated fluoropolymer, the contamination being in the form of dark spots 10–80 μm in size. Both images in Fig. 2.10 are in the C 1*s* photoelectron peak, but of different chemical states, i. e. different binding energies. The upper is in that contribution arising from carbon bound to fluorine, and the lower from carbon bound to carbon (i. e. graphitic). The complementarity of the images indicates that the contamination is graphitic in nature.

If the analyzer is set to accept electrons of an energy characteristic of a particular element, and if the incident X-ray beam is rastered over the surface to be analyzed, a visual display the intensity of which is modulated by the peak intensity will correspond to the distribution of that element over the surface. The result is also an image and this technique is realized with the Quantum 2000.

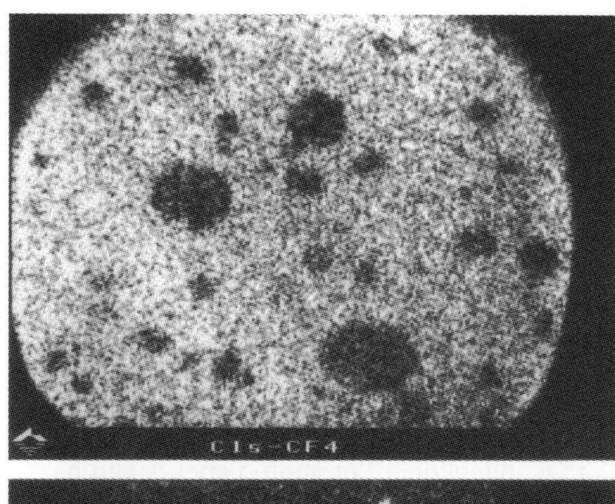

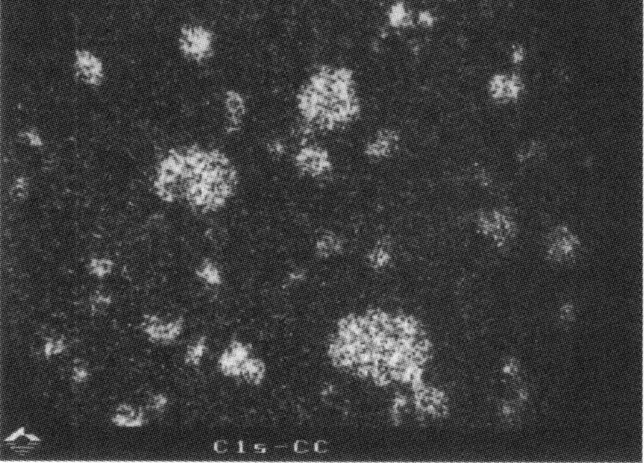

Fig. 2.10. Chemical state images obtained with the Escascope in Fig. 2.9, from a contaminated fluoropolymer [2.29]. (A) image in contribution to C 1*s* from C–F bonding, (B) image in contribution to C 1*s* from C–C bonding.

When the whole sample surface is irradiated by the exciting X-rays, an image can be obtained in a different way: The spot accepted by the transferring lens system in front of the input of the CHA is rastered by introducing deflector plates in front of the lens system. Again, only electrons of a characteristic energy can pass the analyzer. This technique is realized with the Axis series.

Tonner et al. have taken scanning XPS microscopies at the Advanced Light Source Synchrotron Radiation Center of Lawrence Berkeley National Laboratory [2.6]. They investigated a polished and sputter-cleaned surface of mineral ilmenite with the nominal composition $FeTiO_3$, and used the Fe $3p$ and Ti $3p$ lines for imaging. Using synchrotron radiation they demonstrated spatial resolution of approximately 0.25 μm.

Care must be taken in interpreting the intensity distribution, because the electron intensity depends not only on the local concentration of the element but on the topography also, because surface roughness can affect the inelastic background underneath the line. Therefore elemental maps are customarily presented as variations of the ratio of peak intensity divided by the magnitude of the background on both or one side of the line; this can easily be performed by computer.

2.1.5 The Auger Parameter

Section 2.1.3 shows that in an XPS spectrum, X-ray excited Auger peaks are often as prominent as the photoelectron peaks themselves. For many elements, the chemical shifts in Auger peaks are actually greater than the shifts in photoelectron peaks. The two shifts can be combined in a very useful quantity called the Auger parameter α^*, first used by Wagner [2.30] and defined in its modified form [2.31] as

$$\alpha^* = E_{\text{kin}}(\text{XYZ}) + E_{\text{B}}(\text{W}) \tag{2.10}$$

where $E_{\text{kin}}(\text{XYZ})$ is the kinetic energy of the Auger transition XYZ, and $E_{\text{B}}(\text{W})$ is the binding energy of an electron in an arbitrary level W. The Auger and photoelectron peak energies must be measured in the same spectrum.

α^* is independent of photon energy and has the great advantage of being independent of any surface charge. It is, therefore, measurable for all types of material from metals to insulators. Extensive tabulations of α^* have appeared, based on a large number of standard materials; the most reliable data have been collected by Wagner [2.32].

As defined in Eq. (2.10), α^* is purely empirical; any prominent and conveniently situated Auger and photoelectron peaks can be used. If α^* is defined slightly more rigorously, by replacing $E_{\text{B}}(\text{W})$ by $E_{\text{B}}(\text{X})$ in Eq. (2.10), it is related closely to a quantity called the extra-atomic relaxation energy , because $\Delta\alpha^*$ can be shown to be equal to $2\Delta E_{\text{R}}^{\text{ea}}(\text{X})$. In other words, the difference between the Auger parameters of two chemical states is equal to twice the change in the extra-atomic relaxation energies of the two states, associated with the hole in the core level X. Thus α^* has an important physical basis as well as being useful analytically.

2.1.6 Applications

XPS has been used in almost every area in which the properties of surfaces are important. The most prominent areas can be deduced from conferences on surface analysis, especially from ECASIA, which is held every two years. These areas are adhesion, biomaterials, catalysis, ceramics and glasses, corrosion, environmental problems, magnetic materials, metals, micro- and optoelectronics, nanomaterials, polymers and composite materials, superconductors, thin films and coatings, and tribology and wear. The contributions to these conferences are also representative of actual surface-analytical problems and studies [2.33 a,b]. A few examples from the areas mentioned above are given below; more comprehensive discussions of the applications of XPS are given elsewhere [1.1, 1.3–1.9, 2.34–2.39].

2.1.6.1 Catalysis

An understanding of the basic mechanisms of catalysis has been one of the major aims of XPS from its inception. To this end the technique has been used in two ways:

(1) in the study of "real" catalysts closely resembling those used industrially; and
(2) in the study of "model" catalysts, for which the number of variables has been reduced by use of well-characterized crystal surfaces.

For industrial catalysts the chemical information obtainable from XPS is all important, because the course and efficiency of the catalyzed reaction will be governed by the states of oxidation of the elements at the surface. Thus XPS is used to analyze both elemental and chemical composition as a function of bulk composition, of surface loading where the active material is deposited on a support, of oxidation and reduction treatments, and of the time in reactor under operating conditions. Figure 2.11 shows the effect of composition on the chemical state of molybdenum in the surface of Mo–Sm–O catalysts [2.40]. For the highest molybdenum concentration (spectrum d) the Mo 3*d* doublet is very similar to that of MoO_3, but for the samarium-rich composition (spectrum a), it is more like that of a molybdate or a mixed oxide. The effect of successive reduction on a cobalt–molybdenum–alumina catalyst is shown by changes in the Mo 3*d* spectra in Fig. 2.12 [2.41]. The air-fired catalyst in spectrum a contains entirely Mo(VI) at the surface, but with repeated hydrogen reduction the spectrum changes radically and can be curve-resolved into contributions from Mo(VI), Mo(V), and Mo(IV). The chemical state of the surface in spectrum f contains approximately equal proportions of these three oxidation states.

In a recent paper Pijpers et al. [2.42] have reviewed the application of XPS in the field of catalysis and polymers. Other recent applications of XPS to catalytic problems deal with the selective catalytic reduction of NO_x using Pt- and Co-loaded zeolites. Although the Al 2*p* line (Al from zeolite) and Pt 4*f* line interfere strongly, the two oxidation states Pt^0 and Pt^{2+} can be distinguished after careful curve-fitting [2.43].

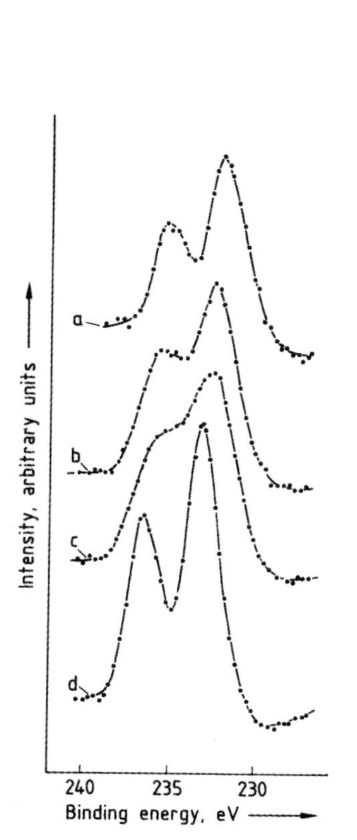

Fig. 2.11. The Mo 3*d* XPS spectra of model catalysts in the Mo–Sm–O system [2.40]. Mo:Sm ratios are (a) 0.25, (b) 1.33, (c) 4.00, (d) 8.00.

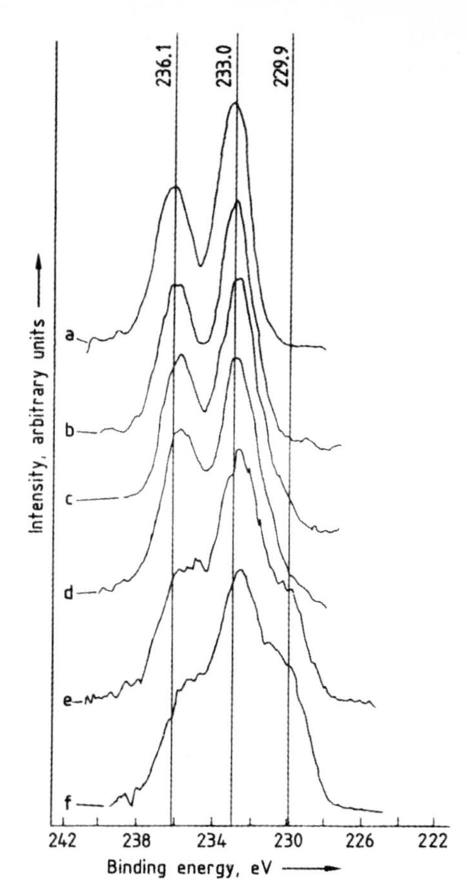

Fig. 2.12. Changes in the Mo 3*d* XPS spectrum from a cobalt–molybdenium–alumina catalyst during successive reduction treatments in hydrogen at 500 °C [2.41]. (a) air-fired catalyst, (b) reduction time 15 min, (c) 50 min, (d) 60 min, (e) 120 min, (f) 200 min.

The influence of Zn-deposition on Cu(111) surfaces on methanol synthesis by hydrogenation of CO_2 shows that Zn creates sites stabilizing the formate intermediate and thus promotes the hydrogenation process [2.44]. Further publications deal with methane oxidation by various layered rock-salt-type oxides [2.45], poisoning of vanadia in VO_x/TiO_2 by K_2O, leading to lower reduction capability of the vanadia, because of the formation of V^{5+} [2.46], and interaction of SO_2 with Cu, Cu_2O, and CuO to show the temperature-dependence of SO_2 absorption or sulfide formation [2.47].

2.1.6.2 **Polymers**

Polymers have been the most prominent objects of investigation since XPS became commercially available at the end of the nineteen-sixties. Nowadays XPS is still the most widely applied method for characterizing polymer surfaces. "Bibles" of high resolution XP spectra of more than hundred different organic polymers have been published by Beamson and Briggs [2.48] and Briggs [2.49]. A crucial problem is the X-ray degradation that occurs in a more or less pronounced manner on almost all polymer surfaces. One must, therefore, exercise caution when investigating, and quantifying, polymer-surface reactions; in practice that means reducing time of irradiation with the exciting X-rays.

Because polymers are typically non-conductive, sample charging can occur and has to be compensated carefully, e.g. by use of a low-energy electron-flood gun, to avoid line-shape distortion and misinterpretation of the measurements.

The photoelectron line of main interest is C1*s*. Different bonding in the environments of the carbon atoms leads to very small chemical shifts of this line. High-resolution XPS is, therefore, required and monochromatic radiation should be used to prevent overlap with satellite lines.

Many publications appear every year dealing with the analysis of polymers; a few examples will be given here. Butyl rubbers are mainly used in tire production and, to study the influence of different polymers on the surface chemistry and to determine the dependence of the adhesion of bromobutyl blends on temperature, valence-band XPS has been applied to a variety of pure polymers and different butyl–elastomer blends. These investigations show the value of valence-band XPS for polymers which are not easy to analyze by core-level XPS [2.50]. In other work dealing with enhancement of the adhesion strength of the Cu–polyimide interface, polyimide surfaces were subjected to plasma treatment, and the modified surfaces were studied by combining TOF–SIMS/XPS [2.51]. In the last five years more than eighty review articles have appeared dealing with special problems in polymer analysis, e.g. polymer membranes [2.52], polymeric biomedical and optoelectronic applications [2.53], or the current status of XPS instrumentation in polymer analysis [2.54].

2.1.6.3 **Corrosion and Passivation**

Apart from the application of XPS in catalysis, the study of corrosion mechanisms and corrosion products is a major area of application. Special attention must be devoted to artifacts arising from X-ray irradiation. For example, reduction of metal oxides (e.g. CuO → Cu_2O) can occur, loosely bound water or hydrates can be desorbed in the spectrometer vacuum, and hydroxides can decompose. Thorough investigations are supported by other surface-analytical and/or microscopic techniques, e.g. AFM, which is becoming increasingly important.

Corrosion products formed as thin layers on metal surfaces in either aqueous or gaseous environments, and the nature and stability of passive and protective films on metals and alloys, have also been major areas of XPS application. XPS has been used in two ways, one in which materials corroded or passivated in the natural environment are analyzed, and another in which well-characterized, usually pure metal surfaces are studied after exposure to controlled conditions.

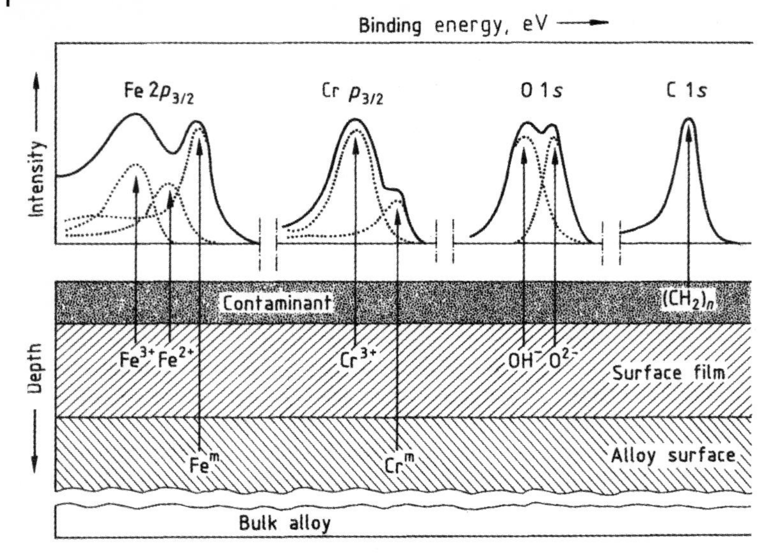

Fig. 2.13. Schematic diagram of the type of information obtainable from XPS spectra from an Fe–Cr alloy with oxide film underneath a contaminant film [2.57].

More effort has probably been devoted to study of the corrosion and passivation properties of Fe–Cr–Ni alloys, e.g. stainless steel and other transition-metal alloys, than to any other metallic system [2.42, 2.44, 2.55, 2.56]. The type of spectral information obtainable from an Fe–Cr alloy of technical origin, carrying an oxide and contaminant film after corrosion, is shown schematically in Fig. 2.13 [2.57].

Corrosion-inhibiting organic films on steel have been studied by XPS [2.58] as have inhibiting films formed in seawater [2.59]; the latter work showed the stability of these films, i.e. no pitting corrosion was observed. Protection layers, frequently TiN, deposited on different materials, are of major interest. Investigation of TiN coatings exposed to different humid SO_2-aggressive environments has shown that the most aggressive atmosphere can penetrate through pores toward the TiN–Fe interface and cause corrosion [2.60]. The corrosion behavior of sandwich-structured TiN/Ti layers was recently a subject of interest [2.61].

Biocorrosion of stainless steel is caused by exopolymer-producing bacteria. It can be shown that Fe is accumulated in the biofilm [2.62]. The effect of bacteria on the corrosion behavior of the Mo metal surface has also been investigated by XPS [2.63]. These last two investigations indicate a new field of research in which XPS can be employed successfully. XPS has also been used to study the corrosion of glasses [2.64], of polymer coatings on steel [2.65], of tooth-filling materials [2.66], and to investigate the role of surface hydroxyls of oxide films on metal [2.67] or other passive films.

2.1.6.4 **Adhesion**

The adhesion of metal and ink to polymers, and the adhesion of paint and other coatings to metal, are of vital importance in several technologies. Aluminum-to-aluminum adhesion is employed in the aircraft industry. The strength and durability of an adhesive bond are completely dependent on the manner in which the adhesive compound interacts with the surfaces to which it is supposed to adhere; this, in turn, often involves pretreatment of the surfaces to render them more reactive. The nature and extent of this reactivity are functions of the chemical states of the adhering surfaces, states that can be monitored by XPS.

In electronic packaging technology, metal–polymer adhesion has been important for some years. Many studies have been devoted to the interactions between metals and polymers; that of Chin-An Chang et al. [2.68] is a good example. They deposited 200-nm-thick films of copper, chromium, titanium, gold, and aluminum on a variety of fluorocarbon polymer films, and found that adhesion, in terms of peel strength, was highest for titanium. Fig. 2.14 shows the C 1*s* spectra after deposition of metal on the perfluoroalkoxy polymer tested; it is apparent from the peak at 282–283 eV that for titanium, and to a lesser extent for chromium, a strong carbide bond is formed. With the other metals, very low adhesion was found, and no carbide-like peaks occurred.

The path of failure of an adhesive joint can give information about the mechanism of failure if analysis of the elemental and chemical composition can be conducted along the path. Several authors have performed such analyses by loading the adhesive joint until it fractures and then using XPS to analyze each side of the fracture.

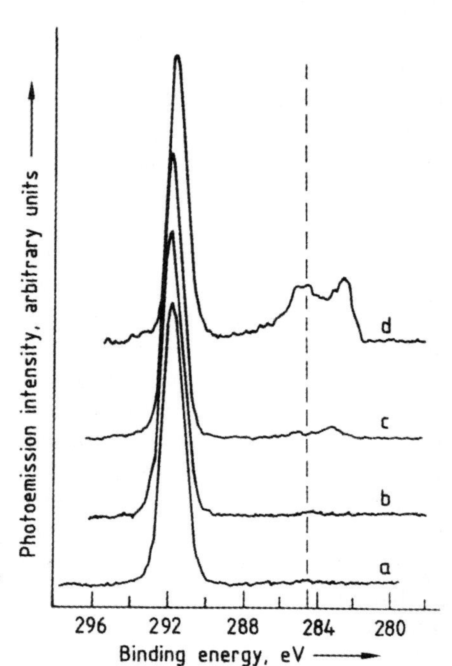

Fig. 2.14. The C 1*s* XPS spectra recorded by Chin–An Chang et al. [2.68] from perfluoroalkoxy polymer (PFA). (a) before deposition, (b) deposition of copper, (c) deposition of chromium, (d) deposition of titanium.

Two recently published works demonstrate the complementary use of XPS and TOF–SIMS. Fitzpatrick et al. and Watts et al. [2.69, 2.70] studied the failure of hot-dipped galvanized steel pretreated with a zinc phosphate conversion coating and joined by means of epoxy resin. They found that electrochemical activity resulting from degradation at the zinc–polymer interface was responsible for the ingress of aggressive substances. Wolany et al. [2.71] investigated the adhesion of untreated and plasma-modified polyimide by XPS and static SIMS. They showed that the number and kind of functional groups available for adhering bonds were enhanced by plasma treatment and that the adhesion strength for Cu deposition was improved. This was also corroborated by SIMS.

Other topics recently studied by XPS include the effects of thermal treatment on the morphology and adhesion of the interface between Au and the polymer trimethylcyclohexane–polycarbonate [2.72]; the composition of the surfaces and interfaces of plasma-modified Cu–PTFE and Au–PTFE, and the surface structure and the improvement of adhesion [2.73]; the influence of excimer laser irradiation of the polymer on the adhesion of metallic overlayers [2.74]; and the behavior of the Co-rich binder phase of WC–Co hard metal and diamond deposition on it [2.75].

2.1.6.5 **Superconductors**

Since the discovery of high-T_c superconducting oxides in 1986 many papers have been published describing the application of XPS to superconductors. One reason for this frenzied activity has been the search for the precise mechanism of superconductivity in new materials. An essential piece of information is the chemical states of the constituent elements, particularly copper, which is common to all of the materials. Some theories predict that small but significant amounts of the Cu^{3+} state should occur in superconductors, other models do not. Comparison by Steiner et al. [2.76] of the Cu $2p_{3/2}$ spectra from copper compounds containing copper as Cu^0 (metal), Cu^+, Cu^{2+}, and Cu^{3+} with the spectra from the superconducting oxide $YBa_2Cu_3O_7$ and from the base material La_2CuO_4, showed conclusively that the contribution of Cu^{3+} to the superconducting mechanism must be negligible. Some of their comparative spectra are shown in Fig. 2.15. The positions of the Cu $2p_{3/2}$ peak in $YBa_2Cu_3O_7$ and La_2CuO_4, along with the satellite structure, are very similar to those in CuO (i. e. Cu^{2+}) rather than those in $NaCuO_2$ (which contains Cu^{3+}).

Recent effort in superconductor research has been devoted mostly to attempts to fabricate thin films with the same superconducting properties as the bulk material, and to prepare suitable electrical contact to the surfaces of superconductors. XPS has been of great use in both areas because of its ability to monitor both composition and chemical state during the processes either of thin superconducting film formation, or of the deposition of thin conducting and semiconducting films on to a superconducting surface. Typical of the latter studies is that of Ziegler et al. [2.76], who deposited increasing amounts of silicon on to epitaxial films of $YBa_2Cu_3O_{7-x}$. Figure 2.16, from their paper, shows the Si $2p$ and Ba $4d$ spectra obtained during deposition of silicon and after heating the films in oxygen at 250 °C when 6 nm of silicon had accumulated. The Ba $4d$ spectrum at the bottom can be resolved into three $4\,d_{5/2,3/2}$ doublets, corresponding – with increasing energy – to barium in the bulk

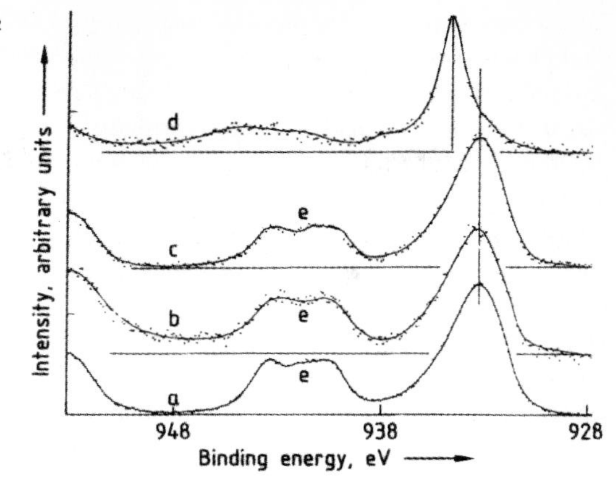

Fig. 2.15. Comparison of the Cu $2p_{3/2}$ and satellite XPS spectra from several copper compounds with the spectrum from the superconducting oxide $YBa_2Cu_3O_7$ [2.76].
(a) CuO, $\Gamma = 3.25$ eV,
(b) La_2CuO_4, $\Gamma = 3.30$ eV,
(c) $YBa_2Cu_3O_7$, $\Gamma = 3.20$ eV,
(d) $NaCuO_2$, $\Gamma = 1.60$ eV,
(e) Cu^{2+} satellites.

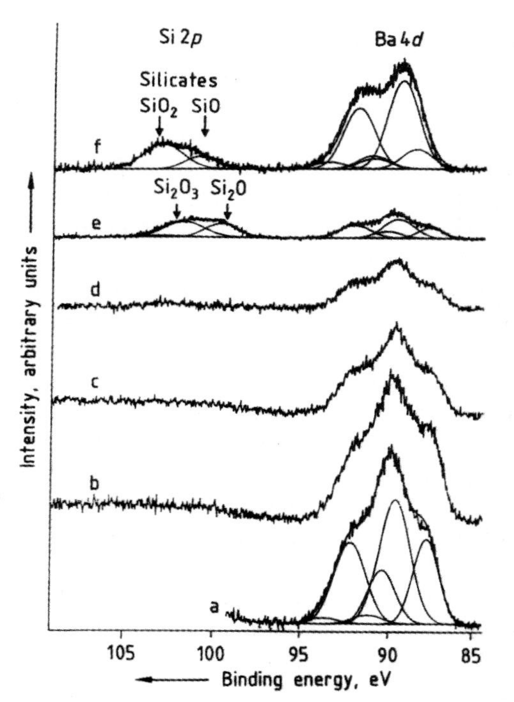

Fig. 2.16. The Si $2p$ and Ba $4d$ XPS spectra [2.77] from: (a) the clean surface of the superconducting oxide $YBa_2Cu_3O_{7-x}$, (b)–(e) the same surface after increasing deposition of (b) 0.8 nm Si, (c) 2.8 nm Si, (d) 4.4 nm Si, (e) 6.0 nm Si, (f) the deposited surface after heating to 250 °C in oxygen (0.8 Pa).

superconductor, in the surface of the superconductor, and as the carbonate. With increasing silicon thickness the barium spectrum attenuates but does not change in character, but silicon does not appear in the spectrum until nearly 5 nm has been deposited, suggesting migration into the superconductor at room temperature. When the thickness is 6 nm, the Si $2p$ spectrum can be separated into suboxide contributions, indicating reaction of silicon with the $YBa_2Cu_3O_{7-x}$, surface. After heating in oxygen the Si $2p$ spectrum changes to that characteristic of SiO_2 and silicates; the

Ba 4*d* spectrum also changes by increasing in intensity and conforming mostly to that expected of a barium silicate. As a result of the latter changes the superconducting properties of the film were destroyed. The Y 3*d* and Cu 2*p* spectra establish that yttrium and copper oxides are also formed.

Other surface reactions studied have been those of gold [2.78] and silver [2.79] with $Bi_2Sr_2CaCu_2O_{8+y}$, of niobium with $YBa_2Cu_3O_{7-x}$ [2.80], and of lead with $YBa_2Cu_3O_{7-x}$ [2.81, 2.82]. Characteristics of superconductors in the form of thin films prepared in different ways have been recorded by XPS for $YBa_2Cu_3O_{7-x}$ [2.83–2.87], $GdBa_2Cu_3O_{7-x}$ [2.88], $Nd_{2-x}Ce_xCuO_{4-y}$ [2.89], $Rb_xBa_{1-x}BiO_3$ [2.90], $(Bi_{0.9}Pb_{0.1})_{2.3}$ $Sr_2Ca_{1.9}Cu_3O_{10+x}$ [2.91], and $Tl_2Ca_2BaCu_3O_{10}$ [2.92]. Detailed electronic studies have been made of the bulk superconductors $Bi_2Sr_2Ca_{1-x}Nd_xCu_2O_y$ [2.93, 2.94], $Nd_{2-x}Ce_xCuO_{4-y}$ [2.95–2.99], Er–$Ba_2Cu_4O_8$ [2.100], and $Tl_2Ba_2CaCu_2O_8$ [2.101].

In several applications depth profiles were recorded to check the extent of reaction of various materials with superconductors, or to analyze uniformity and concentration during preparation of superconductors. Such investigations have been performed by both XPS and AES.

2.1.6.6 **Interfaces**

The chemical and electronic properties of elements at the interfaces between very thin films and bulk substrates are important in several technological areas, particularly microelectronics, sensors, catalysis, metal protection, and solar cells. To study conditions at an interface, depth profiling by ion bombardment is inadvisable, because both composition and chemical state can be altered by interaction with energetic positive ions. The normal procedure is, therefore, to start with a clean or other well-characterized substrate and deposit the thin film on to it slowly at a chosen temperature while XPS is used to monitor the composition and chemical state by recording selected characteristic spectra. The procedure continues until no further spectral changes occur, as a function of film thickness, of time elapsed since deposition, or of changes in substrate temperature.

A good example is the study of the reaction at the interface of the rare-earth element thulium with the silicon (111) surface performed by Gokhale et al. [2.102]. Figure 2.17 shows the Si 2*p* spectrum from the substrate as thulium is deposited on it at room temperature. With the Mg Kα, radiation used for the analysis, the kinetic energy of the Si 2*p* electrons is approximately 1155 eV, corresponding to an inelastic mean free path of approximately 2 nm. The Si 2*p* peak is still plainly detectable after 12.5 nm of thulium have been deposited, and in addition a new feature appears on the low binding-energy side of the peak, increasing in magnitude and shifting to lower energy with thickness. Increasing the temperature, in stages, of the thickest film and use of X-ray diffraction established that the energetic final position, labeled R_1, corresponded to Tm_5Si_3, and the initial energetic position, found at a thickness of 0.5–1.0 nm, to $TmSi_2$. The magnitude of the residual Si 2*p* peak was, however, too great at all stages to be accounted for by compound formation alone, indicating the formation of thulium clusters followed by their coalescence into islands. Measurement of the Tm $4d_{5/2}$ binding energy during the interaction also showed that the valence of thulium was 3+ at all temperatures and coverages.

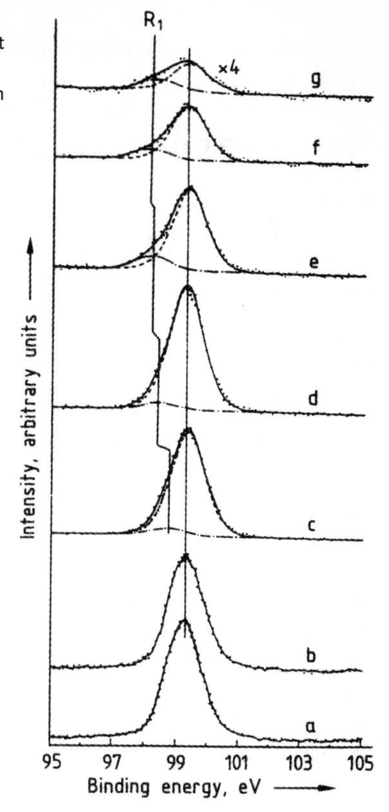

Fig. 2.17. The Si 2p XPS spectra with Mg Kα excitation during deposition of thulium to a thickness of 12.5 nm at room temperature [2.102]. (a) clean silicon, (b) 0.2 nm Tm, (c) 0.5 nm Tm, (d) 2 nm Tm, (e) 4 nm Tm, (f) 6.5 nm Tm, (g) 12.5 nm Tm.

Recent XPS studies of interfaces have been performed on U/Si(111)–7 × 7 face [2.103], on interfaces forming during the oxidation of $Ti_{1-x}Al_xN$ (x = 0.2–0.65) [2.104], on ZrO_2–Y_2O_3 on steel [2.105], on naphthalocyanine on Si(111)–7 × 7 and Si(100)–2 × 1 faces [2.106], and on (phenylacetylene–p-nitrophenylacetylene) copolymer on Cr [2.107]. The five examples given are only to show that in the different areas of application discussed so far interfaces play a dominant role, because they determine the chemical and electronic properties of the effectiveness of adhesive, the adhesion of protective or decorative coatings, the functionality of microelectronics and sensors, and the behavior of catalysis.

Other important areas of application of XPS have included lubricating films [2.108, 2.109], glasses [2.110–2.114], ion implantation [2.115–2.119], and cleaning [2.120–2.122] and passivation [2.123, 2.124] of semiconductor surfaces.

2.1.7
Ultraviolet Photoelectron Spectroscopy (UPS)

In its principle of operation, UPS is similar to XPS, in that a surface is irradiated with photons and the energies of the ejected photoelectrons are analyzed. The physical relationship involved is the same as that of Eq. (2.1) for XPS, but in UPS the energy $h\nu$ of the exciting photons is much lower, because the photons are derived from a gaseous discharge that produces hard UV radiation. The gas normally used is helium, which, depending on pressure conditions in the discharge, will provide line sources of energy 21.21 eV (He I) or 40.82 eV (He II) with very narrow line-widths (~20 meV). Because of these low exciting energies, the binding energies E_B that appear in Eq. (2.1) do not refer to core levels, as in XPS, but to shallow valence-band levels and to other shallow levels, such as those for adsorbates, near the valence band. For energy analysis, the CHA is again used, but the pass energies employed for UPS are much lower than in XPS, because better energy resolution is required.

UPS is not an analytical technique in the sense that it can provide quantified elemental concentrations on a surface, as can XPS and AES, but it is a powerful tool in basic research. It is used in two ways. In its angle-integrated form it has been very extensively employed to study the interactions of gaseous molecules with surfaces, usually in combination with other techniques, e.g. XPS, AES, LEED, and TDS. The other way in which it has been used is in *angle-resolved UPS* (ARUPS). In this form the energy analyzer, a CHA of small radius, is mounted mechanically in such a way that it can traverse the space around the sample in both polar and azimuthal directions, while the angle of photon incidence is kept constant. The angle of incidence is then changed by rotation of the sample. If the sample is a single crystal with a clean surface, the dispersion of spectral features (i.e. their movement along the energy axis and their increase or decrease in magnitude) can be used theoretically to plot the band structure in the surface region. The changes in surface-associated band structure during reaction with gaseous molecules can then be followed, leading to additional information about the nature of the reaction.

2.2
Auger Electron Spectroscopy (AES)

Henning Bubert and John C. Rivière

After XPS, AES is the next most widely used surface-analytical technique. As an accepted surface technique AES actually predates XPS by two to three years, because the potential of XPS as a surface-specific technique was not recognized immediately by the surface-science community. Pioneering work was performed by Harris [2.125] and by Weber and Peria [2.126], but the technique as it is known today is basically the same as that established by Palmberg et al. [2.127].

2.2.1 Principles

The surface to be analyzed is irradiated with a beam of electrons of sufficient energy, typically in the range 2–10 keV, to ionize one or more core levels in surface atoms. After ionization the atom can relax by either of the two processes described in Sect. 2.1.1 for XPS – ejection of a characteristic X-ray photon (fluorescence) or ejection of an Auger electron. Although these are competing processes, for shallow core levels ($E_B < 2$ keV) the probability of the Auger process is far higher. The Auger process is described schematically in Fig. 2.1 (left side), which points out that the final state of the atom is doubly ionized.

Eq. (2.2) shows that the Auger energy is a function of atomic energy levels only. Because no two elements have the same set of atomic binding energies, analysis of Auger energies provides elemental identification. Even if levels L_1 and L_{23} are in the valence band of the solid, analysis is still possible because the dominant term in Eq. (2.2) is always the binding energy of K, the initially ionized level.

Considering that heavy elements have more levels than just K and L, Eq. (2.2) also indicates that the heavier the element, the more numerous are the possible Auger transitions. Fortunately, there are large differences between the probabilities of different Auger transitions, so that even for the heaviest elements, only a few intense transitions occur, and analysis is still possible.

In principle, chemical and elemental information should be available in AES, because the binding energies appearing in Eq. (2.2) are subject to the same chemical shifts as measured in XPS. In practice, because the binding energies of three levels are involved, extracting chemical information from Auger spectra is usually difficult, although significant progress has been made in data processing methods (Sect. 2.1.4.2), which yield valuable chemical information.

The reasons AES is a surface-specific technique have been given in Sect. 2.1.1, with reference to Fig. 2.2. The normal range of kinetic energies recorded in an AES spectrum would typically be from 20 to 1000 eV, corresponding to inelastic mean free path values of 2 to 6 monolayers.

The nomenclature used in AES has also been mentioned in Sect. 2.1.1. The Auger transition in which initial ionization occurs in level X, followed by the filling of X by an electron from Y and ejection of an electron from Z, would therefore be labeled XYZ. In this rather restricted scheme, one would thus find in the KLL series the six possible transitions KL_1L_1, KL_1L_2, KL_1L_3, KL_2L_2, KL_2L_3, and KL_3L_3. Other combinations could be written for other series such as the LMM, MNN, etc.

2.2.2 Instrumentation

2.2.2.1 Vacuum Requirements

The same considerations discussed in Sect. 2.1.2.1 for XPS apply to AES.

2.2.2.2 Electron Sources

The energy of the exciting electrons does not enter into Eq. (2.2) for the Auger energy, unlike that of the exciting X-ray photon in XPS; thus the energy spread in the electron beam is irrelevant. But the actual energy of the primary electrons is relevant because of the dependence on that energy of the cross-section for ionization by electron impact. For an electron in a core level of binding energy E_B, this dependence increases steeply with primary energy above E_B, passes through a maximum of approximately 3–5 times E_B, and then decreases to a fairly constant plateau. This is illustrated in Fig. 2.18, from Casnati et al. [2.128], in which various theoretical and experimental cross-sections are shown for ionization of the Ni K shell. The implication of this dependence is that for efficient ionization of a particular core level, the primary energy should be approximately five times the binding energy of an electron in that level. Because, in most samples for analysis, several elements are found in the surface region with core-level binding energies extending over a large range, choice of a primary energy sufficiently high to ionize all the core levels efficiently is advisable. Hence for conventional AES, primary energies are typically in the range 3–10 keV. In *scanning* Auger *microscopy* (SAM), much higher primary energies are used, in the range of 25–50 keV, because in that version of AES the electron spot on the sample must be as small as possible, and the focusing required can be effected only at such energies.

For the primary energy range 3–10 keV used in conventional AES, the electron emitter is thermionic, usually an LaB_6 or tungsten cathode, and focusing of the electron beam is performed electrostatically. Typically, such an electron source would be able to provide a spot size on the specimen of approximately 0.5 μm at 10 keV and a beam current of approximately 10^{-8} Å. The beam can normally be rastered over the specimen surface, but such a source would not be regarded as adequate for SAM. Sources for SAM can be of either the thermionic or the field-emission type, the latter being particularly advantageous in achieving minimum spot sizes, because it is a high-brightness source. Focusing is performed electromagnetically, and for optimum performance a modern electron gun for SAM would be capable of a spot size of 15 nm at 30 keV and a beam current of approximately 10^{-10} Å. The beam is rastered over the surfaces at scan rates variable up to TV rates, with the dimensions of the scanned area variable also.

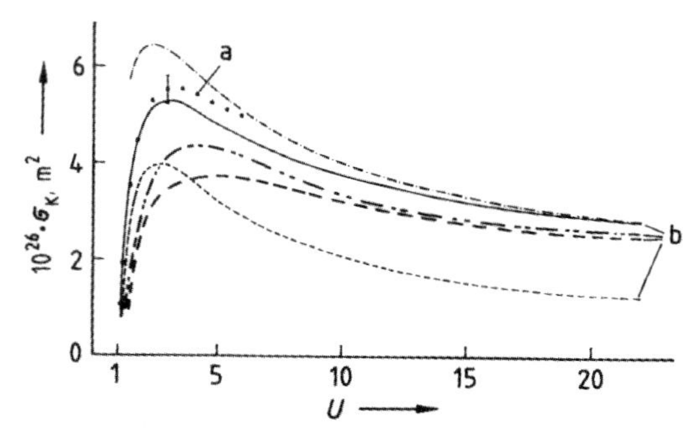

Fig. 2.18. Electron-impact ionization cross-section for the Ni K shell, as a function of reduced electron energy U [2.128] $U = E_p/E_K$, where E_p is the primary electron energy and E_K the binding energy of the K shell. (a) experimental points, (b) semi-empirical or theoretical curves.

2.2.2.3 Electron-energy Analyzers

An electron-energy analyzer still often applied in AES is the cylindrical mirror analyzer (CMA). In the CMA shown schematically in Fig. 2.19, two cylinders of radii r_1 (inner) and r_2 (outer) are accurately positioned coaxially. Annular entrance and exit apertures are cut in the inner cylinder, and a deflection potential $-V$ is applied between the cylinders. Electrons leaving the sample surface at the source S with a particular energy E on passing into the CMA via the entrance aperture are then deflected back through the exit aperture to the focus F. For the special case in which the acceptance angle $\alpha = 42.3°$, the first-order aberrations vanish, and the CMA becomes a second-order focusing device. The relationship between the electron energy E and the deflection potential is then:

$$E = 1.31\,\mathrm{eV}/\ln\,(r_2/r_1) \tag{2.11}$$

As in the CHA, therefore, scanning the deflection potential $-V$ and recording the signal as a function of electron energy provide the distribution in energy of electrons leaving the sample surface.

If the angular spread of the acceptance angle is $\Delta\alpha$, the relative energy resolution of a CMA is:

$$\Delta E/E = 0.18\,w/r_1 + 1.39\,(\Delta\alpha)^3 \tag{2.12}$$

where w is the effective slit width. For typical angular spread $\Delta\alpha \approx 6°$, w can be replaced by the source size (i.e. the area on the sample from which electrons are accepted). The source size is very small, because of the focused primary beam, and the important property of the CMA is, therefore, its very high transmission, arising from the large solid angle of acceptance; for $\Delta\alpha \approx 6°$, transmission is approximately 14%.

Conventionally, Auger spectra are presented in the differentiated energy distribution form $\mathrm{d}N(E)/\mathrm{d}E$, rather than in the undifferentiated form $N(E)$ used in XPS. Differentiating enhances the visibility of Auger features, as is demonstrated in Fig. 2.20, in which the differentiated spectrum from boron is shown above the corresponding $N(E)$ spectrum. The $N(E)$ distribution is nowadays recorded digitally, and differentiation is performed by computer. With the rapid improvement in the signal-to-noise

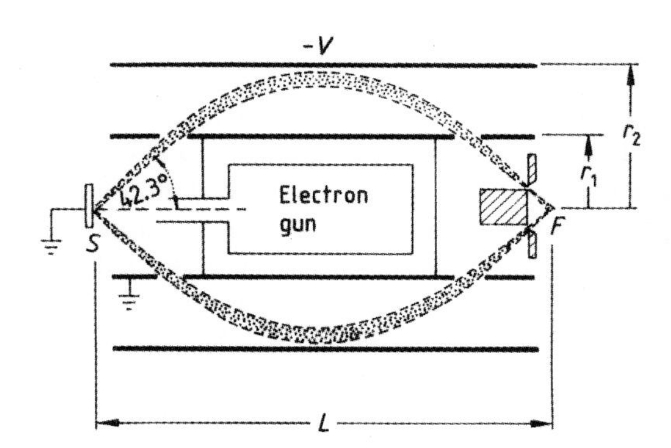

Fig. 2.19. Schematic diagram of a cylindrical mirror analyzer (CMA).

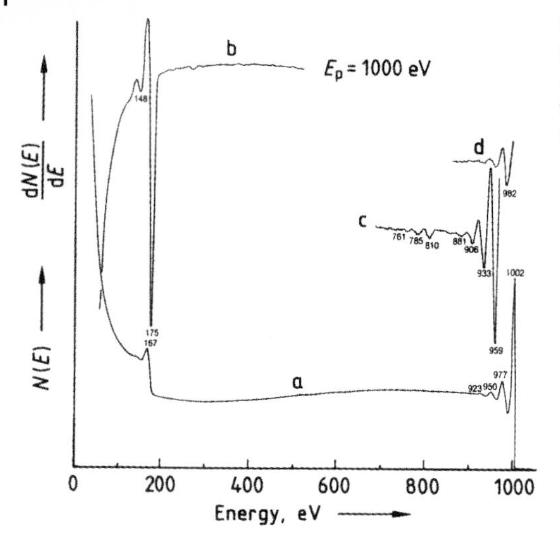

Fig. 2.20. Secondary electron distribution from boron (a) in the $N(E)$ mode, (b)–(d) in the $dN(E)/dE$ mode, where the differentiation reveals the plasmon satellites.

characteristics of detection equipment, however, the trend in presenting Auger spectra is increasingly toward the undifferentiated $N(E)$ distribution.

Note from Fig. 2.20 that although the true position of the boron KLL Auger peak in the $N(E)$ spectrum is at 167 eV, the position in the $dN(E)/dE$ spectrum is taken for purely conventional reasons to be that of the negative minimum, i.e. at 175 eV.

2.2.3
Spectral Information

As stated above, the range of primary energies typically used in conventional AES is 3–10 keV, and the ionization cross-section passes through a maximum at 3–5 times the binding energy of an electron in the ionized core level. The core levels that can be ionized efficiently are, therefore, limited. Thus K-shell ionization efficiency is adequate up to $Z \approx 14$ (silicon), L_3 up to $Z \approx 38$ (strontium), M_5 up to $Z \approx 76$ (osmium), and so on (for the much higher primary energies used in SAM, the ranges are correspondingly extended). This means that in various regions of the periodic table, characteristic Auger transitions occur that are most prominent under typical operating conditions.

Figures 2.21–2.23 show some of these prominent Auger features, all recorded using a CMA, in the differential distribution [2.129]. Examples of the KLL Auger series are shown in Fig. 2.21, of the LMM series in Fig. 2.22, and of the MNN series in Fig. 2.23. In the KLL series the most prominent feature is a single peak arising from the $KL_{2,3}L_{2,3}$ transition, with minor features from other Auger transitions to lower energy. The LMM series is characterized by a triplet arising from $L_3M_{2,3}M_{2,3}$, $L_3M_{2,3}M_{4,5}$, and $L_3M_{4,5}M_{4,5}$, transitions in ascending order of kinetic energy, with another intense peak, the $M_{2,3}M_{4,5}M_{4,5}$, appearing at very low energy (e.g. 50 eV for iron). Note that AES transitions involving electrons in valence band levels are often written with a V rather than the full symbol of the level. Thus $L_3M_{2,3}M_{4,5}$ and

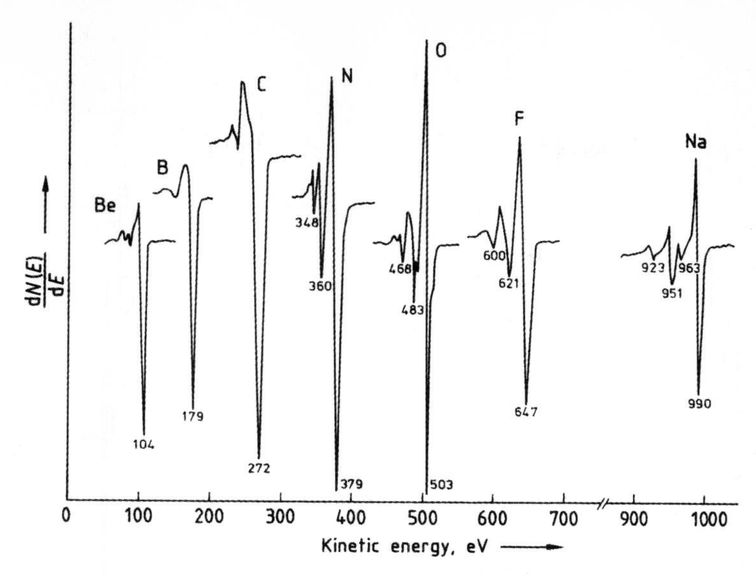

Fig. 2.21. KLL Auger series characteristic of the light elements [2.129].

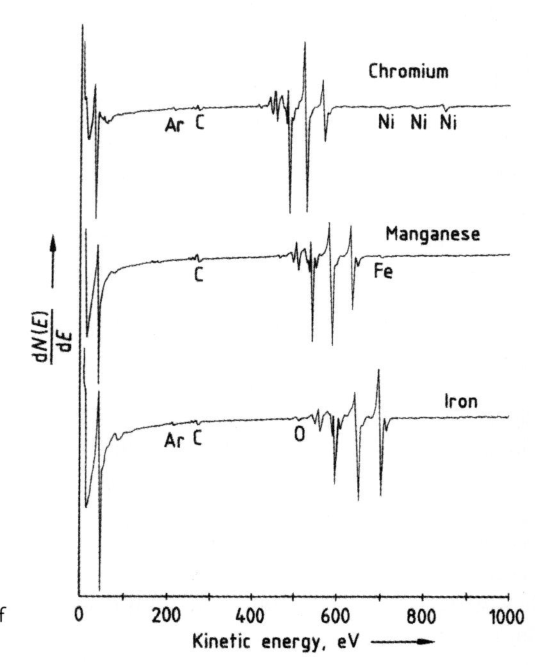

Fig. 2.22. LMM and MMM Auger series in the middle of the first series of transition elements [2.129].

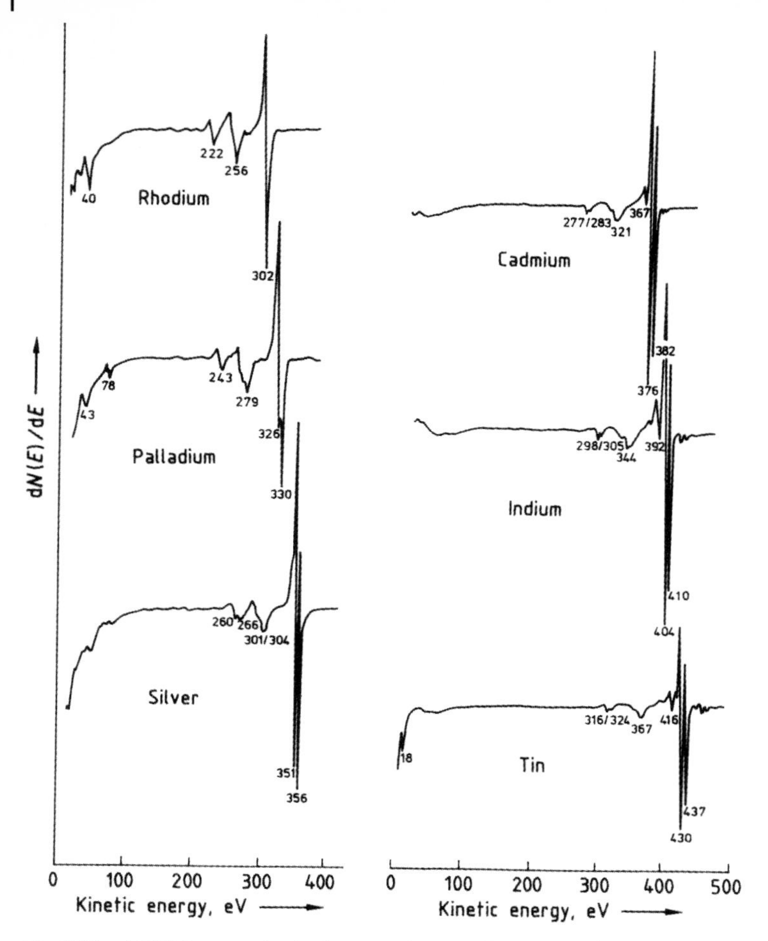

Fig. 2.23. MNN Auger series in the second series of transition elements [2.129].

$L_3M_{4,5}M_{4,5}$ would often appear as $L_3M_{2,3}V$ and L_3VV, respectively, and similarly $M_{2,3}M_{4,5}M_{4,5}$, as $M_{2,3}VV$. In Fig. 2.22 the increase in the intensity of the L_3VV peak relative to the other two, upon going from chromium to iron, is because of the progressive increase in the electron density in the valence band. The characteristic doublet seen in the MNN series arises from the $M_{4,5}N_{4,5}N_{4,5}$ transitions, in which the doublet separation is that of the core levels M_4 and M_5.

Chemical effects are quite commonly observed in Auger spectra, but are difficult to interpret compared with those in XPS, because additional core levels are involved in the Auger process. Some examples of the changes to be seen in the KLL spectrum of carbon in different chemical environments are given in Fig. 2.24 [2.130]. Such spectra are typical components of data matrices (see Sect. 2.1.4.2) derived from AES depth profiles (see below).

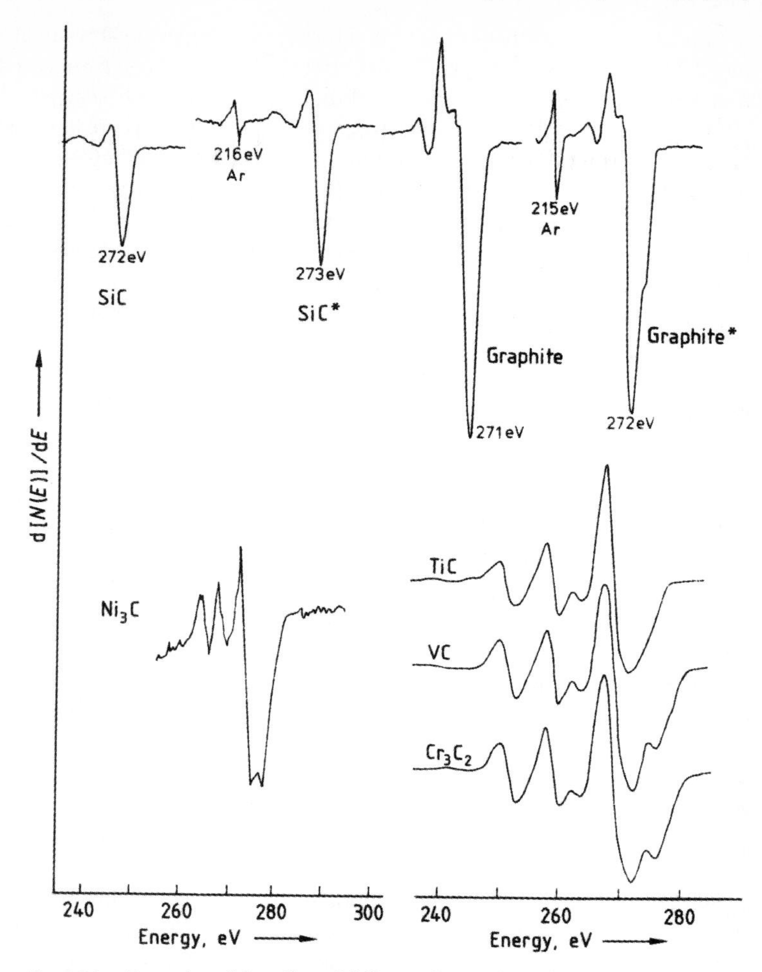

Fig. 2.24. Examples of the effect of different chemical states on the KLL Auger spectrum of carbon [2.130] (SiC* and graphite* denote Ar^+-bombarded surfaces of SiC and graphite, respectively).

Associated with prominent features in an Auger spectrum are plasmon energy loss features that are also found associated with photoelectron peaks in an XPS spectrum. Plasmon energy losses arise from excitation of modes of collective oscillation of the conduction electrons by outgoing secondary electrons of sufficient energy. Successive plasmon losses suffered by the back-scattered primary electrons can be seen in both the $N(E)$ and the $\mathrm{d}N(E)/\mathrm{d}E$ spectra of Fig. 2.20, from boron. The same magnitude of plasmon loss, ~27 eV, is also associated with the KLL Auger peak of boron. Plasmon losses are also present in Figs 2.21–2.24, but they are difficult to disentangle from minor Auger features.

Another type of loss feature not often recorded in Auger spectra can also be seen in Fig. 2.20, in the $dN(E)/dE$ spectrum at 810 eV. This type arises from core-level ionization and forms the basis for the technique of *core electron energy-loss spectroscopy* (CEELS). A primary electron interacting with an electron in a core level can cause excitation either to an unoccupied continuum state, leading to complete ionization, or to localized or partly localized final states, if available. Because the primary electron need not give up all its energy in the interaction, the loss feature appears as a step in the secondary electron spectrum, with a tail decreasing progressively to lower energies. Differentiation, as in Fig. 2.20, accentuates the visibility. In that figure, the loss at 810 eV corresponds to the K-shell ionization edge of boron at approximately 190 eV.

2.2.4
Quantification and Depth Profiling

2.2.4.1 Quantification

If ionization of a core level X in an atom A in a solid matrix M by a primary electron of energy E_p gives rise to the current $I_A(XYZ)$ of electrons produced by the Auger transition XYZ, then the Auger current from A is

$$I_A(XYZ) = K\sigma(E_p, X)\,[1 + r_M(E, \alpha)]\,\bar{N}_A\,\lambda_M(E_A)\cos\theta \tag{2.13}$$

where $\sigma(E_p, X)$ is the cross-section for ionization of level X by electrons of energy E_p; r_M the back scattering factor that takes account of the additional ionization of X (binding energy E_B) by inelastic electrons with energies E between E_p and E_B; α the angle of the incident electron beam to the surface normal; $\bar{N}_A$ the atomic density of A averaged over depth of analysis; $\lambda_M(E_A)$ the inelastic mean free path in matrix M containing A, at the Auger kinetic energy E_A; and θ the angle of Auger electron emission to surface normal.

Similarly to Eq. (2.6), K is a proportionality constant containing fixed operating conditions, for example incident electron current density, transmission of the analyzer at the kinetic energy E_A, efficiency of the detector at the kinetic energy E_A, and the probability of the Auger transition XYZ.

The cross-section for electron impact ionization has already been mentioned in Sect. 2.2.2.2 in connection with electron sources, and a variety of experimental and theoretical cross-sections have been shown in Fig. 2.18 for the particular case of the K-shell of nickel. The expression for the cross-section derived by Casnati et al. [2.128] gives reasonably good agreement with experiment; the earlier expression of Gryzinski [2.131] is also useful.

For the inelastic mean free path, Eq. (2.10) given for XPS also applies.

The new factor in Eq. (2.13), compared with Eq. (2.6) for XPS, is the back scattering factor r_M. Ichimura and Shimizu [2.132] have performed extensive Monte Carlo calculations of the back scattering factor as a function of primary beam energy and angle of incidence, of atomic number and of binding energy; a selection of their results for $E_p = 10$ keV at normal incidence is shown in Fig. 2.25. The best fit of their results to experiment gives the following relationship

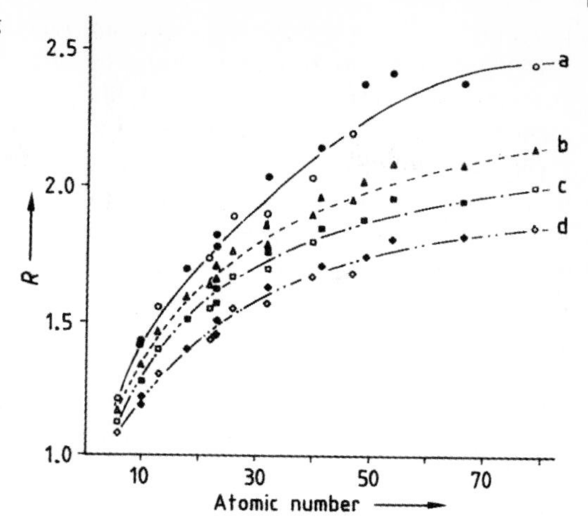

Fig. 2.25. Values of the back-scattering factor R (i.e., $1 + r$) calculated as a function of atomic number for an electron beam of 10 keV at normal incidence by Ichimura et al. [2.132]. (a) $E_B = 0.1$ keV, (b) $E_B = 0.5$ keV, (c) $E_B = 1.0$ keV, (d) $E_B = 2.0$ keV.

$$r = (2.34 - 2.1\,Z^{0.14})\,U^{-0.35} + (2.58Z^{0.14} - 2.98) \qquad \text{for } \alpha = 0^\circ$$

$$r = (0.462 - 0.777\,Z^{0.20})\,U^{-0.32} + (1.5\,Z^{0.20} - 1.05) \qquad \text{for } \alpha = 30^\circ$$

$$r = (1.21 - 1.39\,Z^{0.13})\,U^{-0.33} + (1.94\,Z^{0.13} - 1.88) \qquad \text{for } \alpha = 45^\circ \qquad (2.14)$$

where Z is the atomic number and $U = E_p/E_B$.

Thus, as for XPS, the average surface concentration $\bar{N}_A$ can, in principle, be calculated by measurement of the Auger current, according to Eq. (2.13). Again, as in XPS, relative sensitivity factors are generally used. The Auger current I_A^∞ for the same transition XYZ in a standard of pure A is measured under the same experimental conditions as in the analysis of A in M, whereupon the ratio of the atomic concentrations is

$$\frac{\bar{N}_A}{N_A^\infty} = \frac{I_A\,(1 + r_A)\,\lambda_A\,R_A^\infty}{I_A^\infty\,(1 + r_M)\,\lambda_M\,R_M} \qquad (2.15)$$

because the ionization cross-section, σ, is the same for both standard and sample. Then, as in XPS, assuming nearly equal roughness factors $R_A^\infty \approx R_M$, Eq. (2.15) can be reduced to:

$$\bar{N}_A = S_A\,I_A \qquad (2.16)$$

where:

$$S_A = \frac{\bar{N}_A^\infty\,(1 + r_A)\,\lambda_A}{I_A^\infty\,(1 + r_M)\,\lambda_M} \qquad (2.17)$$

S_A is the relative sensitivity factor. Normally, values of S_A are derived empirically or semi-empirically. Tables of such sensitivity factors have been published by Payling

[2.133] and by Mroczkowski and Lichtman [2.134] for differential Auger spectra, and for the direct, undifferentiated, spectra by Sato et al. [2.135]. A comprehensive discussion of the assumptions and simplifications involved and the corrections that should be applied is given elsewhere [2.9, 2.136]. Although modern Auger spectrometers detect the spectra in the undifferentiated mode, many workers still evaluate the spectra in the subsequently electronically differentiated mode, i.e. using peak-to-peak heights, so that changes in peak shape as a result of changes in chemical bonding are not taken into account and quantification in AES is still not as accurate as that in XPS.

2.2.4.2 Depth Profiling

As in XPS, elemental distributions near the surface, but to depths greater than the analytical depth resolution, are frequently required when using AES. Again, the universally employed technique is that of progressive removal of the surface by ion sputtering; again one must be aware that the electron spectroscopic sampling depth corresponds roughly to the depth of the layer modified by preferential sputtering effects, and again the edges of the sputter crater should not contribute to the signal.

A variation on depth profiling that can be performed by modern scanning Auger instruments (see Sect. 2.2.6) is to program the incident electron beam to jump from one pre-selected position on a surface to each of many others in turn, with multiplexing at each position. This is called multiple point analysis. Sets of elemental maps acquired after each sputtering step or each period of continuous sputtering can be related to each other in a computer frame-store system to derive a three-dimensional analysis of a selected micro volume.

2.2.5 Applications

Like XPS, the application of AES has been very widespread, particularly in the earlier years of its existence; more recently, the technique has been applied increasingly to those problem areas that need the high spatial resolution that AES can provide and XPS, currently, cannot. Because data acquisition in AES is faster than in XPS, it is also employed widely in routine quality control by surface analysis of random samples from production lines of, for example, integrated circuits. In the semiconductor industry, in particular, SIMS is a competing method. Note that AES and XPS on the one hand and SIMS/SNMS on the other, both in depth-profiling mode, are complementary, the former gaining signal from the sputter-modified surface and the latter from the flux of sputtered particles.

2.2.5.1 Grain Boundary Segregation

One of the original applications of AES, and still one of the most important, is the analysis of grain boundaries in metals and ceramics. Very small amounts of impurity or dopant elements in the bulk material can migrate under appropriate temperature conditions to the boundaries of the grain structure and accumulate there. In that way the concentration of minor elements at the grain boundaries can become much higher than in the bulk, and the cohesive energy of the boundaries can be so altered that the material becomes brittle. Knowledge of the nature of the segregating elements and their grain boundary concentrations as a function of temperature can be used to modify fabrication conditions and thereby improve the strength performance of the material in service.

Goretzki [2.137] has discussed the importance and analysis of internal surfaces such as grain boundaries and phase boundaries, and given several examples. He emphasizes that to avoid ambiguity in interpretation of the analysis, the internal surfaces must be exposed by fracturing the material under UHV conditions inside the electron spectrometer. If this is not done, atmospheric contamination would immediately change the surface condition irrevocably. One of the examples from Goretzki's paper is given in Fig. 2.26, in which Auger spectra in the differential mode from fracture surfaces in the embrittled (Fig. 2.26A) and unembrittled (Fig. 2.26B) states of a 12% Cr steel containing phosphorus as an impurity, are compared.

Embrittlement occurs when the steel is held at 400–600 °C for a long time. Note in the spectrum from the embrittled material not only is the phosphorus concentration greatly enhanced at the grain boundary, but the amounts of chromium and nickel, relative to iron, are also increased. By setting the analyzer energy first to that of the phosphorus peak, and then to that of the principal chromium peak, Auger maps could be recorded showing the distribution of phosphorus and chromium over the fracture surface. The two maps showed high correlation of the two elements, indicating association of phosphorus and chromium at the grain boundary, possibly as a compound. Other recently published studies of the same 12% steel have reported enhanced phosphorus levels at the grain boundaries of the embrittled material [2.138, 2.139]; similar results have been obtained for other high-chromium steel alloys [2.140, 2.141]. Reichl et al. [2.142] investigated the temperature-dependence of surface segregation of S and N in α-Fe. In an initial phase, S diffusion occurred at the grain boundaries and displaced N at the surface at higher temperatures. Other

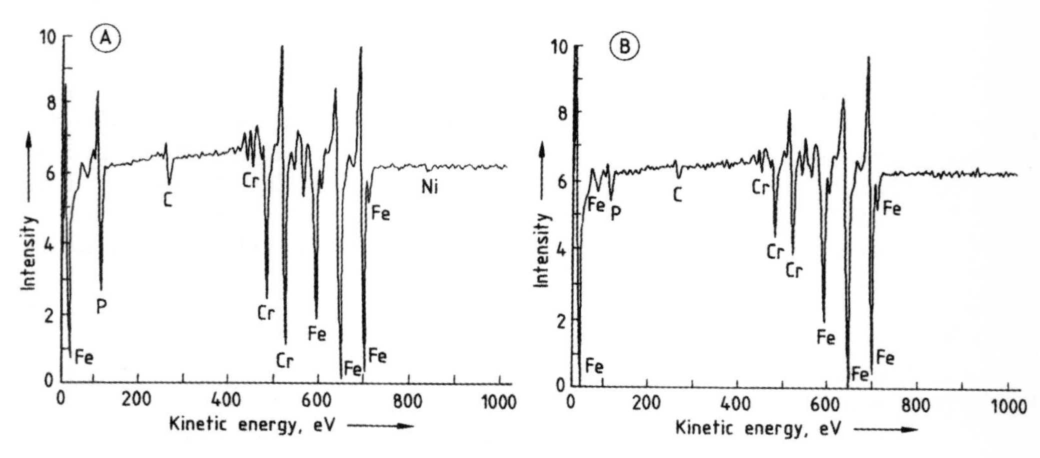

Fig. 2.26. Auger spectra from fracture surfaces of a 125 Cr steel [2.137]. (A) embrittled state, (B) unembrittled state.

applications in the study of grain boundary segregation in metals have included the effects of boron, zirconium, and aluminum on Ni_3Al intermetallics [2.143], the segregation of phosphorus and molybdenum to grain boundaries in the superalloy Nimonic PE 16 as a function of aging treatment [2.144], and the enrichment of chromium carbide and chromium–iron carbide at the grain boundaries in steel [2.145].

Application of AES to zirconia ceramics has been reported by Moser et al. [2.146]. Elemental maps of Al and Si demonstrate the grain boundary segregation of small impurities of silica and alumina in these ceramics.

2.2.5.2 **Semiconductor Technology**

With ever-increasing miniaturization of integrated circuits goes a need for the ability to analyze ever smaller areas so that the integrity of fabrication at any point on a circuit can be checked. This checking process includes not only microdetails such as continuity of components in the form of thin films and establishment of the correct elemental proportions in contacts, diffusion barriers, and Schottky barriers, but also the effectiveness of macro treatments such as surface cleaning. In all these types of analysis, depth profiling is used extensively because, in general, elemental compositional information is required, not information about chemical state. As mentioned in Sect. 2.1.4.2, chemical state information tends to become blurred or destroyed by ion bombardment.

Several materials have been prescribed as diffusion barriers to prevent one circuit material from diffusing into another (often Si) at the temperatures needed for preparation of certain components such as oxide films in integrated circuits. Zalar et al. [2.147] studied interdiffusion at an Al_2O_3/Ti interface. They covered Si (111) with a thin film of TiN as a diffusion barrier, then with a 45 nm layer of polycrystalline Ti, and finally with 55 nm amorphous Al_2O_3. The same samples were heated to different temperatures at a rate of 40 °C min^{-1} and the interdiffusion processes were investigated by AES depth profiling. Figs 2.27 A–C show that oxygen from Al_2O_3, and to a lesser extent Al also, migrate into the Ti layer. (The "N + Ti" intensity in the Ti layer is that of Ti only, whereas in the neighboring TiN layer the N + Ti intensity is that of Ti and N, because of peak overlapping.) Control measurements indicate that oxygen does not arise from the outer gas phase. In the interface region a new phase consisting of α-Ti_3Al is formed and detected by selected-area electron diffraction. Such AES investigations are normally performed in combination with other surface-analytical techniques – in this instance with XPS, XRD, TEM, and electron diffraction also. Other work dealing with the analysis of CVD W_xN films as barriers for Cu metallization has recently been published by Hua Li et al. [2.148], who used AES and RBS, AFM, XRD, and TEM.

The nature of the metallic contacts to compound semiconductors has also been an interesting problem. Ohmic contacts must be made to semiconducting surfaces, otherwise circuits could not be fabricated. During fabrication, however, excessive interdiffusion must not occur, or the electrical characteristics of the semiconductor cannot be maintained. Among the many studies of the reactions of metallizing alloys with compound semiconductors is that of Roberts et al. [2.149], who investigated the W/TiN/Ti/Si contact structure using a multi-spectral Auger microscope (MUL-

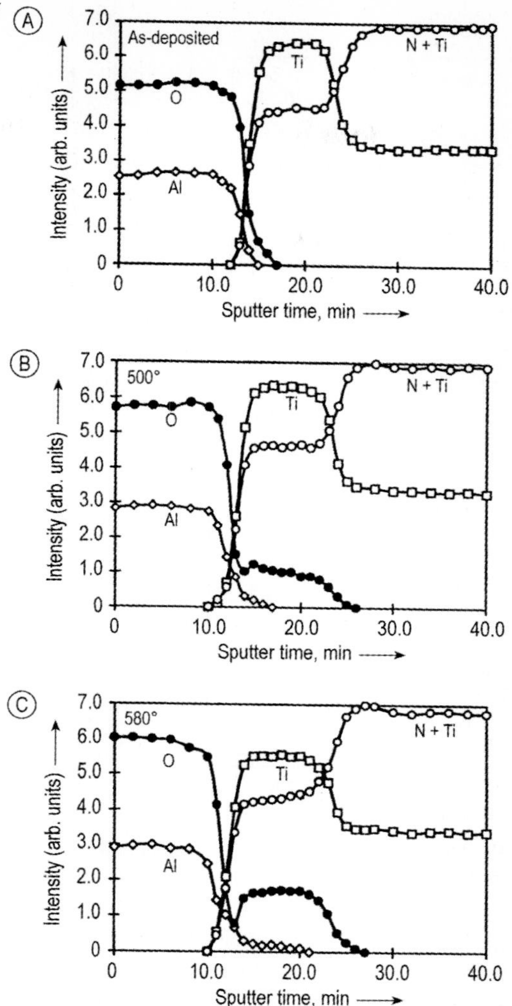

Fig. 2.27. AES sputter-depth profiles of the α-Al_2O_3–Ti thin-film structure on a smooth Si substrate covered with a TiN thin-film diffusion barrier, (A) as-deposited, (B) after heating to 500 °C, (C) after heating to 580 °C [2.147].

SAM) and TEM. An Auger electron spectrometer with high spatial resolution imaging capability was developed especially for the detection of small particles and defects which might be present in the ULSI regime; this enabled the inspection of wafers up to 200 mm in diameter [2.150].

AES has also been applied to study preferential sputtering of $TiSi_x$ forming for low-resistivity conductor films in ULSI devices [2.151], or the electromigration behavior of Au–Ag films on SiO_2 using AES, XPS and AFM [2.152].

2.2.5.3 **Thin Films and Interfaces**

The nature of the interface formed between very thin metallic films and substrates of various types has been studied extensively by AES, just as it has by XPS

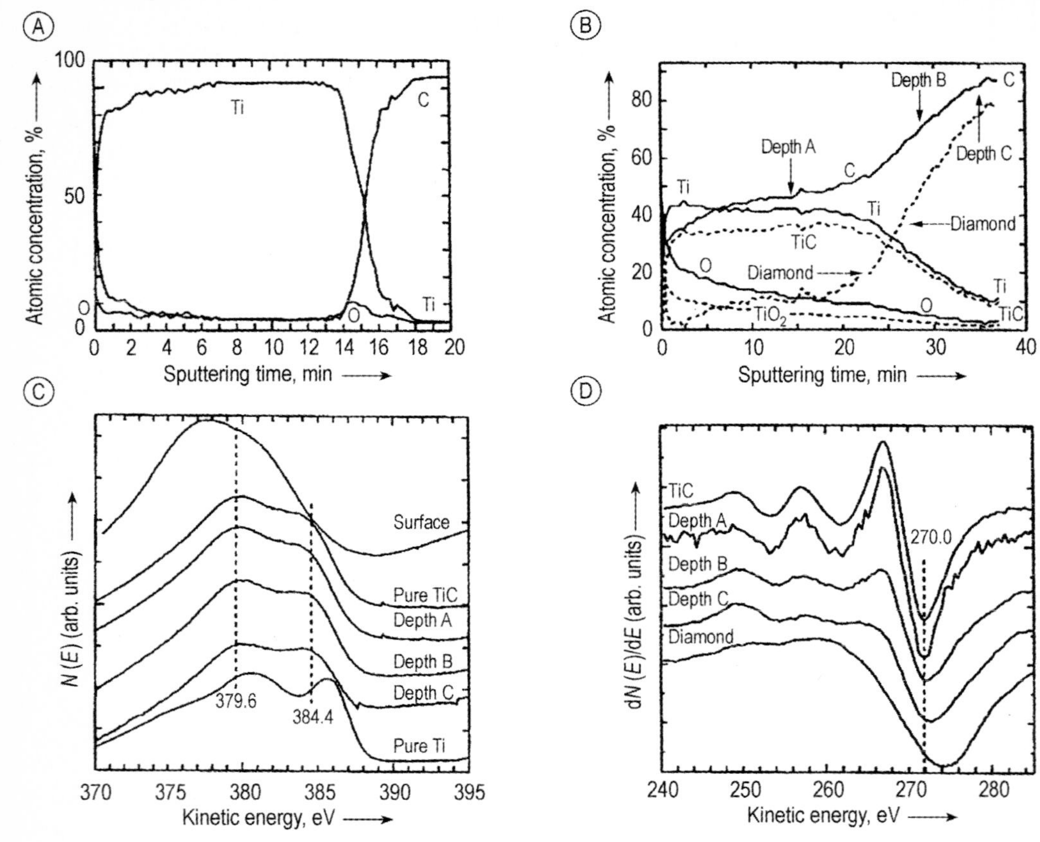

Fig. 2.28. Depth profiles of a Ti–diamond layer (A) before and (B) after annealing at 600 °C for 4 h [2.153]. Auger line shapes of (C) Ti LM_1M_4 and (D) C KLL at various depths.

(Sect. 2.1.6.6). The extent of interaction and interdiffusion can be established by AES with depth profiling, although chemical information is normally absent, but the development of the interface is usually studied by continuous recording of Auger spectra during film deposition.

Zhu et al. [2.153] deposited Ti or Cr layers on diamond samples. The Ti–diamond or Cr–diamond specimens were annealed at 600 °C for 4 h during metallization. Figs 2.28 A and B show the depth profiles before and after annealing of a Ti/diamond specimen, and Figs 2.28 C and D show the Ti LM_1M_4 Auger line shapes in the $N(E)$ mode and the C KLL line shape in the differentiated $dN(E)/dE$ mode at different depths. The unstructured Ti signal at the surface can be attributed to TiO_2, because of smooth oxidation of the surface. But the greatest part of the layer consists of TiC. The formation of TiC is also recognizable by the strongly structured carbon line which changes its shape with depth and changes gradually into that of diamond. All these features are consistent with the presence of TiC, indicating complete metallization.

Other interesting thin-film studies using AES have included the growth of platinum on TiO_2- and SrO-terminated (100) $SrTiO_3$ single-crystal substrates [2.154], of epitaxial niobium films on (110) TiO_2 [2.155], the interaction of copper with a (0001) rhenium surface [2.156], and the characterization of radio-frequency (rf) sputtered TiN films on stainless steel [2.157].

A multilayer-type structure probably due to cords in the molten zone between single arc sprayed (0.25 MPa) Ni droplets and steel substrate were found in AES point depth profiles [2.158]. That particular arc spraying condition turned out to yield the best adhesion. Plasma-sprayed Al_2O_3 layers separated from pre-oxidized Ni Substrate had a micrometer-thick NiO layer on the substrate-sided face and micrometer-deep oxide interdiffusion [2.159]. In this work also, AES point depth profiling substantiated technological assumptions about adhesion mechanisms.

2.2.5.4 **Surface Segregation**

The surface composition of alloys, compounds, and intermetallic materials can change profoundly during heat treatment as a result of segregation of one or more constituents and impurity elements. Surface enhancement can also occur as a result of chemical forces during oxidation or other reactions. The mechanism of segregation is similar to that of grain boundary segregation already discussed, but, of course, at the free surface material is much more likely to be lost through reaction, evaporation, etc., so that excessive continued segregation can change the bulk properties as well as those of the surface. Nearly all such studies have been performed on metals and alloys; AES and LEED were often commonly applied to surface segregation.

Sometimes the segregation of metallic and non-metallic elements on surfaces can result in the formation of surface compounds, as demonstrated by Uebing [2.160, 2.161]. He heated a series of Fe–15%Cr alloys containing 30 ppm nitrogen, 20 ppm carbon, or 20 ppm sulfur to 630–800 °C and obtained the typical Auger spectra shown in Fig. 2.29. Each spectrum shows that the amounts of both chromium and the non-metal are greater than would be expected if the bulk concentration were maintained. The fine structure of the nitrogen and carbon KLL spectra is characteristic of nitride and carbide, respectively. Thus the cosegregation of chromium with nitrogen and with carbon has led to the formation of surface nitrides and carbides.

Song et al. [2.162] have investigated surface segregation in $Ni_{0.65}Nb_{0.35}$ up to 500 °C and 1 atm O_2 atmosphere. Below 100 °C, Nb migrated to the surface, with formation of Nb_2O_5. At temperatures $\geq$ 300 °C, Ni migrated to the surface as Ni and NiO. This inverse segregation can be explained by the different diffusion speeds of O and Ni in the Nb_2O_5 layer formed at the surface.

Other recent investigations involving AES, often with depth profiling, deal with the surface segregation of Ag in Al–4.2% Ag [2.163], of Sn in Cu and formation of superficial Sn–Cu alloy [2.164], of Mg in Al–Mg alloy [2.165], and of Sb in Fe–4% Sb alloy [2.166]. Note the need to differentiate between, particularly, segregation, i. e. original sample properties, from the artifact of preferential sputtering.

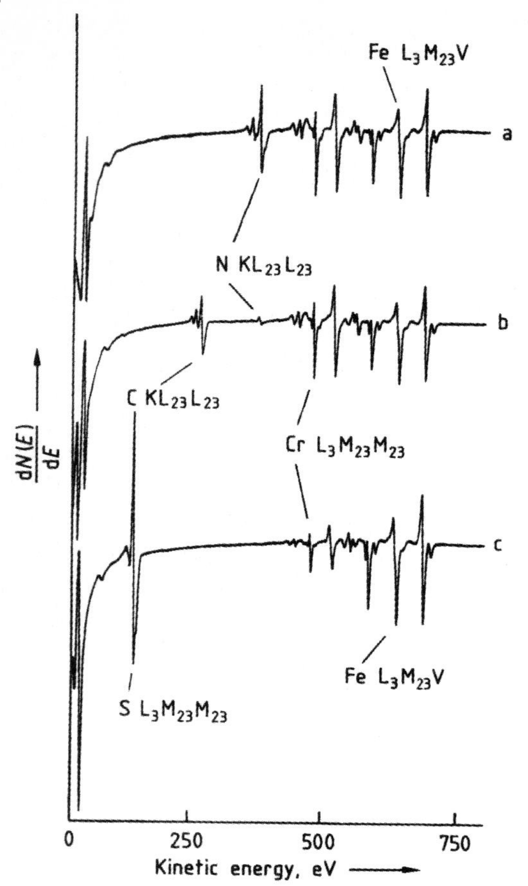

Fig. 2.29. Formation of surface compounds on Fe–15%Cr alloys by cosegregation of chromium and a non-metallic element [2.160]. (a) nitride, (b) carbide, (c) sulfide.

2.2.6 Scanning Auger Microscopy (SAM)

If an incident electron beam of sufficient energy for AES is rastered over a surface in a manner similar to that in a scanning electron microscope (SEM), and if the analyzer is set to accept electrons of Auger energies characteristic of a particular element, then an elemental map or image is again obtained, similar to XPS for the Quantum 2000 (Sect. 2.1.2.5).

This is illustrated in Fig. 2.30 [2.167]. The surface was that of a fractured compact of SiC to which boron and carbon had been added to aid the sintering process. The aim of the analysis was to establish the uniformity of distribution of the additives and the presence or absence of impurities. The Auger maps show not only very non-uniform distribution of boron (Fig. 2.30 a) but also strong correlation of boron with sodium (Fig. 2.30 c), and weaker correlation of boron with potassium (Fig. 2.30 b). Point analyses for points A and B marked on the images reveal the presence of sulfur and cal-

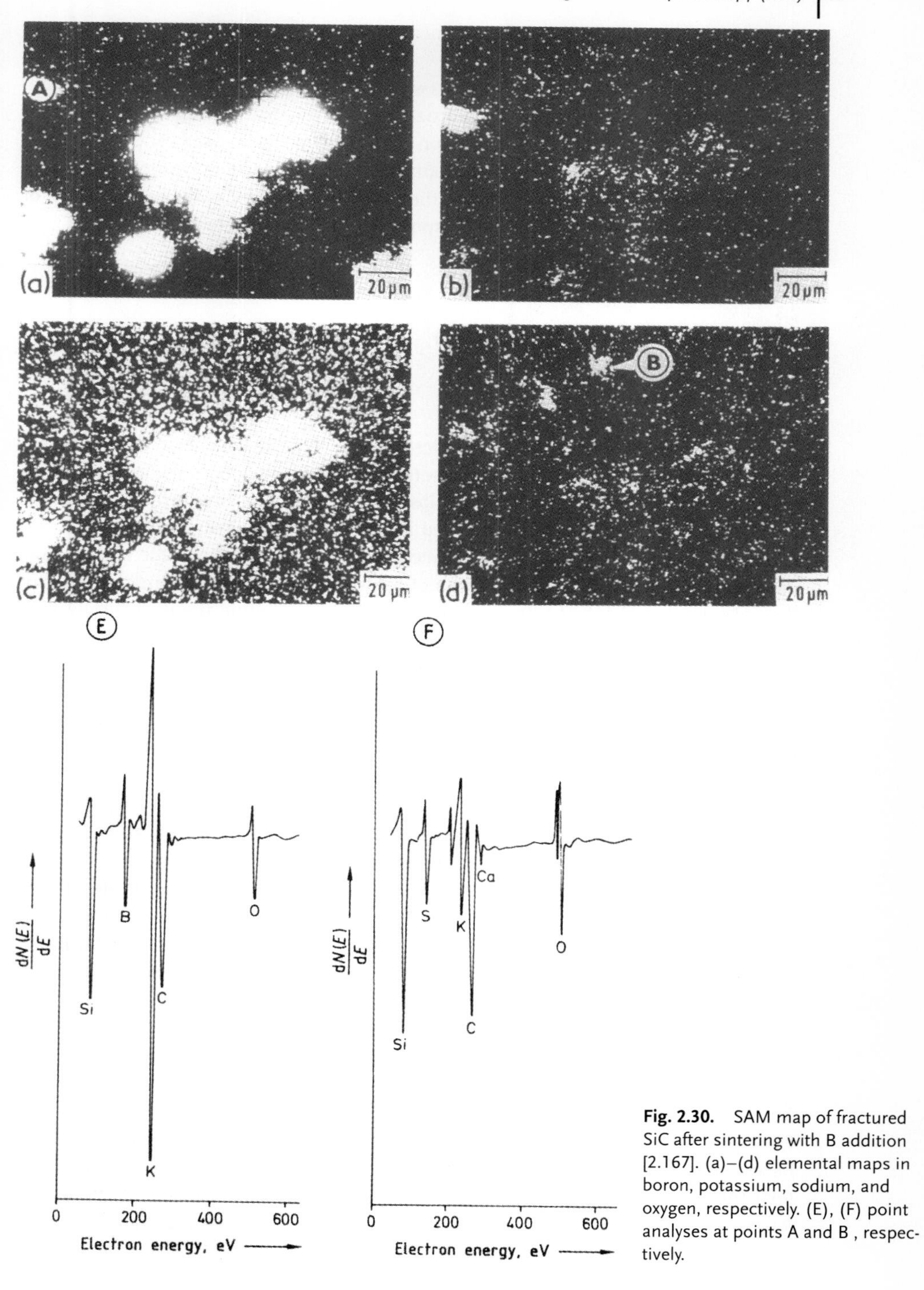

Fig. 2.30. SAM map of fractured SiC after sintering with B addition [2.167]. (a)–(d) elemental maps in boron, potassium, sodium, and oxygen, respectively. (E), (F) point analyses at points A and B , respectively.

cium in some areas also. In this instance the sintering process had not been optimized.

Dedicated scanning auger microscopes with ultimate spatial resolution have been developed towards smallest possible spot size and still acceptable Auger signal-to-noise ratio obtained within tolerable acquisition time. Electron guns used for SAM employ electromagnetic focusing. Because focusing electrons of low energy into a small spot is difficult, energies used in SAM are in the range of 25–50 keV, with beam currents of the order of 1 nA or less and minimum spot size of approximately 15 nm.

2.3
Electron Energy-loss Spectroscopy (EELS)

Reinhard Schneider

When, in a transmission electron microscope (TEM), high-energy electrons are transmitted through a thin specimen there is a probability of both inelastic and elastic scattering, i. e. electrons can always lose energy when passing through a sample. This probability is given by the so-called inelastic scattering cross-section which essentially depends on the energy of the electrons (and, therefore, on the high voltage used to accelerate them), the thickness of the specimen, and the atomic number of the atom excited [2.168–2.170]. In conventional TEM imaging the inelastically scattered electrons are not wanted, because they strike the image plane away from the correct image point (defocusing); this results in a diffuse background and reduced image contrast. These electrons, having lost energy, can, however, be used to obtain detailed information about the chemical composition of the material through which they are transmitted – its electronic structure, the chemical bonding of the elements inside, and their nearest-neighbor distances.

The physical method used to measure the energy loss of electrons scattered into a specific angular range is electron energy-loss spectroscopy (EELS). It can be used in analytical transmission electron microscopy (AEM) for chemical analysis at high lateral resolution, but also in reflection arrangements in surface-science physics. In the latter a specimen surface is bombarded with electrons of relatively low energy (only some 100 eV up to approximately 3 keV) and the energy and angular distribution of the reflected beam are analyzed. The instrumentation necessary for reflection EELS (REELS) is usually combined with Auger electron spectroscopy (Sect. 2.2), because an Auger spectrometer can usually also be operated as an energy-loss spectrometer.

The following discussion is restricted to EELS in combination with TEM and scanning transmission electron microscopy (STEM), and only a brief overview on the fundamentals, experimental technique, and the information gained from EEL spectra is given. For a thorough understanding of the theoretical background and peculiarities of the EELS experiment, and its application, further reading of appropriate textbooks [2.171–2.176] and original papers is recommended. Also, to simplify matters some experience of electron microscopy is assumed, and such details can be read elsewhere [2.173–2.177].

It should be noted that for TEM at accelerating voltages of 100–400 keV the specimen thickness must be of the order of 10–100 nm which requires dedicated preparation techniques [2.173, 2.176, 2.178].

2.3.1 Principles

In electron-optical instruments, e.g. the scanning electron microscope (SEM), the electron-probe microanalyzer (EPMA), and the transmission electron microscope there is always a wealth of signals, caused by the interaction between the primary electrons and the target, which can be used for materials characterization via imaging, diffraction, and chemical analysis. The different interaction processes for an electron-transparent crystalline specimen inside a TEM are sketched in Fig. 2.31.

Often, the specimen is illuminated by means of a focused electron probe to obtain the signals from different regions. Owing to the crystalline structure diffracted beams also accompany the directly transmitted beam. The energy of the transmitted electrons can be analyzed by means of an electron energy-loss spectrometer and the electrons which have suffered losses yield chemical information. Likewise, Auger electrons generated as secondary particles of the inelastically scattered primary electrons are important for microanalysis (Sect. 2.2). Secondary electrons that escape from the surface after elastic or inelastic scattering enable the imaging of the topography if the electron probe is scanned across the specimen, and elastically back scattered electrons yield images showing materials contrast. Also photonic signals are excited, including X-rays, in a process complementary to the emission of Auger electrons, and cathodoluminescence (CL), because of corresponding electronic transitions.

In a closer view (Fig. 2.32), the following interactions occur on an atomic scale when a material is hit by electrons. Firstly, in addition to elastic scattering, inelastic scatter-

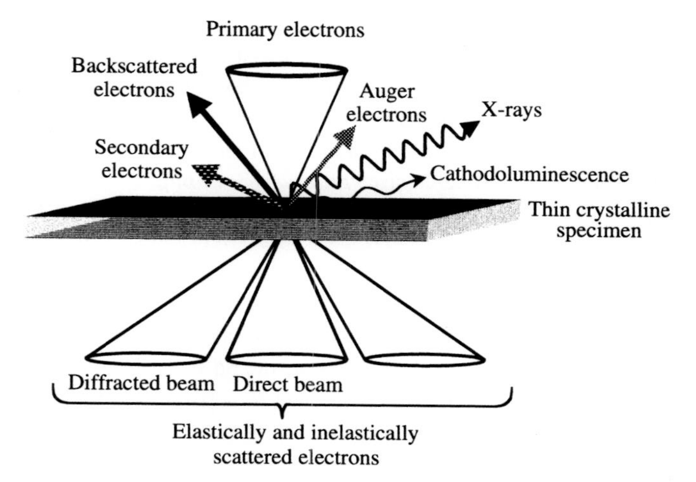

Fig. 2.31. Schematic diagram of the interactions between high-energy electrons and matter in TEM.

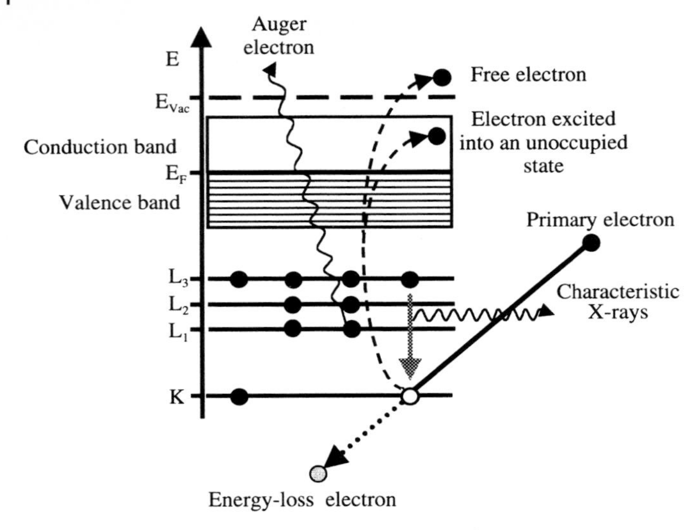

Fig. 2.32. Atomistic view of interaction processes between incident high-energy electrons and electrons of an individual atom.

ing can also be observed as a result of direct interaction between the primary electrons of energy E_0 and electrons in inner shells and the valence or conduction band of the atoms in a solid. The energy transferred to the excited electron can be measured as an energy loss, E, of the transmitted electron, which reduces its kinetic energy to $E_0 - E$. The tightly bound core electrons or the more loosely bound valence electrons can be excited to higher unoccupied energy states in the conduction band, or even be set free into the vacuum. Thus, a hole is generated and the atom becomes ionized. After a certain dwell time of excitation the system relaxes and the hole state in the originally excited level is filled with an electron from an outer shell. This process is accompanied by secondary processes – the emission either of Auger electrons or X-rays, which yield additional information about the chemical composition of the specimen under investigation. A full description of all the imaginable interaction processes is possible only by quantum-mechanical physics, and will not presented here (details are available elsewhere [2.168, 2.169]).

2.3.2
Instrumentation

Entirely different electron spectrometers have been constructed to be used as energy analyzers combined with TEM. But, owing to chromatic aberration, even an electron-optical round lens could be used to separate electrons passing through it. Because of its small dispersion this is of no practical importance. Well-known types of energy analyzers are the electrostatic Möllenstedt energy analyzer [2.179], the Wien filter [2.180, 2.181], based on the combined use of crossed electrostatic and magnetic fields, single magnetic sector fields or prisms [2.182–2.187], resp., magnetic prisms combined with an electron mirror (Castaing–Henry filter) [2.188–2.191], and systems employing magnetic prisms, e.g. the Ω filter [2.192–2.196] and the so-called Mandoline filter [2.197].

Electrostatic filter systems, including those with electron mirrors, have disadvantages compared with the magnetic variety, predominantly because of the need for high-voltage connections and the possible contamination of electrodes, both of which lead to instability. In contrast with that, a current of amperes only is needed to build up a magnetic field acting as energy analyzer, because of the Lorentz force. In addition, the electron-optical properties of magnetic sector fields have been studied intensively [2.174, 2.198], and their imaging aberration can often be corrected without great difficulty; they are also simpler to construct. Magnetic energy analyzers are, therefore, commonly applied to EELS in combination with TEM/STEM.

The literature contains many publications on self-made magnetic EEL spectrometers [2.182–2.196], and such systems have also been commercially available for many years. Several companies have developed such spectrometers, but the 90° magnetic prisms of the firm Gatan (parallel-detection EELS model 666 and the new Enfina model) are the only ones recently offered for TEM/STEM. Because they can easily be attached to a transmission microscope below its camera chamber, they are widely used.

In general, a magnetic sector field has electron-optical properties comparable with those of a round lens, i.e. first-order imaging of a spectrometer object point into a point in the image plane is realized. The spectrometer object can be a cross-over of the incident beam or an intermediate image of the ray-path in the TEM. In the image plane, however, electrons differing in their energies are separated owing to the dispersive action of the homogeneous magnetic field. In former times, the EEL spectra generated in this way were sequentially recorded by scanning them across a slit. This can be done by continuously increasing the excitation of the magnet or using an additional magnetic dipole field. To correct imaging aberrations, which must not neglected because they particularly determine the energy resolution achievable, the best solution is to increase the accelerating voltage of the microscope. Hence, when the electron optics is well aligned for the primary electrons the path of the energy-loss electrons is also optimized. With the development of detectors with local sensitivity, e.g. photodiode arrays or charge-coupled devices (CCD), parallel recording of spectra has become feasible and respective spectrometers are called parallel-detection EELS (PEELS). Because a PEELS gathers the whole energy-loss spectrum simultaneously it is much more efficient than serial-detection EELS (SEELS). Thus, either spectra of better signal-to-noise ratio (SNR) can be recorded or the acquisition time can be kept short, which is of particular importance for electron-sensitive materials, e.g. polymers and organic samples.

An example of a serial-recording EEL spectrometer is shown in Fig. 2.33; it features a magnetic prism system which was constructed for a TEM/STEM of the type JEOL JEM 100S [2.199, 2.200]. Its second-order aberrations are corrected by curved pole-piece boundaries, an additional field clamp, and two extra hexapoles acting as stigmators. The electron beam can be adjusted relative to the optical axis by use of several deflection coils. A magnetic round lens is positioned just in front of the prism to

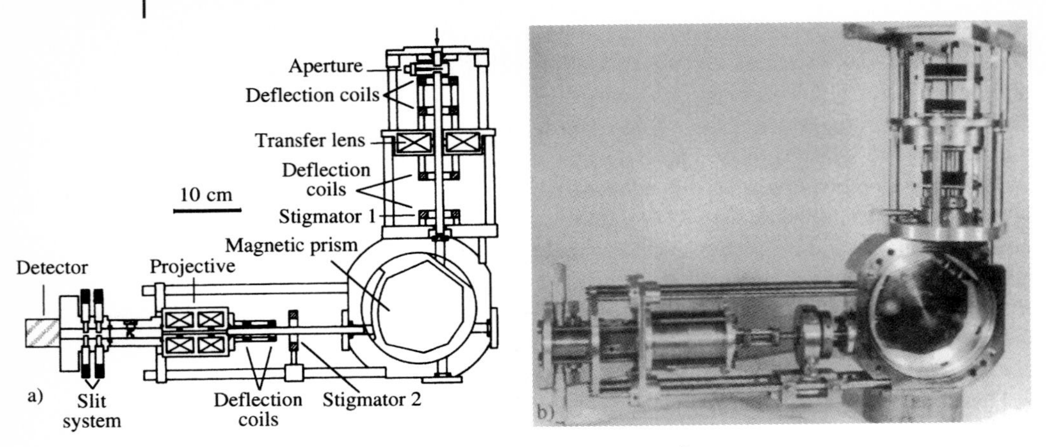

Fig. 2.33. Self-constructed magnetic-prism spectrometer for a TEM/STEM: (a) schematic diagram of set-up; (b) photograph of the system with the prism opened.

adapt to the different imaging and diffraction modes of the microscope. The energy analyzer is followed by a double-projective lens providing a rotation-free magnified image of the spectrum. A scintillator–photomultiplier combination is used as electron detector and EEL spectra are recorded serially by increasing the high voltage of the TEM stepwise. The detector output signal is recorded by means of a multichannel analyzer or personal computer for subsequent processing. Operation of a thermionic electron source results in an energy resolution of approximately 1.5 eV.

In principle, energy-analyzer systems can be designed such that their electron-optical properties do not limit the energy resolution attainable, i. e. their intrinsic energy resolution is much better than the energy width of the primary electron beam, which is of the order of approximately 1.5–2.5 eV for a tungsten hairpin cathode, approximately 1 eV for a LaB_6 cathode, approximately 0.7 eV for a Schottky field emitter, and 0.3–0.5 eV for a pure cold-field emitter.

With regard to the instrumentation, some of the energy-analyzer systems mentioned above, viz. the Castaing–Henry filter, the Ω filter and the Mandoline filter, are not only spectrometers, but also image-forming systems enabling energy-filtered imaging or diffraction on a TEM, i. e. with stationary illumination of the specimen. Thus, they enable the formation of a two-dimensional image or diffraction pattern using energy-loss electrons within a specified energy range. In addition to the contrast enhancement obtained by filtering out all the inelastically scattered electrons, this also enables imaging of element distribution (cf. Sect. 2.5.5). Nevertheless, for a STEM equipped solely with an EEL spectrometer, energy-filtered imaging is possible simply by recording the electrons of a certain energy loss passing the slit while scanning the electron probe over the specimen in a rectangular field.

All the energy-filter systems referred to above are incorporated into the electron-microscopic column, usually between the diffraction lens and the following intermedium lens. They are, therefore, known as in-column filters, which are also commer-

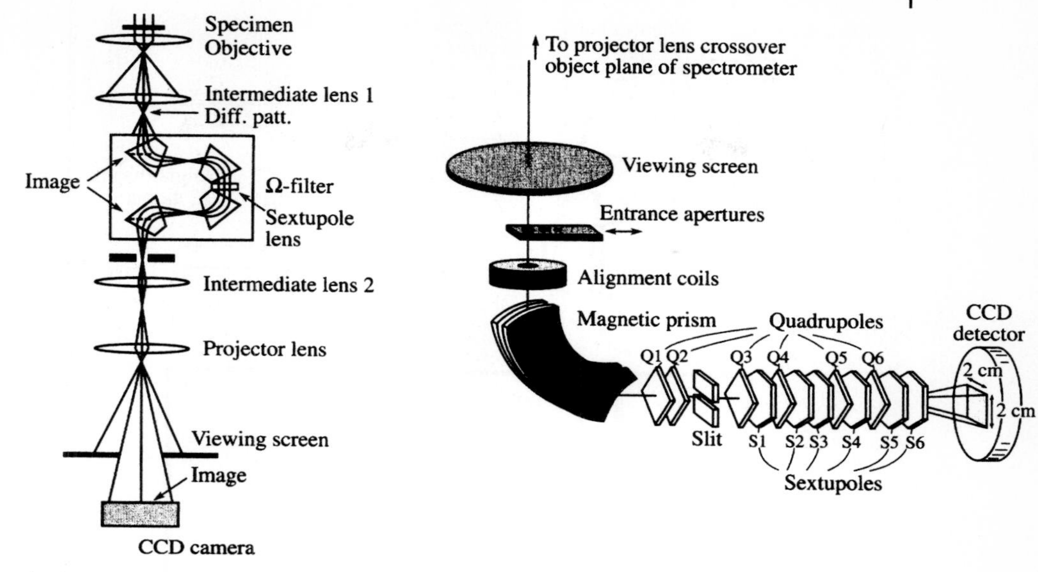

Fig. 2.34. Arrangement of post-column and in-column imaging energy filters [2.173].

cially available in the form of different TEMs, e.g. as the Castaing–Henry filter in the Zeiss EM 902 and as Ω filters in the LEO EM 912, EM 922 (cf., e.g. Ref. [2.201]) and in the Jeol JEM 2010 FEF and 3010 FEF [2.202]. The operation of only a single magnetic prism, which is not part of the column, can, however, also be extended to energy-filtered TEM. A post-column filter is offered by Gatan; it is the well-known and widespread Gatan imaging filter (GIF) attachable to any TEM/STEM. Both concepts, i.e. the in-column and the post-column filters, are compared schematically in Fig. 2.34. It should be noted that the post-spectrometer optics of the GIF consists of multipoles only, viz. quadrupoles, preferably, for imaging and hexapoles for correcting image aberrations. The in-column filter drawn in Fig. 2.34 could be that of the EM 922 Omega or JEM 2010 FEF, respectively.

2.3.3 Qualitative Spectral Information

The EEL spectrum measured is the plot of the intensity of the inelastically scattered electrons as a function of energy loss at a certain spectrometer collection angle, which is defined by an aperture in front of the energy analyzer. Typically, the spectrum can be divided into two regions – the low-loss region ranging from 0 eV to approximately 50 eV and the high-energy region lying beyond that (50 to approximately 2000 eV). The upper limit of energy loss is determined by the signal intensity, which generally decreases exponentially with increasing energy loss. It can be much higher than 2 keV, particularly when a PEELS is used, because of its better detection effi-

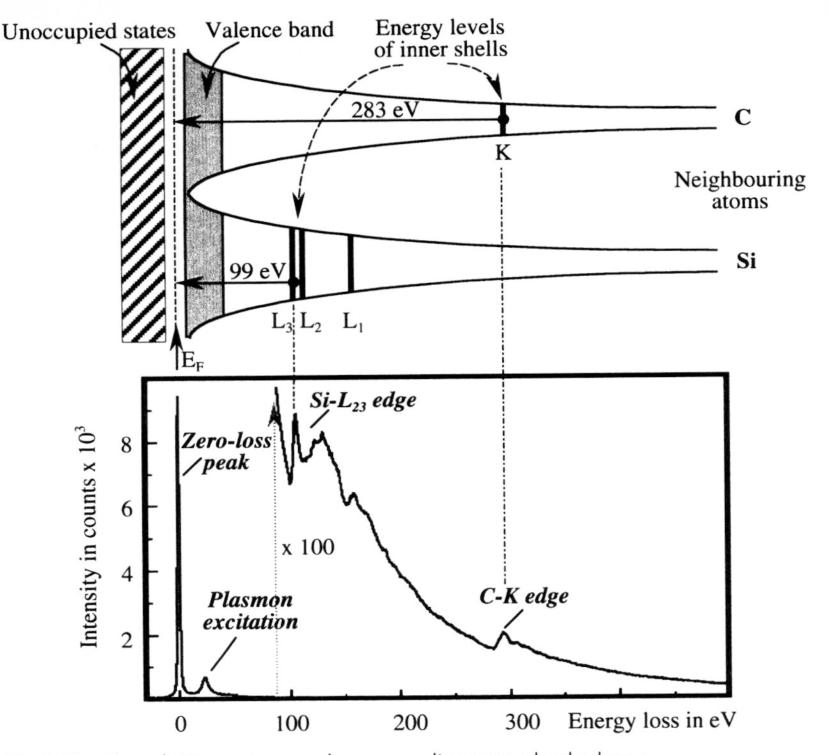

Fig. 2.35. Typical EEL spectrum and corresponding energy-level scheme.

ciency. The characteristic losses giving information on the elemental composition arise from ionization of inner shells. They appear as edges superimposed on a background caused by low-loss excitations (plasmons, single-valence electron excitations) or by the tail of an ionization edge of lower energy. The reason for the appearance of edges instead of peaks is that usually not only the threshold energy is transferred to the inner-shell electrons but also some additional kinetic energy.

In Fig. 2.35 a spectrum of silicon carbide is shown as an example of a typical EEL spectrum; the spectral features are correlated to the electronic structure of this semiconductor material. In the low-loss region besides the zero-loss peak of highest intensity there is only one relatively sharp plasmon peak at approximately 21 eV. On its high-energy side the signal intensity drops dramatically. To make visible the characteristic ionization edges following in the region of higher losses, viz. the $Si–L_{23}$ and C–K edges, the vertical scale must be expanded by a factor of a few hundred. Owing to the high energy resolution of approximately 1 eV fine structures can be observed near the edge onset, especially for the $Si–L_{23}$ edge. Such energy-loss near-edge structures (ELNES) occur because the individual inner-shell electron can be excited from its ground state into unoccupied states in the conduction band (cf. Sect. 2.5.3.3.1).

2.3.3.1 Low-loss Excitations

In the low-loss region the predominant feature is the zero-loss peak containing itself no useful analytical information. It is also termed the elastic peak, because mainly elastically scattered electrons contribute to it, and those transmitted by the thin specimen without interaction. Strictly speaking, however, the term zero-loss peak is not correct, because electrons of very low energy loss (ca. 100 meV) caused by the excitation of phonon vibrations are also part of that peak. On account of the energy width of the primary electrons and of the finite energy resolution of the spectrometer system used they are not resolved on the high-energy tail of the peak. The only way to detect them is to adapt a monochromator to the illumination system of the TEM for reducing the primary energy width. Commonly, the full width at half maximum (FWHM) of the zero-loss peak is accepted as a measure of the energy resolution of an EEL spectrometer.

The region from approximately 5 to 50 eV contains mostly Gaussian-shaped peaks, which might be of different origin [2.203]. For materials with free electrons, e.g. metals, they clearly come from collective oscillations (plasmons) of these electrons. If, however, the specimen is composed of a material in which free electrons are not abundant, there is an alternative mechanism involving the transfer of energy to single valence electrons. Thus, this effect of single-electron excitation is rather important for solids such as semiconductors and insulators, which also generate plasmon-like peaks in the 20–30 eV range. In addition, interband transitions can also cause extra resonances in that energy region.

Plasmon oscillations are not only excited in the bulk of the specimen but also on the surface. The proportion of surface plasmons increases when the specimen becomes thinner and their excitation energy is approximately half that of the corresponding volume plasmon. Analyzing the plasmon energy, E_{pl}, is a local probe of the density of free electrons, N_e, because of the dependence $E_{pl} \propto N_e^{1/2}$ [2.204]. In special cases this phenomenon can be applied to materials analysis. This has been demonstrated, e.g., on Al–Mg and Al–Zn alloys [2.205, 2.206] for which the displacement of plasmon energy enables phase identification. More recently, the region close to the right of the zero loss peak has become of increasing interest in so-called band-gap spectroscopy [2.207], which directly yields data on electronic properties as, e.g., shown for (In,Ga,Al)N compounds [2.208].

When inelastic electron scattering is treated in terms of particle collisions the individual events are independent and obey Poisson statistics. Because the number of scattering processes increases with the specimen thickness, t, the incident electron can undergo multiple losses. Thus, the thicker the specimen, the more interactions generating plasmon oscillations the incident electron will have, which results in the occurrence of several equidistant peaks of multiples of the plasmon energy. The dependence of plural scattering on specimen thickness can be used to measure the thickness in the region illuminated by the electron beam. It can be shown that the relative thickness t/λ is given by

$$\frac{t}{\lambda} = \ln\left(\frac{I_{tot}}{I_0}\right) \tag{2.18}$$

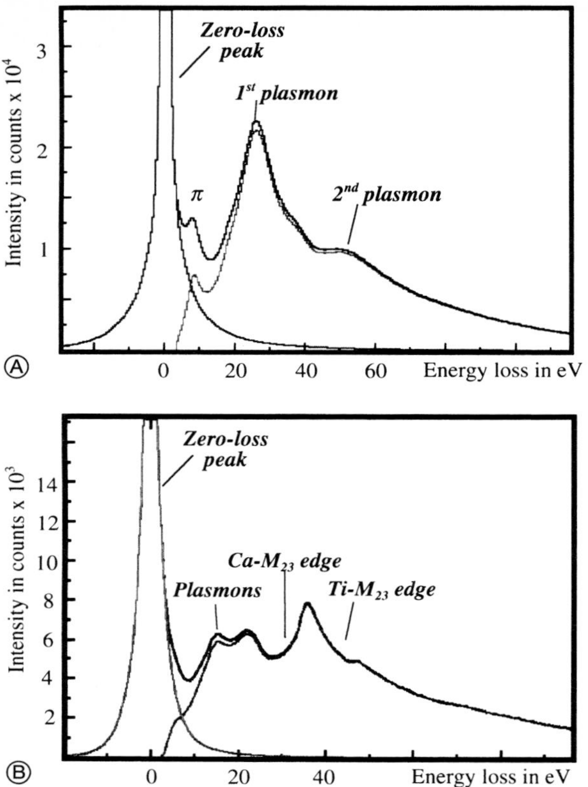

Fig. 2.36. Low-loss spectra of: (A) (BN), and thickness determination, and (B) $CaTiSiO_5$ with outer-shell ionization edges.

where λ is a total mean free path for all inelastic scattering, I_0 is the area integrated under the zero-loss peak, and I_{tot} that under the whole spectrum recorded up to approximately 200 eV. The total inelastic mean free path, λ, must be calculated to determine the absolute thickness, t. Corresponding λ values can be computed when the composition of the material is known [2.171]. A rough estimate is otherwise possible via the formula λ (in nm) = 0.8 E_0, where E_0 is the primary electron energy in keV.

Figure 2.36 A shows a typical low-loss spectrum taken from boron nitride (BN). The structure of BN is similar to that of graphite, i. e. sp^2-hybridized carbon. For this reason the low-loss features are quite similar and comprise a distinct plasmon peak at approximately 27 eV attributed to collective excitations of both π and σ electrons, whereas the small peak at 7 eV comes from π electrons only. Besides the original spectrum the zero-loss peak and the low-loss part derived by deconvolution are also drawn. By calculating the ratio of the signal intensities I_{tot} and I_0 a relative specimen thickness t/λ_{Pl} of approximately unity was found. Owing to this specimen thickness there is slight indication of a second plasmon.

That even low-loss spectra can be complex in appearance is recognizable from that of the silicate $CaTiSiO_5$ given in Fig. 2.36 B. Here the problem arises of superposition of regular plasmon or single-valence electron excitations, and excitation of tightly bound outer-shell electrons with extremely low binding energy. Respective ionization edges are the Ca–M_{23} edge at 25 eV and the Ti–M_{23} edge at 35 eV which appear partly overlapped. The peaks at about 15 and 22 eV are presumably caused by interband transitions and plasmons, respectively.

It should be noted that low-loss spectra are basically connected to optical properties of materials. This is because for small scattering angles the energy-differential cross-section $d\sigma/dE$, in other words the intensity of the EEL spectrum measured, is directly proportional to $\mathrm{Im}\{-1/\varepsilon(E,\mathbf{q})\}$ [2.171]. Here $\varepsilon = \varepsilon_1 + i\varepsilon_2$ is the complex dielectric function, E the energy loss, and $\mathbf{q}$ the momentum vector. Owing to the comparison to optics ($|\mathbf{q}| = 0$) the above quoted proportionality is fulfilled if the spectrum has been recorded with a reasonably small collection aperture. When $\mathrm{Im}\{-1/\varepsilon\}$ is gathered its real part can be determined, by the Kramers–Kronig transformation, and subsequently such optical quantities as refraction index, absorption coefficient, and reflectivity.

2.3.3.2 **Ionization Losses**

In general, EELS is particularly sensitive to light elements, because the ionization cross-sections increase with decreasing atomic number. Because of this energy-loss spectroscopy and energy-dispersive X-ray spectroscopy (EDXS) complement each other reasonably well, since the detection sensitivity of EDXS behaves vice versa (Sect. 4.2). Most of the transmitted electrons, having suffered energy losses via ionization and plasmon excitations, can, moreover, be readily collected owing to their strong forward scattering [2.173]. They are scattered into an angular range of typically 10^{-2} rad.

Thus, the features most important of materials analysis are the inner-shell edges, because the ionization energies are characteristic of the chemical elements. Nearly all elements can principally be detected and in the energy range up to 2 keV there appear series of K edges for Li to Si, of L_{23} edges for Al to Sr, and of M_{45} edges for Rb to Os [2.209]. The specific shape of the edge varies with the edge type (K, L, M, etc.) and strongly depends on the electronic structure and chemical bonding. Usually, K shells arising from 1s to 2p transitions have a so-called hydrogenic-like shape which is characterized by a steep signal increase at the threshold energy. L-shell edges ($p \rightarrow s$ and $p \rightarrow d$ transitions) occur whether as rounded profiles or nearly hydrogenic-like with sharp, so-called white lines at the edge onset. The latter can be observed for elements of atomic number $19 \leq Z \leq 28$ and $38 \leq Z \leq 46$. In EELS the term "white line" is used by analogy with X-ray absorption spectroscopy, where the features were historically first observed as severe bright lines on photographs. M edges tend to appear as delayed signals, or they also show up as white lines when empty d or f states are present.

An example of a high-energy loss spectrum is given in Fig. 2.37; it is an EEL spectrum in the range from 400 to 900 eV taken from a thin region of a $BaTiO_3$ particle at 200 keV primary energy. Ionization edges of all three constituting elements are visible, viz. the Ti–L_{23}, the O–K, and the Ba–M_{45} edges. The K edge of oxygen is saw-tooth-shaped, but the two others have more or less well resolved peaks near the edge onsets, indicating individual energy levels (L_3 and L_2 for Ti, M_5 and M_4 for Ba). Parti-

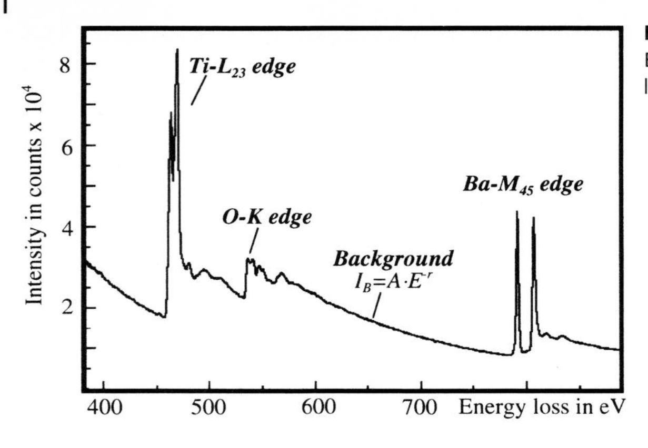

Fig. 2.37. EEL spectrum of $BaTiO_3$ exhibiting ionization losses.

cularly the Ba–M_{45} edge demonstrates the effect of white lines caused here by the excitation from initial 3 d states to unfilled 4 f states. In general, besides the gross edge profiles pronounced fine structures can also occur in the region of the very edge onset up to approximately 30 eV beyond it. These so–called energy-loss near-edge structures (ELNES) are caused by solid-state effects (cf. Sect. 2.3.3.3). In Fig. 2.37 the O–K edge, in particular, shows such ELNES details, and additional weak modulations of the signal intensity reaching some 100 eV above the edge threshold can also be seen. These oscillations are extended energy-loss fine structures (EXELFS) giving information on the number of nearest neighbors and their distances.

Each inner-shell edge has a high-energy tail which is the background signal on which the next edge is superimposed. In front of the first edge occurring in an EEL spectrum the background contribution comes from low-loss excitations of relatively high intensity. Thus, several different interaction phenomena give rise to the background signal and it is, therefore, impossible to calculate it *ab initio*. It was, however, experimentally proved that for a limited energy range the background intensity, I_B, can be fitted by a power law:

$$I_B = A\,E^{-r} \tag{2.19}$$

where E is the energy loss and A and r are constants. Both fit parameters vary with the mass thickness of the material investigated, the accelerating voltage, and the acceptance angle of the spectrometer. Typical values of the exponent r range from 2 to 5, whereas A can vary dramatically.

EELS analyses can be complicated or even fail if the specimen is so thick that intense plural scattering occurs. For spectra recorded from extremely thick regions (some 100 nm) the characteristic ionization edges vanish in the high background. Thus, the inner-shell edge measured often does not represent single but multiple scattering resulting from combined excitations of inner-shell electrons and valence electrons. This behavior is illustrated schematically in Fig. 2.38, where the effect of the convolution of a hydrogenic-like edge with the low-loss spectrum is shown. The single-scattering curve of an edge can be gained from the EEL spectrum measured

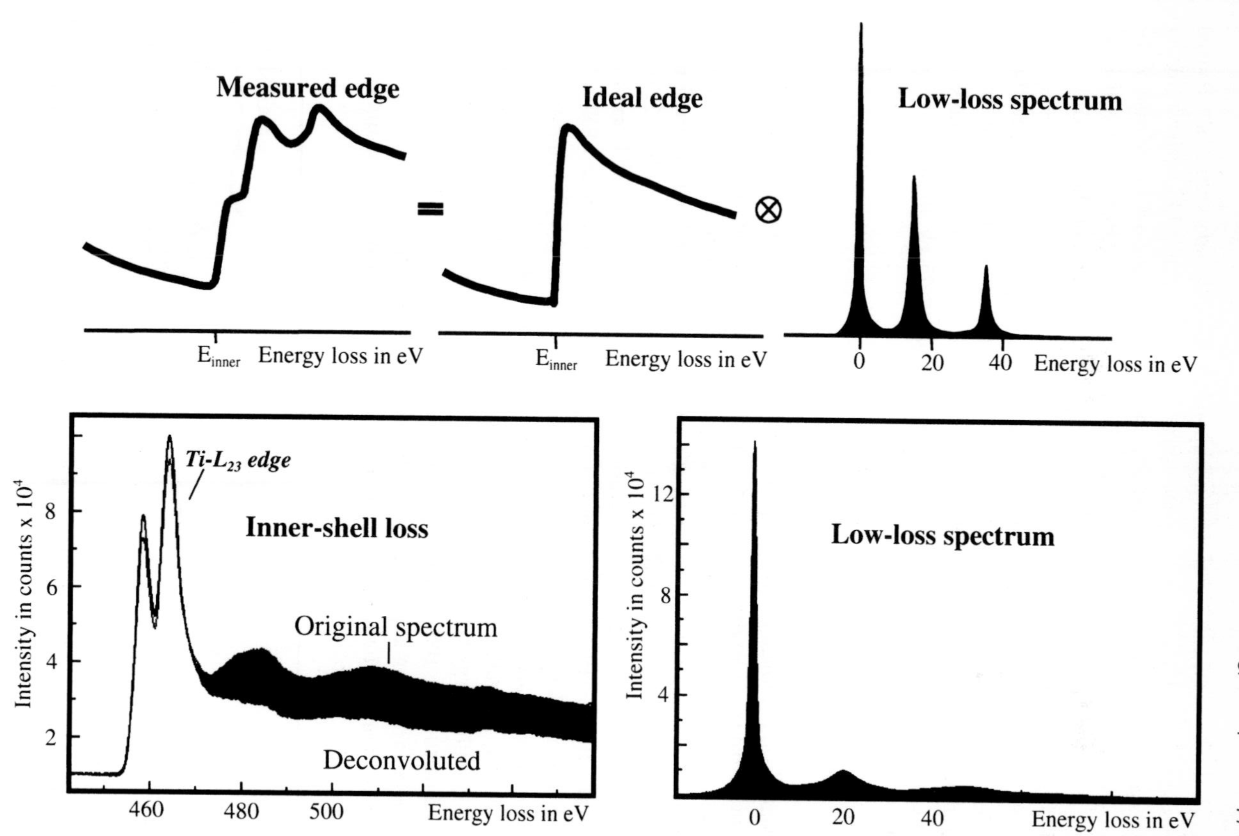

Fig. 2.38. Convolution of inner-shell losses and low-loss features, schematically (above) and as observed experimentally (below) for the Ti-L_{23} edge.

by deconvoluting it by means of the corresponding low-loss spectrum. Fast Fourier routines are preferably applied to spectrum processing, because EEL spectra involve many data points equally spaced in energy. Deconvolution is possible by use of the Fourier–log or Fourier–ratio method, and by multiple least-squares fitting which are described in more detail elsewhere [2.171, 2.173]. An example of the determination of the single-scattering profile via the Fourier–ratio method, viz. for the Ti–L_{23} edge taken from a thin Ti film, is given in the lower part of Fig. 2.38. On the left the original measured and deconvoluted spectra are compared. It is obvious that the near-edge region remains almost unaffected by multiple scattering. There is, however, a strong contribution of plural scattering starting at about 15 eV above the edge onset.

2.3.3.3 Fine Structures

The occurrence of fine structures has already been noted in the sections on spectral information and ionization losses (Sects. 2.5.3 and 2.5.3.2). In the following text some principal considerations are made about the physical background and possible applications of both types of feature, i.e. near-edge and extended energy-loss fine structures (ELNES/EXELFS). A wealth of more detailed information on their usage is available, especially in textbooks [2.171, 2.173] and monographs [2.210–2.212].

In Fig. 2.39 A the Al–K edge (onset at approximately 1560 eV) recorded from Al_4C_3 is shown, and the specific energy ranges of the appearance of ELNES and EXELFS are marked. In the ELNES region reaching from the edge onset up to approximately 30 eV above there are only two distinct peaks, visible at 1566 and 1578 eV. EXELFS which are much weaker in intensity start at approximately 1590 eV and extend over the whole high-energy tail of the edge shown. Usually, EXELFS modulations can be observed up to some 100 eV beyond the ionization threshold.

Near-edge structures superimposed on the general atomic edge profile are caused by transitions of inner-shell electrons of a particular atom to the lowest unoccupied

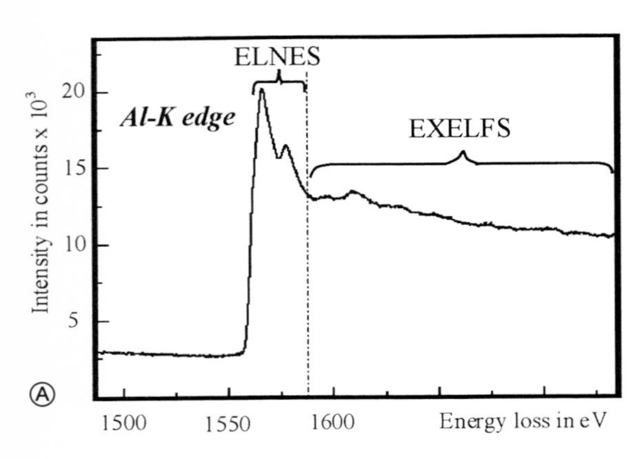

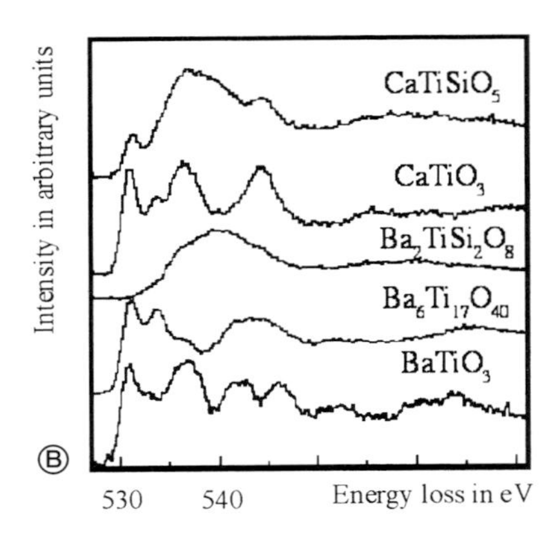

Fig. 2.39. Fine structures of inner-shell edges: (A) Al-K edge of Al_4C_3 exhibiting ELNES and EXELFS, (B) O–K ELNES of different titanates and Ti-containing silicates.

states in the solid. Thus, ELNES probes the density of empty states above the Fermi level at the site of the excited atom. This local density of states (DOS) is related to the atom excited and also to the arrangement, number, and types of atoms in its environment determining the charge transfer and chemical bonding. All these peculiarities influence such ELNES details as the number, energy width, and intensity of individual peaks, and also the threshold energy of the ionization edge. It is mainly the first coordination shell which affects the ELNES features observed. Hence, near-edge structures are a so-called coordination fingerprint which for combination of EELS and STEM/TEM enables analysis of the chemical bonding of a particular compound with high lateral resolution. A practical way of phase identification by ELNES analysis is to compare the fine-structure features recorded from the unknown phase with those measured from reference materials of known structure and stoichiometry (fingerprint method). The literature contains a multitude of data on the several materials systems available; the EEL spectra published in reference books [2.209, 2.213] are also helpful in this context.

It should be noted that a comprehensive ELNES study is possible only by comparing experimentally observed structures with those calculated [2.210–2.212]. This is an extra field of investigation and different procedures based on molecular orbital approaches [2.214–2.216], multiple-scattering theory [2.217, 2.218], or band structure calculations [2.219, 2.220] can be used to compute the densities of electronic states in the valence and conduction bands.

Because near-edge structures reflect the nearest chemical neighborhood of an atomic species in a solid for single-crystalline materials they can also depend on the orientation of the specimen. A classical example is graphite, for which the height of the prepeak of the C–K edge caused by transitions from 1s states into unoccupied π^* states is strongly orientation-dependent [2.221]. The π^* peak intensity reaches a maximum when the incident electron beam is parallel to the *c*-axis of the graphite. Because of the structural similarity of boron nitride to graphite its B–K edge behaves in the same manner (cf. Fig. 2.41 a). Thus, such orientation dependence yields additional information on the anisotropy of chemical bonding and band structure. If, however, these effects must be avoided it is recommended that ELNES data be collected from polycrystalline, or even amorphous, samples, if available, of the material under investigation.

Neglecting theoretical treatment of ELNES analyses, practical application result is presented in Fig. 2.39 in which the O–K ELNES of different titanates and Ti-containing silicates are compared [2.222]. The fine structures shown were acquired from reference materials at an energy resolution of approximately 0.8 eV using a PEELS (Gatan model 666) attached to a TEM/STEM Philips CM20 FEG running at 200 keV with a thermally assisted field-emission gun. All ELNES features were obtained from the raw data after sharpening the resolution (deconvolution with the zero-loss peak), subtracting the background (power-law fit), and subsequently deconvoluting with the low-loss spectrum to obtain the single-scattering profile. O–K ELNES is extremely sensitive to nearest-neighbors surroundings, which is also known from other titanates [2.223]. The $Ba_6Ti_{17}O_{40}$ phase can be distinguished from $BaTiO_3$ by its near-edge details at the very edge onset and by those lying in the range from 540 to 550 eV. The O–K ELNES of the silicate fresnoite ($Ba_2TiSi_2O_8$) is most different – its onset is shifted to higher energies and delayed. There is, moreover, only one broad maximum at approximately 542 eV. Such pronounced differences in the ELNES can be used for phase identification, as is demonstrated in Fig. 2.40. Fig. 2.40 a shows a TEM bright-field image of a $BaTiO_3$ ceramic processed via liquid phase sintering at approximately 1350 °C with an excess of 1 mol% TiO_2. Between two $BaTiO_3$ grains an unknown phase was found, where EELS (cf. Fig. 2.40 b) clearly indicates the presence of $Ba_2TiSi_2O_8$. In the series of spectra recorded along a line of approximately 50 nm length in the STEM mode the local transition from $BaTiO_3$ to $Ba_2TiSi_2O_8$ is clearly visible from the change of the O–K ELNES. Differences in the Ti–L_{23} edge are also apparent. The $Ba_2TiSi_2O_8$ detected is presumably a reaction product introduced by the impurity SiO_2.

Extended energy-loss fine structures (EXELFS) are analogous to the EXAFS effect observed in X-ray absorption [2.224, 2.225]. These weak modulations (cf. Fig. 2.39 a), still observable ca. 100 eV away from the edge onset, occur because the excited elec-

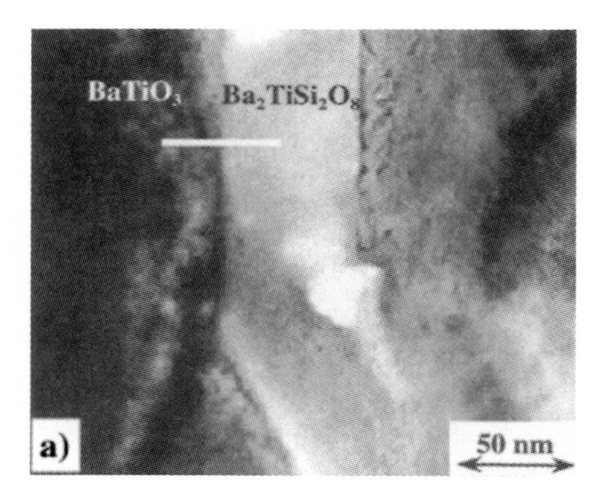

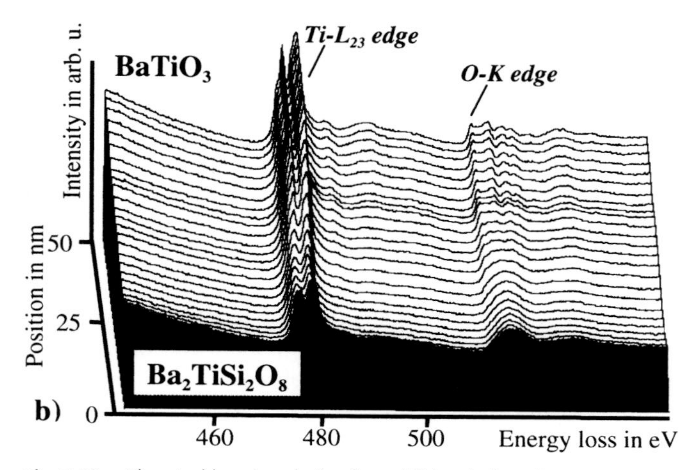

Fig. 2.40. Chemical-bond analysis of an additional phase in a $BaTiO_3$ ceramic using differences in the O-K ELNES.

tron wave has approximately the same wavelength as the atomic spacing of the substance transmitted. Consequently, the wave can be diffracted by neighboring atoms and can return to interfere with the outgoing wave. The interference can be constructive or destructive yielding an oscillating part of the intensity following the ionization edge. Standard EXAFS theory can be used to interpret extended fine structures in EELS [2.226]. Thus, EXELFS analysis enables interatomic distances between the first-neighbor atoms to be measured, yielding the so-called radial distribution function (RDF). For an ideal single crystal the RDF would consist of a series of delta functions which can be attributed to discrete values of shell radii given by the crystal structure. Even bond lengths in a particular direction can be determined when the collection angle of the transmitted electrons and the orientation of the specimen are precisely chosen.

2.3.4 Quantification

In addition to qualitative analysis of nearly all the elements of the periodic table, EEL spectra also enable determination of the concentration of a single element which is part of the transmitted volume and hence gives rise to a corresponding ionization edge. As in all comparable spectroscopic techniques, for quantification the net edge signal, which is related to the number N of excited atoms, must be extracted from the raw data measured. The net intensity I_k of the kth ionization shell of an individual element is directly connected to this number, N, multiplied by the partial cross-section of ionization $\sigma_k(\Delta, \alpha)$ and the intensity I_0 of the incident electron beam, i. e.:

$$I_k(\Delta, \alpha) = N\sigma_k(\Delta, \alpha)\, I_0 \tag{2.20}$$

where Δ is the energy window for integrating the signal intensity and α the collection semi-angle of the spectrometer. Different approaches are used to calculate partial cross-sections. Calculations making use of atomic wave functions are usually based either on a hydrogenic model [2.227, 2.228] or a Hartree–Slater model [2.229]. Neither model, however, can be used to predict effects resulting from electronic transitions to bound states. The Hartree–Slater cross-sections are in very good agreement with experimental findings, although substantial computational effort is required. The results of calculations by use of the hydrogenic model agree with the Hartree–Slater values to an accuracy of better than 20 % which is within the experimental error.

The background signal, I_B, contributing to the kth inner-shell edge must be subtracted from the total energy-loss intensity to obtain the signal, I_k, of the ionization loss itself. As already mentioned in Sect. 2.3.3.2 the background often follows a power law ($I_B = AE^{-r}$). This fit can be used to extrapolate the background in the higher-loss region; for inner-shell losses of approximately 100 eV and below the use of a polynomial fit is sometimes more suitable. The procedure of background extrapolation and subtraction is demonstrated for the B–K edge of boron nitride in Fig. 2.41 a where, in addition to the recorded edge profile, the extrapolated background and the net edge are shown.

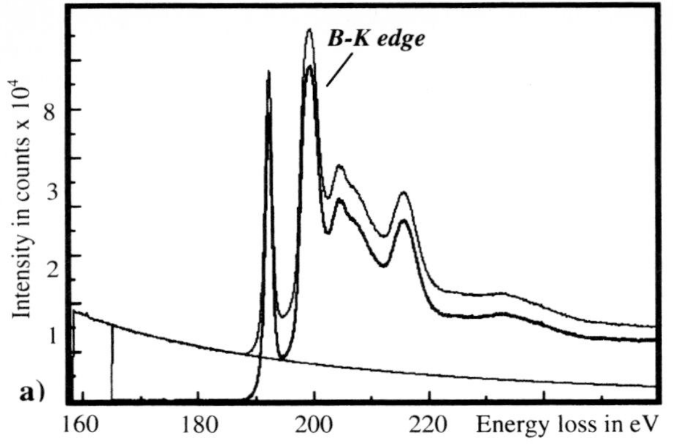

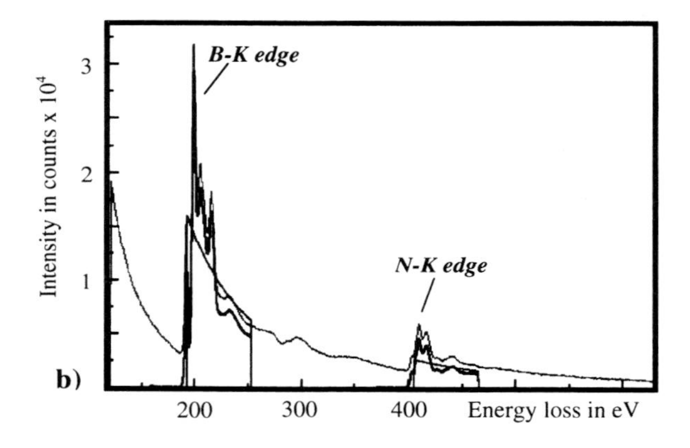

Fig. 2.41. Basic procedure for EELS quantification exercised for BN: (a) background extrapolation and subtraction; (b) determination of net counts in a certain energy window and calculation of corresponding partial cross-sections.

The absolute number of atoms per unit area making the *k*th ionization edge occur is given by:

$$N = \frac{I_k(\Delta, \alpha)}{\sigma_k(\Delta, \alpha) I_0} \,. \tag{2.21}$$

Often only the concentration ratio of two elements A and B is of interest; this is, therefore:

$$\frac{N_A}{N_B} = \frac{I_A(\Delta, \alpha)\,\sigma_B(\Delta, \alpha)}{I_B(\Delta, \alpha)\,\sigma_B(\Delta, \alpha)} \,. \tag{2.22}$$

This kind of estimation of the relative concentration is the most widely used method for quantitative EELS analysis. It is advantageous because the dependence on the primary electron current, I_0, is cancelled out; this is not easily determined in a transmission electron microscope under suitable analytical conditions. Furthermore, in comparison with other methods, e.g. Auger electron spectroscopy and energy-dispersive X-ray spectroscopy, there is no need for additional terms to correct for atomic number effects, specific yield of the interaction process, absorption, etc. Fig. 2.41 b shows the graphical results from a complete quantification procedure for BN, including the determination of the net B–K and N–K edges and the calculation of their partial cross-sections (Hartree–Slater model), yielding a B/N ratio of 0.97 ± 0.16 at $\Delta = 70$ eV and $\alpha = 10$ mrad.

The accuracy achievable by the ratio method amounts to approximately ±5–10 atom% when ionization edges of the same type are used, i.e. only K edges or only L edges, whereas the error in quantification increases to ±15–20 atom% for the use of dissimilar edges. Improvement of the quantification accuracy up to approximately 1 atom% is possible if standards are used.

It is of no use denying that energy-loss spectroscopy cannot usually be used to detect traces of elements in a matrix. The attainable detection sensitivity is highly dependent on the specific elemental composition of the material to be analyzed. For example, the minimum detectable mass fraction amounts to atom% levels for identification of B, N, O, etc., in a thin Si specimen, because the corresponding ionization edges ride on the high-energy tail of the Si–L_{23} edge which limits the sensitivity [2.230]. In contrast with that, it is possible to find approximately 0.1 atom% Li (onset at approximately 55 eV) or Al (approximately 73 eV) in silicon; this corresponds to an absolute number of few thousand atoms. In special circumstances (high primary current of the order of 1.6×10^5 A s cm^{-2}, parallel-recording spectrometer) EELS seems, nevertheless, to be sensitive to a few single atoms – e.g. the detection of Fe atoms in a 10 nm thick carbon film [2.231, 2.232].

2.3.5
Imaging of Element Distribution

The combination of EELS and transmission electron microscopy affords different experimental facilities for imaging of element distributions (see also Sect. 2.3.2) depending on the electron optics of the microscope and the particular type of spectrometer. When the electron microscope can be operated in the STEM mode two-dimensional energy-selective images can easily be gathered by applying the signal of a serial-detection spectrometer to the brightness control of a cathode-ray tube. In general, for an element map three individual images have to be taken – two background images (also termed pre-edge images) and one post-edge image. This procedure is, therefore, called the three-window method [2.171, 2.173, 2.174] and enables determination of the net contribution of an inner-shell edge to image brightness after background extrapolation and subtraction. The image processing necessary is quite similar to the handling of an ionization edge in EEL spectroscopy to obtain the net edge profile. Local differences between the low-loss spectra can, however, also be used for element-specific imaging, e.g. for Al layers on SiO_2 where in cross-section the Al-containing regions appear bright because of a plasmon energy of approximately 15 eV compared with 23 eV for silicon oxide. When a PEELS is run on a STEM the element distribution can be visualized along a line by recording series of spectra (Fig. 2.40 b). For a STEM the lateral resolution is essentially fixed by the diameter of the electron probe used. Thus, sub-nanometer or even atomic resolution can be attained for scanning microscopes equipped with field-emitter electron sources, making the combination of EELS and STEM into a sophisticated means of interface analysis [2.233].

Attachment of an imaging filter to a transmission microscope enables energy-filtered TEM (EFTEM) to be performed; in this technique the method used to generate an elemental map is the same as described above (three-window technique). The only difference is that each of the three individual images is recorded at once while the specimen is illuminated with a parallel stationary electron beam. A corresponding example for mapping the Cr distribution in a Ni-based superalloy SC16 (Ni–68 Cr–18 Mo–2 Ti–4 Al–7 Ta–1) is given in Fig. 2.42. This superalloy is composed of a Cr-rich γ matrix in which cuboidal γ' precipitates (containing approximately 3 atom% Cr) of some 100 nm in size are coherently embedded. In the Cr–L_{23} map the enrichment of chromium amounting to approximately 28 atom% in the matrix region is clearly visible; this can still be recognized by comparing the post-edge and the pre-edge images exhibiting reverse contrast.

EELS line profiles can also be obtained from EFTEM images *a posteriori* by digitally defining a line of interest and representing the element-specific signal along the line chosen. For materials systems having layers of homogeneous thickness the signal can be integrated in the direction parallel to the interfaces, thus yielding an improved signal-to-noise ratio. This is demonstrated in Fig. 2.43 for a cross-section specimen of AlAs/(Al,Ga)As multilayers deposited on GaAs substrate. Here, especially,

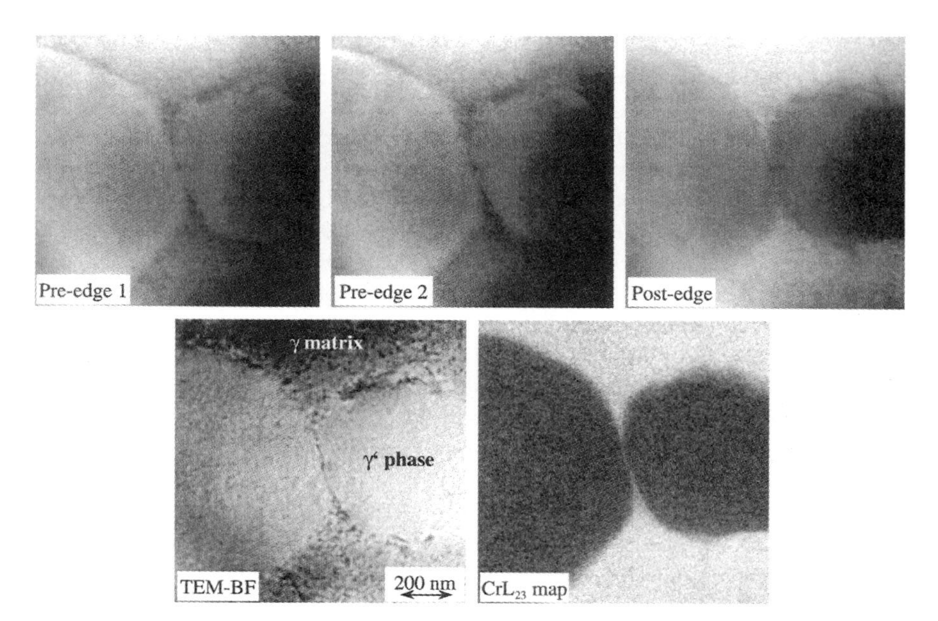

Fig. 2.42. Energy-filtered TEM (three-window method), imaging of the Cr distribution in the Ni-base superalloy SC16.

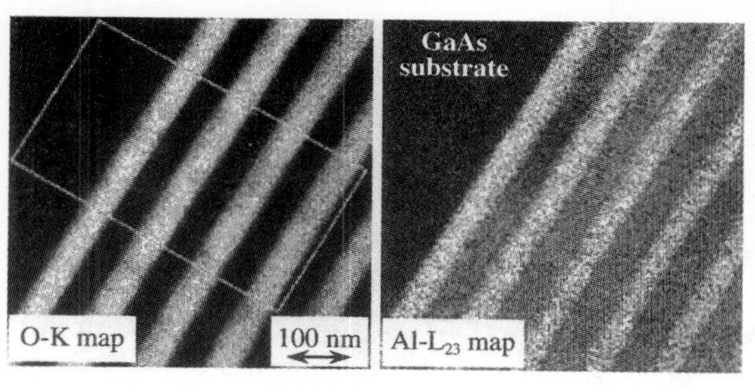

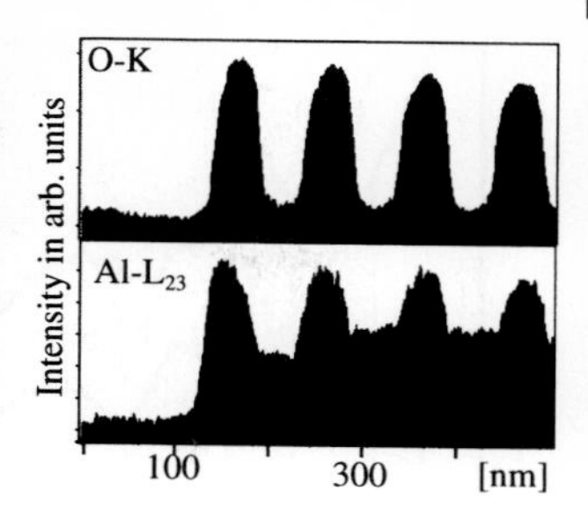

Fig. 2.43. EFTEM images and corresponding line profiles of oxidized AlAs/(Al,Ga)As multilayers.

the AlAs layers are oxidized by exposure to air which is imaged in the elemental maps and corresponding line profiles.

An especially dedicated technique is the so-called spectrum-image method [2.234], which comprises a series of spectra measured in a STEM not along a line but in a rectangular scanned field. This results in a data cube in which the upper plane is the scanned x–y area and the third axis is the energy-loss spectrum. Complementarily, image spectrum means a series of energy-filtered TEM images taken at such small energy increments that for every point in the x–y plane of view a full spectrum can be reconstructed afterwards. Although the spectrum-image and the image-spectrum methods give a comprehensive insight into the specimen, both are limited by the large data capacity needed and the time for data acquisition and off-line processing.

Energy filtering not only enables the imaging of the distribution of an element but even of its chemical bonding when ELNES features are used. Corresponding chemical-bond maps can be obtained in both experimental arrangements, i.e. the combined SEELS/STEM or imaging filter and TEM. In each case the slit width of the spectrometer system must be so small that a single fine-structure detail contributes to the signal. To obtain reliable results the three-window method must usually be used.

Chemical-bond mapping has been practiced on an SiC (Nicalon)-fiber reinforced borosilicate glass (Duran) of high SiO_2 content, as illustrated in Fig. 2.44 [2.235]. The differences in Si–L_{23} ELNES (Fig. 2.44a) were used to image oxidized silicon in the interfacial region between the fiber and the matrix, where an additional C-rich reaction layer is present. Individual STEM images were taken with a 3 eV slit width at approximately 95 eV (background B), at 102 eV (first dominant peak of the Si–L_{23} ELNES of SiC), and at 107 eV (first dominant peak of the Si–L_{23} ELNES of SiO_2). The 3D plot of the image intensity in Fig. 2.44b is obtained by subtraction of the partial image recorded at 102 eV from that at 107 eV, i.e. the value of the resulting signal is higher for regions with strongly oxidized silicon. In Fig. 2.44b the zone of minimum signal in the center corresponds to the carbon layer. On the fiber side of the carbon layer a region is visible where the signal has a local maximum hinting at oxidic-bound Si. The mean signal level is higher in the matrix (right) owing to the presence of SiO_2.

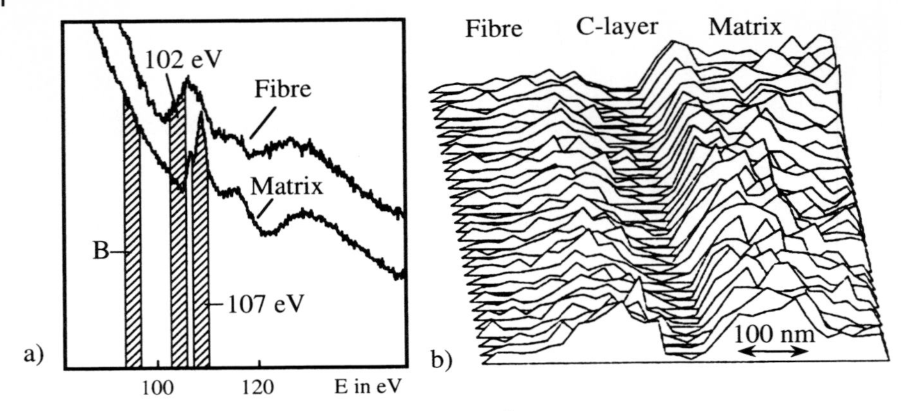

Fig. 2.44. Chemical-bond mapping: (a) comparison of EEL spectra recorded from matrix and SiC-fibre with the window for energy filtering shown; (b) map of oxidic-bound Si.

2.3.6 Summary

EELS in conjunction with TEM/STEM has become a powerful tool for analyzing materials, particularly interlayers and interfaces in structured materials, at a lateral resolution of approximately 1 nm or even better. It yields both qualitative and quantitative information on specimen composition, for which its thickness must be of the order of 10 nm, requiring a high effort of sample preparation. In addition, the chemical bonding of an element detected within the transmitted volume can be characterized by energy-loss near-edge fine structures (ELNES). Investigating low-loss spectra in more detail at high energy resolution (band-gap spectroscopy) enables determination of the electronic and optical properties of materials. The element distribution can be imaged along a line, or two-dimensionally by energy-filtered (S)TEM, where sub-nanometer resolution is achievable. In the future the limits of both energy resolution and lateral resolution will be pushed further. A project named SESAME [2.236, 2.237] which is in progress aims to develop an in-column energy filter TEM with resolution values below 1 Å and 1 eV.

2.4 Low-energy Electron Diffraction (LEED)

Georg Held

2.4.1 Principles and History

When Davisson and Germer reported in 1927 that the elastic scattering of low-energy electrons from well ordered surfaces leads to diffraction spots similar to those observed in X-ray diffraction [2.238–2.240], this was the first experimental proof of the wave nature of electrons. A few years before, in 1923, De Broglie had postulated that electrons have a wavelength, given in Å, of:

$$\lambda_e = h/m_e v = (150/E_{kin})^{1/2} \tag{2.23}$$

and a corresponding wave vector of length:

$$k = 2\pi/\lambda_e \tag{2.24}$$

where h is Planck's constant, m_e the electron mass, v the velocity, and E_{kin} the kinetic energy of the electron, given in eV. Davisson and Germer realized that the diffraction of low-energy electrons (LEED) in the energy range between 40 and 500 eV, where their wavelength ranges between 0.5 and 2 Å, could be used to determine the structure of single crystal surfaces, by analogy to X-ray diffraction. Because of their small inelastic mean free path (IMFP) of only a few Angstrom (typically less than 10 Å) electrons in this energy range sample only the upper-most atomic layers of a surface and are, therefore, better suited to the analysis of surface geometries than X-ray photons, which have a much larger mean free path (typically a few micron). Unlike for photon diffraction, however, multiple scattering plays an important role in the diffraction process of electrons at solid surfaces. Therefore, the analysis of LEED data in respect of the exact positions of atoms at the surface is somewhat more complicated and requires fully dynamic quantum mechanical scattering calculations.

The use of LEED for surface analysis became important when large single crystals became available for surface studies. It was first used solely for a qualitative characterization of surface ordering and the quantitative determination of the two-dimensional surface lattice parameters (e.g. superstructures, see below). The information about the positions of the atoms in the surface is hidden in the energy-dependence of the diffraction spot intensities, the so-called LEED *I–V*, or $I(E)$, curves. In the late nineteen-sixties computer programs became available which could perform fully dynamic scattering calculations for simple surface geometries. Comparison of such theoretical *I–V* curves for a set of model geometries with experimental data enables determination of the atomic positions within the surface by trial and error. With the immense growth of available computer power and speed since then, LEED *I–V* structure determination could be applied to a large number of increasingly complex surface geometries; this has made LEED the standard technique of modern surface crystallography.

For further details of the history, experimental set-up, and theoretical approaches of LEED please refer to books by Pendry [2.241], van Hove and Tong [2.242], van Hove, Weinberg, and Chan [2.243], and Clarke [2.244]. This article relies extensively on these works.

2.4.2
Instrumentation

Standard modern LEED optics are of the "rear view" type, and are shown schematically in Fig. 2.45. The incident electron beam, accelerated by the potential V_0, is emitted from the electron gun behind the hemispherical fluorescent glass screen and hits the sample through a hole in the screen. The surface is at the center of the hemisphere so that all back-diffracted electrons travel towards the LEED screen on radial trajectories. Before the electrons hit the screen they must pass a retarding field energy analyzer. It typically consists of four (or three) hemispherical grids concentric with the screen, each containing a central hole through which the electron gun is inserted. The first grid (nearest to the sample) is connected to earth ground, as is the sample, to provide an essentially field-free region between the sample and the first grid. This minimizes undesirable electrostatic deflection of diffracted electrons. A suitable negative potential $-(V_0 - \Delta V)$ is applied to the second and third (only second) grids, the so-called suppressor grids, to enable a narrow energy range $e\Delta V$ of elastically scattered electrons to be transmitted to the fluorescent screen. The fourth

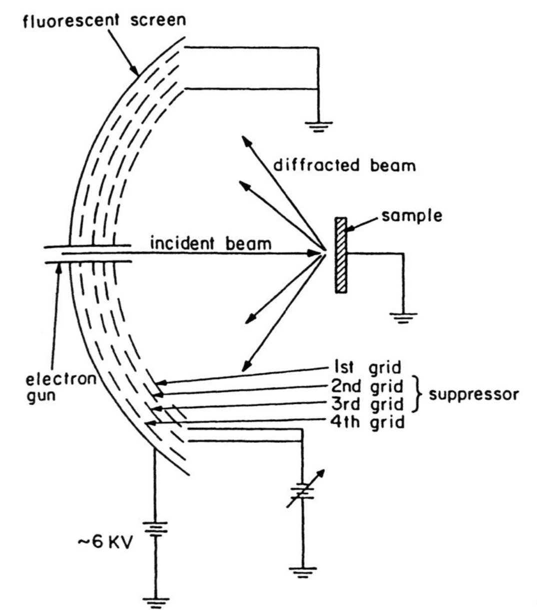

Fig. 2.45. Schematic diagram of a four-grid LEED display system [2.243].

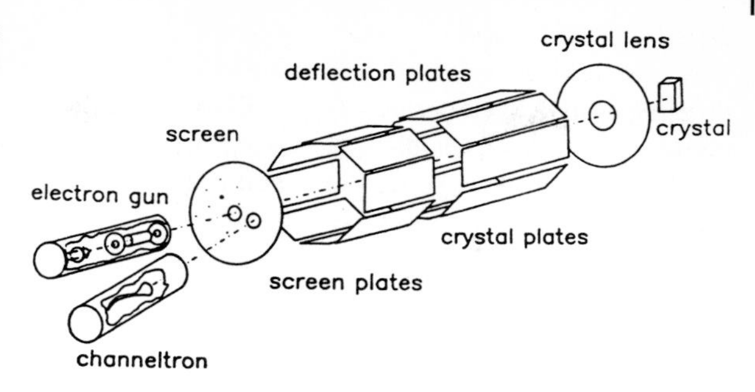

Fig. 2.46. Schematic diagram of the set-up of the SPA-LEED system [2.245].

(third) grid is usually grounded to reduce field penetration of the suppressor grids by the screen voltage, which is of the order of a few kilovolts, to make the diffraction spots visible. Because the fluorescent screen is transparent, the spots can be observed through a view-port behind the screen without being shadowed by the sample holder. Only the electron gun assembly (diameter <15 mm) obstructs the view slightly. Older designs have opaque screens which allow observation from behind the sample only. A typical diameter of the complete optics is approximately 140 mm, so it fits into a 150 mm (i.d.) flange.

To date, the usual way of recording the LEED pattern is a light-sensitive video camera with a suitable image-processing system. In older systems movable Faraday cups (FC) were used which detected the electron current directly. Because of long data acquisition times and the problems of transferring motion into UHV, these systems are mostly out of use nowadays.

Only one widely used type of LEED optics uses a Faraday cup arrangement as electron detector, and this is specially designed for accurate spot profile analysis (SPA–LEED, cf. Fig. 2.46) [2.245]. In this design the FC is at a fixed position and the grids are replaced by round apertures defining the lateral resolution of the system. Instead of the position of the electron detector, the angle of the incoming electron beam is varied by means of an electrostatic octupole lens, thereby deflecting the desired part of the LEED pattern towards the detector without the need for moving parts. This type of LEED optics is specially designed to have a large resolution in k-space. The better the k resolution, the larger are the typical distances on the surface for which effects can be observed in the LEED pattern (spot position, profile). The largest resolvable length on the surface is called the transfer width and characterizes the LEED instrument [2.246]. Typical values are around 150 Å for conventional rear-view LEED systems and up to 1000 Å for SPA–LEED systems.

2.4.3
Qualitative Information

The most direct information obtained from LEED is the periodicity and intermediate range order within the transfer width of the surface under investigation. This can be gathered by visual inspection of the diffraction pattern and/or by relatively simple mathematical transformations of the spot profiles.

2.4.3.1 LEED Pattern

Because the electrons do not penetrate into the crystal bulk far enough to experience its three-dimensional periodicity, the diffraction pattern is determined by the two-dimensional surface periodicity described by the lattice vectors $\mathbf{a}_1$ and $\mathbf{a}_2$, which are parallel to the surface plane. A general lattice point within the surface is an integer multiple of these lattice vectors:

$$\mathbf{R} = n_1\,\mathbf{a}_1 + n_2\,\mathbf{a}_2 \tag{2.25}$$

The two-dimensional Bragg condition leads to the definition of reciprocal lattice vectors $\mathbf{a}_1^*$ and $\mathbf{a}_2^*$ which fulfil the set of equations:

$$\mathbf{a}_1\,\mathbf{a}_1^* = \mathbf{a}_2\,\mathbf{a}_2^* = 2\,\pi \tag{2.26 a}$$
$$\mathbf{a}_1\,\mathbf{a}_2^* = \mathbf{a}_2\,\mathbf{a}_1^* = 0 \tag{2.26 b}$$

These reciprocal lattice vectors, which have units of Å^{-1} and are also parallel to the surface, define the LEED pattern in k-space. Each diffraction spot corresponds to the sum of integer multiples of $\mathbf{a}_1^*$ and $\mathbf{a}_2^*$.

$$\mathbf{g}_{(n1,n2)} = n_1\,\mathbf{a}_1^* + n_2\,\mathbf{a}_2^* \tag{2.27}$$

The integer numbers (n_1,n_2) are used as indices to label the spot. The parallel component of the corresponding wave vector is:

$$k_{||,(n1,n2)} = k_{||,0} + g_{(n1,n2)} \tag{2.28}$$

where $k_{||,0}$ is the parallel component of the wave vector of the incoming electron beam. The vertical component, k_z, of the back-diffracted electrons is defined by energy conservation:

$$k_{z,(n1,n2)} = \left(2\,m_e\,E_{kin}/\,h^2 - k^2_{||,(n1,n2)}\right)^{1/2} \tag{2.29}$$

This equation also limits the set of observable LEED spots by the condition that the expression inside the brackets must be greater than zero. With increasing electron energy the number of LEED spots increases while the polar emission angle relative to the surface normal, $\theta = \arctan\,(k_{||}/k_z)$, decreases for each spot except for the specular spot (0,0) which does not change. Fig. 2.47 shows examples of common surface unit cells and the corresponding LEED patterns.

Often (adsorption, reconstruction) the periodicity at the surface is larger than expected for a bulk-truncated surface of the given crystal; this leads to additional (superstructure) spots in the LEED pattern for which fractional indices are used. The lattice vectors $\mathbf{b}_1$ and $\mathbf{b}_2$ of such superstructures can be expressed as multiples of the (1×1) lattice vectors $\mathbf{a}_1$ and $\mathbf{a}_2$:

$$\mathbf{b}_1 = m_{11}\,\mathbf{a}_1 + m_{12}\,\mathbf{a}_2 \tag{2.30 a}$$

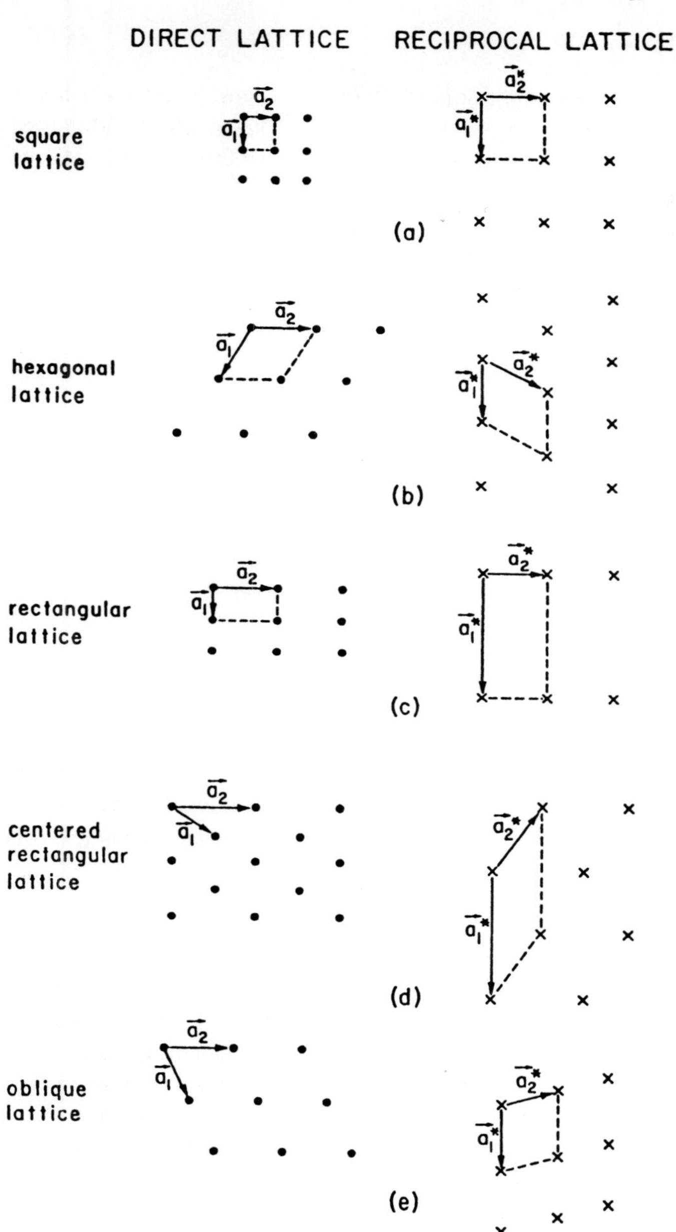

Fig. 2.47. Direct (left) and reciprocal (right) lattices for the five two-dimensional Bravais lattices [2.243].

$$\mathbf{b}_2 = m_{21}\,\mathbf{a}_1 + m_{22}\,\mathbf{a}_2 \tag{2.30 b}$$

where the numbers m_{ij} are the coefficients of the superstructure matrix, which is a straightforward way of characterizing the superstructure. The positions and indices of the additional LEED spots can be calculated directly from this matrix [2.243] by use of the formulas:

$$\mathbf{b}_1^* = (m_{11}\,m_{22} - m_{12}\,m_{21})^{-1}\,(m_{22}\,\mathbf{a}_1^* - m_{21}\,\mathbf{a}_2^*) \tag{2.31 a}$$

$$\mathbf{b}_2^* = (m_{11}\,m_{22} - m_{12}\,m_{21})^{-1}\,(-m_{12}\,\mathbf{a}_1^* - m_{11}\,\mathbf{a}_2^*) \tag{2.31 b}$$

The area of the superstructure unit cell, A, in units of the (1×1) unit cell area can also easily be calculated from the coefficients of the superstructure matrix:

$$A = (m_{11}\,m_{22} - m_{12}\,m_{21}) \tag{2.32}$$

Another, less general method uses the notation according to Wood [2.247], where the lengths of the vectors $\mathbf{b}_1$ and $\mathbf{b}_2$ are specified in units of $\mathbf{a}_1$ and $\mathbf{a}_2$ together with the rotation angle, α, between $\mathbf{b}_1$ and $\mathbf{a}_1$ (only specified if non-zero):

$$\mathrm{p/c} = \left(\frac{|\mathbf{b}_1|}{|\mathbf{a}_1|} \times \frac{|\mathbf{b}_2|}{|\mathbf{a}_2|}\right) \tag{2.33}$$

"p" indicates a "primitive" and "c" a "centered" surface unit cell. Examples are "p(2×2)", "p$(\sqrt{3} \times \sqrt{3})$ R30°", and "c$(\sqrt{2} \times \sqrt{2})$". This notation is not always applicable but it is more frequently used than the matrix notation because it is shorter. Fig. 2.48 shows examples of common superstructures with the corresponding matrix and Wood notations.

2.4.3.2 **Spot-profile Analysis**

Whereas the spot positions carry information about the size of the surface unit cell, the shapes and widths of the spots, i.e. the spot profiles, are influenced by the long range arrangement and order of the unit cells at the surface. If vertical displacements (steps, facets) of the surface unit cells are involved, the spot profiles change as a function of electron energy. If all surface unit cells are in the same plane (within the transfer width of the LEED optics), the spot profile is constant with energy.

A periodic arrangement of equal steps at the surface leads to a spot splitting at certain energies from which the step height can be determined directly. For a statistical arrangement of steps the analysis of energy-dependent changes in the spot profiles often enables the determination of the mean step height and characterization of the step distribution [2.248]. Facets lead to extra spots which move in $k_{||}$ upon changes of the kinetic energy.

Point defects, static disorder, and thermally induced displacements lead to an increase of the background intensity between the spots. Depending on the correlation between the scatters, the background is either homogeneous (no correlation) or

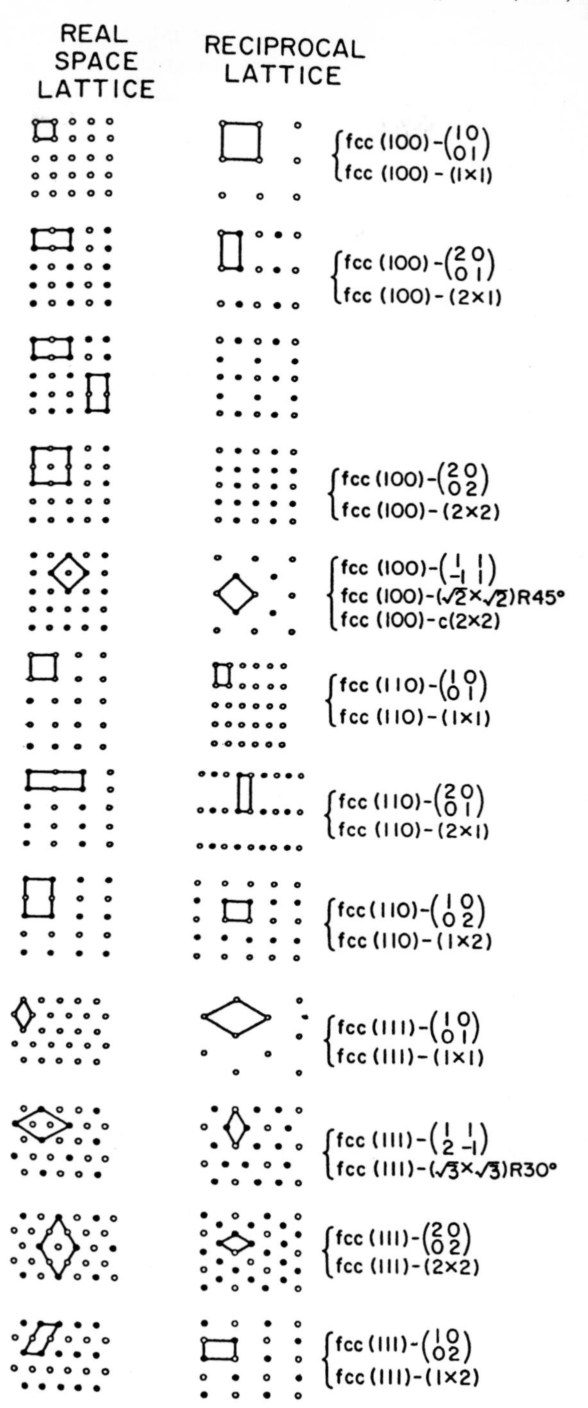

Fig. 2.48. Direct and reciprocal lattices of a series of commonly occurring two-dimensional superlattices. Open circles correspond to the ideal (1 × 1) surface structure, whereas filled circles represent adatoms in the direct lattice and fractional-order spots in the reciprocal lattice [2.243].

structured (correlation). If the coherently ordered surface areas (islands, domains) are smaller than the transfer width of the LEED system and at the same vertical height, the width of these areas, Δw, is directly related to the width of the LEED spots in k-space, $\Delta k_{||}$:

$$\Delta k_{||} = 2\pi/\Delta w \tag{2.34}$$

This relationship is true for each direction parallel to the surface independently. It is particularly useful for determining the size of adsorbate islands which lead to extra superstructure spots. A good introduction (in German) to spot profile analysis is given by Henzler and Göpel [2.249].

2.4.3.3 Applications and Restrictions

Visual inspection of the LEED pattern is a very good, rapid method for qualitative characterization of the surface under investigation and, therefore, LEED optics are part of most UHV systems dedicated to the analysis of single crystal surfaces. Consequently, many LEED patterns with and without superstructures are reported in the literature [2.250]. Sharp LEED spots are indicative of long-range ordering in the uppermost layers. Their absence is often a sign of large surface roughness (e.g. after ion bombardment) or a thick amorphous layer (e.g. water, hydrocarbons) covering the surface. The appearance of a superstructure is also very useful for characterizing the surface after adsorption or heat-treatment. The literature contains numerous examples of how superstructures can be used to identify certain adsorbates, oxidation, carbide formation, etc. Because these effects depend very much on the reactivity and surface structure, such information is best found in the original literature relating to the particular surface. The size of the surface unit cell can be deduced exactly for a given superstructure by use of Eq. (2.32). Thus, the relative coverage of an adsorbate inducing this superstructure can often also be determined by LEED, because there must be an integer number of adsorbate atoms or molecules per unit cell.

Determination the surface morphology (domain size, step distribution, etc.) by LEED spot profile analysis always takes into account the whole surface area hit by the incoming electron beam. It therefore uses a much larger statistical sample than any microscopic method (e.g. STM, SEM, TEM) could deal with and, therefore, enables more reliable determination of average surface properties such as roughness or step density. Spot profile analysis has been applied to flat metal surfaces to study the development of domain sizes during phase transitions in adsorbate layers [2.251, 2.252]. Changes in the step density and/or distribution of stepped or faceting upon adsorption or annealing have also been studied by LEED on stepped and nominally flat metal and semiconductor surfaces [2.253–2.256]. For these experiments either conventional LEED systems with a fluorescent screen or Faraday cup (FC) systems with small apertures (movable FC or SPA–LEED systems) have been used. Another advantage of LEED is that it is a contact-free analysis method which enables access to the surface while data are acquired. This has been used to analyze surfaces in-situ during adsorption, epitaxy, or sublimation, to study dynamic restructuring processes [2.257–2.259].

All LEED data analysis must, however, rely on prior assumptions and model distributions and is, therefore, not really direct. Microscopic methods have the advantage of delivering directly the shape of the surface (domains, terraces, etc.) without any assumptions being made.

2.4.4
Quantitative Information

Analysis of the LEED pattern or of spot profiles does not give any quantitative information about the position of the atoms within the surface unit cell. This type of information is hidden in the energy-dependence of the spot intensities, the so-called LEED *I–V* curves.

2.4.4.1 Principles

In a simplified picture these intensity variations are caused by the interference of electrons scattered from different atomic layers parallel to the surface. For an infinite penetration depth this would impose a third Bragg condition for k_z and, therefore, for E_{kin} (cf. Eq. 2.29) with sharp intensity maxima and zero intensity between them. Because the penetration depth is very short, the back-scattered electrons only interact with a few layers of atoms, which results in broad maxima and non-zero intensities in the intermediate energy regimes. Because of the influence of the atomic scattering potentials and multiple scattering, shifts relative to the Bragg peaks and other peaks are observed in the *I–V* curves. All these effects are included in modern LEED computer programs which perform fully dynamic quantum mechanical scattering calculations. These programs deliver a set of *I–V* curves which would be expected for a given user-specified surface geometry.

Because multiple scattering dominates the electron diffraction process at low energies, there is no easy way of determining the surface directly such as the Patterson function in X-ray crystallography. Instead, *I–V* curves must be calculated for a large number of model geometries and compared with experimental *I–V* curves. Their agreement is then quantified by the means of a reliability factor (*R* factor). There are several ways of defining such *R* factors [2.243] with Pendry's *R* factor, R_P, being the most common [2.260]. By convention, R_P is 0 when the agreement is perfect, 1 for uncorrelated sets of *I–V* curves, and 2 for completely anti-correlated curves (each maximum of one curve coincides with a minimum of the other). Usually, automated search procedures, which modify the model geometries to be tested according to the *R* factor values achieved by the preceding geometries, are used to find a *R* factor minimum within the set of geometrical data to be optimized. The search strategies are either conventional downhill-oriented algorithms (simplex method, Powell's method, Marquard's algorithm [2.261]), which usually find only the nearest local minimum, or stochastic Monte-Carlo methods (simulated annealing [2.261], genetic algorithm [2.262]) which, in principle, always find the global minimum. A set of experimental and theoretical *I–V* curves for benzene on Ru(0001) [2.263], depicted in Fig. 2.49, shows a typical degree of agreement achieved in a structure optimization for organic molecules.

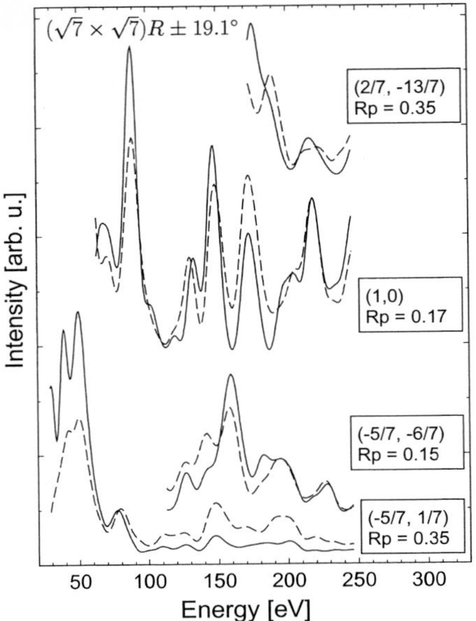

Fig. 2.49. Examples of experimental (solid lines) and calculated (dashed lines) LEED *I–V* curves for the p ($\sqrt{7} \times \sqrt{7}$) R19° superstructure of benzene on Ru(0001). The individual R_P factors are indicated in the figure [2.263].

Usually, the collection of LEED *I–V* curves requires single crystal surfaces with long-range order in the upper-most layers. Structural information can, however, also be obtained in a similar way for certain disordered surfaces, when the energy dependence of the diffusely scattered intensity is analyzed (diffuse LEED [2.264, 2.265]).

2.4.4.2 **Experimental Techniques**

The standard experimental set-up for collecting LEED *I–V* curves uses a video camera recording images of the fluorescent screen for each energy (video LEED) [2.266]. When a conventional video camera is used the rate at which images are collected is fixed at 50 to 60 Hz. In this case several images (typically between 8 and 256) must be averaged to obtain a satisfying signal-to-noise ratio, especially for weak superstructure spots. Newer systems often use slow-scan CCD cameras which can perform the averaging directly on the cooled CCD chip (exposure times up to several seconds) and therefore avoid multiple readout noise [2.267]. The images are either stored on a computer hard disk and analyzed in a second round or the spot intensities are extracted online by a special software. Usually, the intensity is averaged within a given area at the spot position and the averaged local background outside this area is subtracted from this intensity. Commercially available data acquisition software also performs the control of the electron energy and records the electron beam current which is needed to normalize the spot intensities.

Older experimental arrangements used Faraday cups with small apertures mounted on goniometers, which could be moved around the sample to collect the back-scattered electron current directly, or spot photometers, which were directed at one diffraction spot on the fluorescent screen which was then followed by hand while the energy was varied.

Data collection is mostly performed at normal incidence of the primary electron beam. Under these conditions usually several equivalent LEED spots exist because of the surface symmetry. By taking care that the *I–V* curves of equivalent spots are identical, normal incidence conditions can be adjusted to within a few tenths of a degree.

2.4.4.3 Computer Programs

Most computer programs currently used for calculating LEED *I–V* curves are based on the multiple scattering algorithms outlined in the early works by Pendry, Tong, and van Hove [2.241, 2.242]. The most common way of calculating LEED *I–V* curves is to subdivide the crystal into atomic layers parallel to the surface. All possible multiple scattering paths inside each layer are first added then combined in a layer diffraction matrix. The total back-diffracted intensity for each LEED spot is then calculated in a second step by combining these layer diffraction matrices in a way that includes all remaining multiple scattering paths between the layers. The amount of computer time needed for calculating a set of *I–V* curves for one model geometry depends on the number of atoms per surface unit cell and on the number of spots within the LEED pattern. For both the dependence is cubic, so the time requirements vary quite substantially from a few seconds up to several hours depending on the complexity of the surface geometry. The computational effort can be significantly reduced by making use of rotational or mirror symmetries at the surface.

Some more recent software uses the tensor LEED approximation of Rous and Pendry which can save a substantial amount of computer time [2.268–2.270]. In tensor LEED the amplitudes $A_g(0)$ of all escaping electron waves (spots) are first calculated conventionally as described above for a certain reference geometry. Then the derivatives of these amplitudes $\delta A_g/\delta \mathbf{r}_i$ with respect to small displacements of each atom i in this reference geometry are calculated. These derivatives are the constituents of the "tensor". The wave amplitude for a modified model geometry where atom i is displaced by the vector $\Delta \mathbf{r}_i$ is then approximately given by:

$$A_g(\Delta \mathbf{r}_i = A_g(0) + (\delta A_g/\delta \mathbf{r}_i)\,\Delta \mathbf{r}_i \qquad (2.35)$$

This way, a simple summation over all displaced atoms avoids a new multiple scattering calculation and is therefore several orders of magnitude faster than a conventional LEED calculation, especially for large surface unit cells with many atoms. Obviously, tensor LEED requires much more storage space and overhead than a conventional LEED calculation, because the derivatives must be stored for each atom at each energy and the applicability is limited to geometries in the neighborhood of the reference structure with maximum displacements of a few tenths of Angstroms.

When automated searches are implemented, the LEED *I–V* calculation and the *R* factor comparison are called by the master search program either as subroutines or as separate programs through the operating system. The LEED *I–V* calculations are always the most time-consuming parts of the searches and the main criterion for selecting search algorithms is to reduce the number of trial geometries.

There are no commercially available computer programs for LEED *I–V* structure determinations. Most programs and their documentation can, however, be downloaded from the internet or are distributed by the authors [2.271].

2.4.4.4 Applications and Restrictions

Structure analysis by LEED *I–V* is currently the most accurate and reliable way of determining the atomic positions at surfaces. That LEED is sensitive to the atomic positions not only in the upper-most layer but down to several layers below the surface makes it an especially ideal tool for studying spontaneous or adsorbate-induced surface reconstructions. General drawbacks are certainly that LEED usually requires ordered and conducting surfaces, otherwise charging would distort the LEED pattern. The computer time needed to calculate the *I–V* curves and the number of trial geometries are factors limiting the complexity of accessible surface structures on the computational side; the density of LEED spots on the fluorescent screen is a limiting experimental factor.

Because LEED theory was initially developed for close packed clean metal surfaces, these are the most reliably determined surface structures, often leading to R_P factors below 0.1, which is of the order of the agreement between two experimental sets of *I–V* curves. In these circumstances the error bars for the atomic coordinates are as small as 0.01 Å, when the total energy range of *I–V* curves is large enough (>1500 eV). A good overview of state-of-the-art LEED determinations of the structures of clean metal surfaces, and further references, can be found in two recent articles by Heinz et al. [2.272, 2.273].

For more open adsorbate covered and/or reconstructed surfaces certain approximations used in the standard programs are less accurate and lead to higher R_P factor values. For simple superstructures of mono-atomic adsorbates or small molecules on metal surfaces R_P factors between 0.1 and 0.2 can be expected. For large molecules, which often adsorb in complex superstructures, the optimum R_P factor values are often as large as 0.3. Because the error margins scale with the optimum *R* factor value, the accuracy of atomic coordinates is these latter cases is smaller, with error bars up to 0.1 Å (A recent review concerning structure determinations of molecular adsorbates is given by Over [2.274].) In general the accuracy is higher for coordinates perpendicular to the surface than for lateral coordinates.

Because of the minimization of the number of dangling bonds semiconductor surfaces often show large displacements of the surface atoms from their bulk lattice positions. As a consequence these surfaces are also very open and the agreement is more in the range of R_P factor values of approximately 0.2. Determination of the structure of semiconductor surfaces is reviewed in a recent article by Kahn [2.275].

A relatively complete listing of all surface geometries determined by LEED can be found in the NIST surface structure data base [2.250].

2.5 Other Electron-detecting Techniques

John C. Rivière

2.5.1 Auger Electron Appearance Potential Spectroscopy (AEAPS)

Because of the emission of an Auger electron as an alternative to soft X-ray photons, the total secondary electron yield will increase as an ionization threshold is crossed. It is the total secondary yield that is monitored in AEAPS, in a rather simple experimental arrangement. The secondary current arises not just from the Auger electrons themselves, but also from inelastic scattering of Auger electrons produced at greater depths below the surface. Because the yield change is basically a measure of the probability of excitation of a core-level electron to an empty state above the Fermi level, the fine structure above the threshold will be similar to that seen in SXAPS. One of the problems in AEAPS is that of a poorly behaved background, which means that SXAPS is preferred.

2.5.2 Ion (Excited) Auger Electron Spectroscopy (IAES)

Auger emission after creation of a core hole by electron or photon irradiation has been described under AES and XPS. Incident ions can also create core holes, so that Auger emission as a result of ion irradiation can also occur, giving rise to the technique of IAES. The ions used are normally noble gas ions, but protons and α particles also have occasionally been used. In addition to the normal Auger features found in AES and XPS spectra, peaks are found in IAES spectra arising from Auger transitions apparently occurring in atoms or clusters sputtered from the surface. These peaks do not always coincide with those found in gas-phase excitation, and others are found that are not present in gas-phase measurements. Because of the complexity of the spectra, IAES cannot be used directly as an analytical technique, but it is very useful in basic physics experiments that study Auger processes occurring in excited atoms.

2.5.3 Ion Neutralization Spectroscopy (INS)

The technique of INS is probably the least used of those described here, because of experimental difficulties, but it is also one of the physically most interesting. Ions of He^+ of a chosen low energy in the range 5–10 eV approach a metal surface and within an interaction distance of a fraction of a nanometer form ion–atom pairs with the nearest surface atoms. The excited quasi molecule so formed can de-excite by Auger neutralization. If unfilled levels in the ion fall outside the range of filled levels of the solid, as for He^+, an Auger process can occur in which an electron from the valence band of the solid fills the core hole in the ion and the excess energy is given up to another valence electron, which is then ejected. Because either of the valence electrons can come from anywhere within the valence band, the observed energy spectrum reflects the local density of states at the solid surface, but is a self-convolution of the LDOS and of transition probabilities across the valence band. The technique of INS requires complex mathematical for deconvolution of the spectra. Because the technique of STS is also capable of extracting the LDOS, the results from INS and STS should be comparable, but such a comparison does not seem to have been attempted.

2.5.4 Metastable Quenching Spectroscopy (MQS)

MQS is, in a sense, an extension of INS. Instead of He^+ ions, helium atoms in metastable states are used in the incident beam, at the same low energies. The excited singlet state $He^*(2^1S_1)$ has an energy of 20.62 eV and a lifetime of 2×10^{-2} s, and so is suitable. It is produced by expanding helium gas at high pressure through a nozzle into a cold cathode discharge sustained by combined electrostatic and magnetic fields. The high fields prevent ions and fast neutrals from leaving the discharge, and the beam is then nearly all $He^*(2^1S_1)$. As the metastable ion approaches the surface, either of two mechanisms can lead to de-excitation; both result in Auger emission similar to INS. If the excited level in the atom can resonate with empty states at the Fermi level of the surface, electron transfer from the atom to the surface can occur, leading to resonance ionization of the helium atom. Auger neutralization then occurs as in INS, and the resulting spectrum is again a self-convolution of the LDOS. If, however, the excited level cannot resonate with empty surface states, then direct Auger de-excitation can occur, in which the hole in the inner shell of the metastable helium atom is filled from a surface state of the sample, followed by ejection of the excited electron from the helium atom. The process is also called Penning ionization. In the latter process, only one electron is ejected, and the resulting spectrum is thus an unconvoluted reflection of the LDOS.

2.5.5 Inelastic Electron Tunneling Spectroscopy (IETS)

IETS is unique in that it is entirely surface-specific, but does not require a vacuum environment and has almost never been performed in UHV. Its principle is relatively simple. If two metals are separated by a thin (~3 nm) insulating layer, and a voltage is applied across them, electrons can tunnel from one to the other through the insulator, and if no energy is lost by the electrons, the process is known as elastic tunneling. If, however, discrete impurity states occur in the interfaces between either metal and the insulator, or molecules exist in such an interface that have characteristic vibrational energies, then the tunneling electron can give up some of this energy either to the state or to the vibrational mode, before reaching the other metal. This is called inelastic tunneling. Obviously, the applied voltage must be greater than the state or vibrational energies. If the current across the metal–insulator–metal sandwich is recorded as a function of applied voltage, the current increases as the threshold for each state or vibrational mode is crossed. The increases in current are in fact very small, and for improved detectability the current is double-differentiated with respect to voltage, thereby providing, in effect, a vibrational spectrum that can be compared directly with free-molecule IR and Raman spectra.

The metal–insulator–metal sandwich is known as a tunnel junction, and its preparation is all-important. The standard junction consists of an aluminum strip ca. 60–80 nm thick and 0.5–1.0 mm wide deposited in a very good vacuum on to scrupulously clean glass or ceramic; the surface of the strip is then oxidized either thermally or by glow discharge. The resulting oxide layer is extremely uniform and approximately 3 nm thick. Introduction of adsorbed molecules on to the oxide layer, or "doping", as it is called, is then effected either by immersion in a solution then spinning to remove excess fluid, or by coating from the gas phase. The final stage in preparation is deposition of the second metal (invariably lead) of the sandwich; this deposition is performed in a second vacuum system (i. e. not that used for aluminum deposition), the final thickness of lead being approximately 300 nm, and the width of the lead strip being the same as that of the aluminum strip. The reason for using lead is that all IETS measurements are conducted at liquid helium temperature, 4.2 K, to optimize the energy resolution by reducing the contribution of thermal broadening to the line-width, and lead is, of course, superconducting at that temperature.

Normally not one, but several, junctions are fabricated simultaneously, because the likelihood of junction failure always exists. After fabrication the electrical connections are made to the metal electrodes, the junctions are dropped into liquid helium, and measurements are commenced immediately.

Although insulators other than aluminum oxide have been tried, aluminum is still used almost universally because it is easy to evaporate and forms a limiting oxide layer of high uniformity. To be restricted, therefore, to adsorption of molecules on aluminum oxide might seem like a disadvantage of the technique, but aluminum oxide is very important in many technical fields. Many catalysts are supported on alumina in various forms, as are sensors, and in addition the properties of the oxide film on aluminum metal are of the greatest interest in adhesion and protection.

3
Ion Detection

3.1
Static Secondary Ion Mass Spectrometry (SSIMS)

Heinrich F. Arlinghaus

SSIMS emerged as a technique of potential importance for surface analysis as a consequence of the work of Benninghoven and his group in Münster [3.1] during the late nineteen-sixties and early nineteen-seventies (i. e., at roughly the same time as AES). The prefix "static" was added to distinguish the technique from "dynamic" SIMS – the difference between the two lying in the incident (or primary) ion current densities used, of the order of < 1 nA cm^{-2} for SSIMS but much higher for dynamic SIMS. With this low primary ion dose density, spectral data can be generated on a time scale which is very short compared with the lifetime of the surface layer (typically surface destruction of $< 1\%$).

Together with XPS and AES, SSIMS ranks as one of the principal surface analytical techniques. Because its sensitivity for elements greatly exceeds that of the other two techniques and much chemical information is available, its use is rapidly expanding in many fields of application.

3.1.1
Principles

A beam of positive ions bombards a surface, leading to interactions that cause the emission of a variety of types of secondary particle, including secondary electrons, Auger electrons, photons, neutrals, excited neutrals, positive secondary ions, and negative secondary ions. SSIMS is concerned with the last two of these, *positive* and *negative secondary ions*. The emitted ions are analyzed in a mass spectrometer, resulting in positive or negative mass spectra consisting of parent and fragment peaks characteristic of the surface. The peaks can be identifiable as arising from the substrate material itself, from contaminations and impurities on the surface, or from deliberately introduced species adsorbed on the surface.

When a heavy energetic particle such as an argon ion (typically 1 to 15 keV) hits a surface, it will not be stopped short by the first layer of atoms but continues into the surface until it comes to a halt as a result of energy lost in atomic and electronic scattering. Along its way the ion displaces some atoms from their original positions in the solid structure, which as they recoil displace others, which displace additional atoms, and so on, resulting in a complex sequence of collisions. Depending on the energy absorbed in an individual collision, some atoms are permanently displaced from their normal positions, whereas others return elastically after temporary displacement. This sequence is called a collision cascade and is illustrated schematically, along with some other processes, in Fig. 3.1. As the cascade spreads out from the path of the primary ion, its effect can eventually reach atoms in the surface layer, and with enough kinetic energy left, bonds can be broken so that material leaves the surface in form of neutral or ionized atoms or clusters. The resulting secondary particle emission is of low energy (peak of energy distribution between 5 to 10 eV) with over 95% of the secondary particles originating from the top two monolayers of the solid. Detecting these ionized particles directly with a mass spectrometer thus results in very high surface sensitivity.

The theoretical aspect of the collision cascade is given by Sigmund's collision cascade model [3.2] which explains much about the factors involved in the removal of material from a surface by sputtering, although it cannot predict the extent of positive or negative ionization of the material. A unified theory of secondary ion formation in SSIMS does not yet exist, although many models have been proposed for the process. Because the secondary ion yield (i.e. the probabilities of ion formation) can vary by several orders of magnitude for the same element in different matrices, or for different elements across the periodic table, lack of a means of predicting ion yield in a given situation limits both interpretation and quantification. Some theoretical and semi-theoretical models have had some predictive success for a very restricted range of experimental data, but they fail when extended further. The most important theoretical models are briefly addressed in Sect. 3.2.1; further information is given elsewhere [3.3].

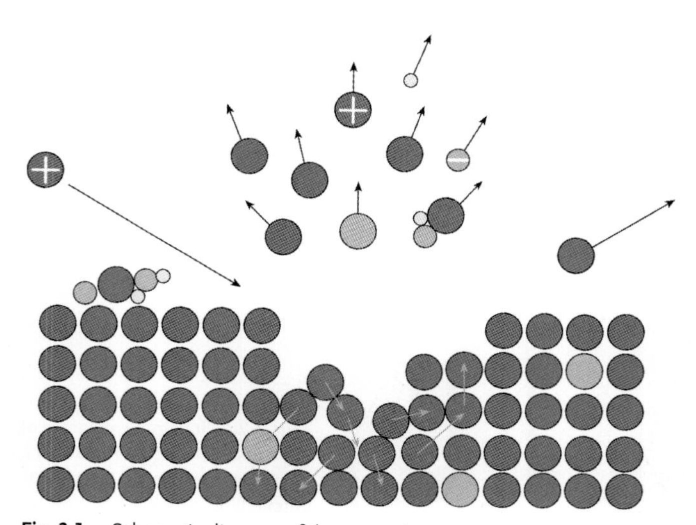

Fig. 3.1. Schematic diagram of the sputtering process.

3.1.2
Instrumentation

The instrumentation for SSIMS can be divided into two parts: (a) the primary ion source in which the primary ions are generated, transported, and focused towards the sample; and (b) the mass analyzer in which sputtered secondary ions are extracted, mass separated, and detected.

3.1.2.1 Ion Sources

Several ion sources are particularly suited for SSIMS. The first produces *positive noble gas ions* (usually argon) either by electron impact (EI) or in a plasma created by a discharge (see Fig. 3.18 in Sect. 3.2.2.). The ions are then extracted from the source region, accelerated to the chosen energy, and focused in an electrostatic ion-optical column. More recently it has been shown that the use of primary polyatomic ions, e.g. SF_5^+, created in EI sources, can enhance the molecular secondary ion yield by several magnitudes [3.4, 3.5].

Argon ion guns (see also Sect. 3.2.2.1) of the EI type operate in the energy range 0.2–10 keV with a 0.1–1 μA dc current, providing optimum spot sizes on the surface of approximately 10 μm. Discharge-type argon ion guns, often using the duoplasmatron arrangement, operate at up to 15 keV, with currents variable up to 20 μA and, depending on the particular focusing lens system used, providing spot sizes down to approximately 1 μm.

Cesium ions are also sometimes used to enhance the secondary-ion yield of negative elemental ions and that of some polymer fragments [3.6]. They are produced by surface ionization with an extraction technique similar to that of EI sources.

Another type of ion gun produces *positive ions from a liquid metal* (almost always gallium) in the manner shown schematically in Fig. 3.2 [3.7]. A fine needle (f, tip radius ~5 μm) of refractory metal passes through a capillary tube (d) into a reservoir of liquid metal (e). The liquid is drawn up through the tube over the needle tip by capil-

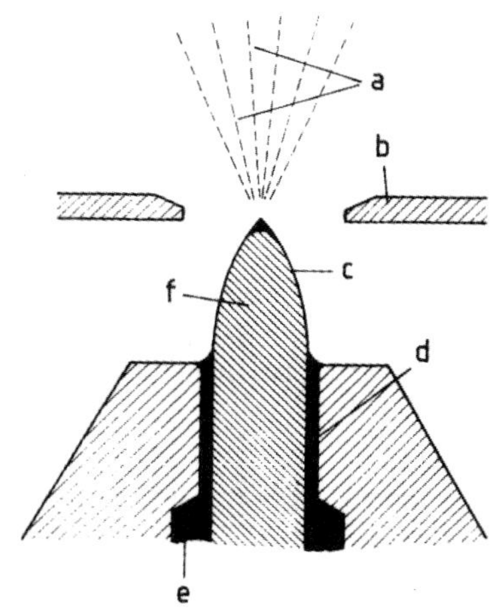

Fig. 3.2. Schematic diagram of the design and operation of a liquid-metal ion source (LMIS) [3.7]; (a) metal ions; (b) extractor; (c) liquid-metal film; (d) capillary tube; (e) liquid metal; (f) needle.

lary action. Application of 10–30 keV between the needle and a nearby extractor electrode (b) draws the liquid up into a cusp (Taylor cone) as a result of the combined forces of surface tension and electrostatic stress. Ions are formed in the region just above the cusp and are then accelerated through an aperture in the extractor and focused through an ion optical column onto the sample surface. Gallium is mainly used as the liquid metal because its melting point is 30 °C and it thus remains liquid at room temperature; other metals such as In and Au have been tried but require heating of the reservoir. Because ion production occurs in a very small volume, *gallium liquid metal ion sources (LMIS)* are very bright (~10^{10} A m^{-2} Sr^{-1}). As a result, the ion beam can be focused to a fine spot, resulting in a spot size of 0.2 μm at 8–10 keV; this is reduced to approximately 20 nm at 30 keV.

All ion gun optical columns are provided with deflection plates for scanning the ion beam over areas adjustable from a few square micrometers to many square millimeters. They have been adapted for pulsing by the introduction of additional deflection plates which rapidly sweep the beam across an aperture. By applying an ion beam bunching technique, ion pulses less than 1 ns wide can be produced.

3.1.2.2 **Mass Analyzers**

To minimize surface damage, static SIMS mass spectrometers should be as efficient as possible for detecting the total yield of secondary ions from a surface. Also, to be able to separate elemental from molecular species, and molecular species from each other, the mass resolution usually given as the mass m divided by the separable mass Δm, should be very high. With this in mind, two types of mass spectrometer have been used – in early work mainly quadrupole mass filters and, more recently, time-of-flight mass spectrometers.

3.1.2.2.1 **Quadrupole Mass Spectrometers**

Quadrupole mass spectrometers have been in use for many years as residual gas analyzers (with an ionizing hot filament) and in desorption studies and SSIMS. They consist of four circular rods, or poles, arranged equally spaced in a rectangular array and exactly coaxial. Figure 3.3 indicates the arrangement, as depicted by Krauss and Gruen [3.8]. Two voltages are applied to the rods, a dc voltage (U_{dc}) and an rf voltage ($U_{rf} \approx U_o \cos \omega t$). When an ion with a certain mass-to-charge ratio, m/q enters the space between the rods it is accelerated by the electrostatic field and for a particular combination of dc and rf voltages the ion has a stable trajectory and passes to a detector. For other combinations of voltages, the trajectory diverges rapidly and the ion is lost either as a result of hitting one of the poles or by passing between them to another part of the system. The mass resolution is governed by the dimensions of the mass spectrometer, the accuracy of its construction, and the stability and reproducibility of the ramped voltage. The quadrupole mass spectrometer is compact, does not require magnets, and is entirely ultra-high vacuum compatible – hence its popularity. It does have disadvantages, however, in that its transmission (typically $<1\%$) is very low and decreases with increasing mass number. It is, furthermore, a scanning instrument enabling only sequential transmission of ions, all other ions being discarded. The information loss is, therefore, very high.

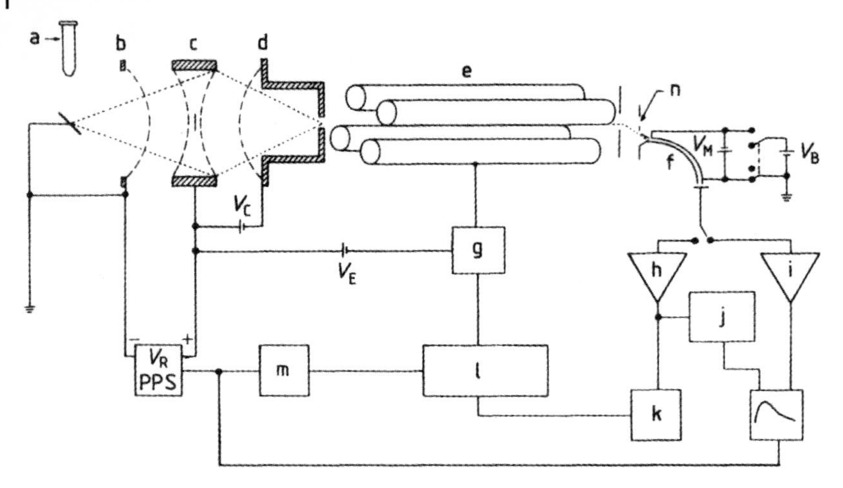

Fig. 3.3. Experimental arrangement used by Krauss and Gruen for SSIMS [3.8]; a quadrupole mass spectrometer was used for mass analysis and a retarding-field analyzer for prior energy selection; (a) ion gun; (b)–(d) lenses 1–3; (e) quadrupole mass spectrometer; (f) charge detecting electron multiplier; (g) quadrupole power supply; (h) pulse amplifier; (i) ISO amplifier; (j) rate meter; (k) multichannel analyzer; (l) mass programmer; (m) sawtooth generator; (n) exit hole for neutrals.

3.1.2.2.2 **Time-of-flight Mass Spectrometers (TOFMS)**

Because the quadrupole mass spectrometer is unsuitable for the analysis of large molecules on surfaces or of heavy metals and alloys, the TOF mass spectrometer was developed for use in SSIMS by Benninghoven and coworkers [3.9]. In a TOF mass analyzer (Fig. 3.4), all sputtered ions are accelerated to a given potential V (2–8 keV), so that all ions have the same kinetic energy. The ions are then allowed to drift through a field-free drift path of a given length L before striking the detector. According to the equation $(mL^2)/(2t^2) = qU_o$, light ions travel the fixed distance through the flight tube more rapidly than identically charged heavy ions. Thus, measurement of the time, t, of ions with mass-to-charge ratio, m/q, provides a simple means of mass analysis with $t^2 = (mL^2)/(2qU_o) \propto m/q$. Because a very well defined start time is required for flight time measurement, the primary ion gun must be operated in a pulsed mode to enable delivery of discrete primary-ion packages [3.10]. Electric fields (e.g. ion mirrors [3.10, 3.11] or electrical sectors [3.12, 3.13]) are used in the drift path to compensate for different incident energies and angular distributions of the secondary ions. For good mass resolution ($m/\Delta m \approx 10\,000$), the flight path must be sufficiently long (1–1.5 m), and very sophisticated high-frequency pulsing and counting systems must be employed to time the flight of the ion to within a tenth nanosecond. One great advantage of TOFMS is its capacity to provide simultaneous detection of all masses of the same polarity. Raw data acquisition [3.14] enables the reconstruction of TOF spectra for any ion species as a function of depth and lateral position.

The original TOF design [3.9] used pulsed beams of argon ions, but commercial development of the LMIS has significantly extended the capabilities of the TOF system.

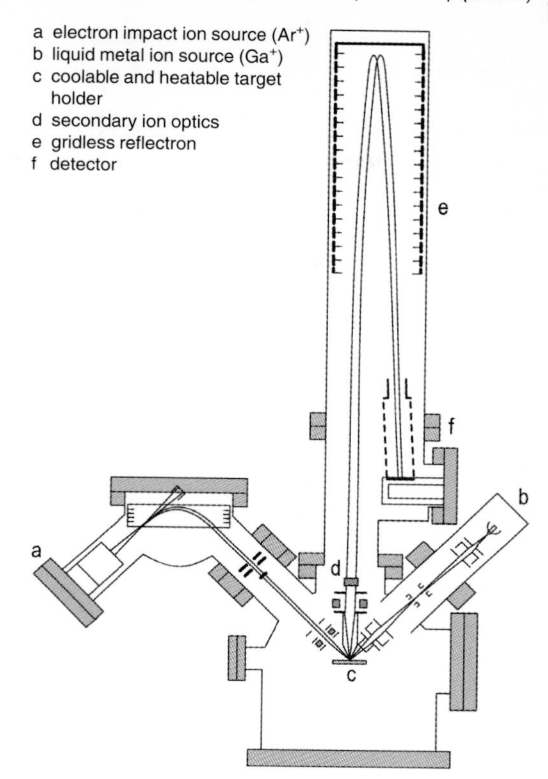

Fig. 3.4. Schematic diagram of the imaging time-of-flight SSIMS system used at the University of Münster, Germany.

The principle of LMIS operation enables the beam of isotopical enriched $^{69}Ga^{+}$ ions to be focused to a probe of 50-nm minimum diameter, while being pulsed at frequencies up to 50 kHz and rastered at the same time. Operating an LMIS at beam currents of 10–100 pA makes it possible to perform scanning SSIMS and to produce secondary ion images from all masses. Pulsing of the ion source is achieved by rapid deflection of the ion beam across a small aperture, the pulse length being variable between 1 and 50 ns. The flight path length is 1.5 m, part of which includes an energy compensator (ion mirror) to ensure that all ions of the same mass, but of different energies, arrive at the same time. Secondary electron detectors (SED) enable generation of topographical images produced by ion-induced secondary electrons; a low-energy ($\leq$20 eV) electron flood gun can be used for charge compensation, if an insulating sample is analyzed; a temperature-controlled sample holder enables acquisition of SSIMS data as a function of surface temperature (temperature-programmed SIMS, TPSIMS).

With such a TOF–imaging SSIMS instrument, the useful mass range is extended beyond 10000 amu; the mass resolution, $m/\Delta m$, is ~10000 with simultaneous detection of all masses; and within each image, all masses can be detected. The number of data generated in a short time is enormous, and very sophisticated data acquisition systems are required to handle and process the data.

3.1.3
Quantification

By bombarding a surface consisting of species A with primary ions, the surface coverage of A is reduced. Particles of A can be removed by desorption, by driving them into a deeper layer or, for molecular species, by fragmentation. The ratio of the number of sputtered particles to the number of primary ions is given by the disappearance yield $Y_D(A)$:

$$Y_D(A) = \frac{\text{No. of sputtered A}}{\text{No. of primary ions}} \tag{3.1}$$

A quantity related to the disappearance yield for a particle of a species which covers a solid surface with a surface density $\vartheta(A)$, is the disappearance cross-section, $\sigma_D(A)$:

$$\sigma_D(A) = \frac{Y_D(A)}{\vartheta(A)} \tag{3.2}$$

$\sigma_D(A)$ corresponds to the average area damaged by one primary ion impact. The change of the surface density $\vartheta(A)$ with increasing primary ion fluence F_{PI} is given by Eq. (3.3):

$$\vartheta(A, F_{PI}) = \vartheta(A)_0 \exp(-\sigma_D(A)\, F_{PI}) \tag{3.3}$$

Some particles sputtered from the surface are neutral whereas others are charged. Molecular particles can be emitted either as intact molecules or fragmented. The probability of the desorption of A into the emission channel X_i^q is given by the transformation probability $P(A \to X_i^q)$:

$$P(A \to X_i^q) = \frac{\text{No. of emitted particles } X_i^q}{\text{No. of sputtered A}} \tag{3.4}$$

where i is an index distinguishing between the different emission channels and q is the charge of the particle X_i with $q = 0, \pm1, \pm2, \pm3, \ldots$ The transformation probability for ionization cannot be measured easily because it is highly dependent on matrix and concentration. This is, in fact, the principal obstacle to achieving proper quantification in SSIMS ("matrix effect").

The average number of emitted particles X_i^q per incident primary ion is given by the secondary yield $Y^{(q)}(X_i^q)$:

$$Y^{(q)}(X_i^q) = \frac{\text{No. of particles } X_i^q}{\text{No. of primary ions}} \tag{3.5}$$

For the detection of sputtered ions, the transmission $T(X_i^q)$ of the mass spectrometer and the detection probability $D(X_i^q)$ must be taken into account. Their product gives the probability of detecting one sputtered particle. For a TOF mass spectrometer, both $T(X_i^q)$ and $D(X_i^q)$ are typically 10–50%.

The number of detected particles per incident primary ion is given by the detected yield $Y(X_i^q)$:

$$Y(X_i^q) = \frac{\text{No. of detected } X_i^q}{\text{No. of primary ions}} \tag{3.6}$$

An important quantity is the so-called useful yield $Y_u(X_i^q(A))$:

$$Y_u(X_i^q(A)) = \frac{\text{No. of detected } X_i^q}{\text{No. of sputtered A}} \tag{3.7}$$

Useful yield provides an overall measure of the extent to which the sputtered material is used for analysis. It is a quantity employed to estimate the sensitivity of the mass spectrometric method. Values of $Y_u(X_i^q(A))$ for elements typically range from 10^{-6} to 10^{-2} in TOF SIMS. The number of sputtered particles A per incident primary ion (*sputtering yield*) can be measured from elemental and multielemental standards under different operational conditions and can, therefore, by judicious interpolation between standards, be estimated with reasonable accuracy for the material being analyzed.

Because measuring A can be problematic, quantification is normally performed by relative sensitivity factor (RSF) methods. If a species A on the surface is detected by the ion X_i^q, the ratio of the detected ion current $I_A(X_i^q)$ to the primary ion current I_{PI} and the surface density $\vartheta(A)$ is called the practical sensitivity factor $S_p(X_i^q(A))$:

$$S_p(X_i^q(A)) = \frac{I_A(X_i^q)}{\vartheta(A) \cdot I_{PI}} \tag{3.8}$$

The ratio of two practical sensitivity factors is called the RSF $S(X_i^q(A), X_i^q(B)$, which is independent of the primary ion current I_{PI}:

$$S(X_i^q(A), X_{i'}^{q'}(B)) = \frac{S(X_i^q(A))}{S(X_{i'}^{q'}(B))} = \frac{I_A(X_i^q)}{I_B(X_{i'}^{q'})} \cdot \frac{\vartheta(B)}{\vartheta(A)} \tag{3.9}$$

By using RSFs it is possible to determine surface densities of other species, if the surface density of the reference species $\vartheta(B)$ is known.

Although the RSF contains matrix-dependent quantities, their variations are damped to some extent by virtue of taking ratios, and in practice the RSF is assumed constant for low concentrations of A (e.g. <1 atom%). It can be evaluated from measurements on a well-characterized set of standards containing A in known dilute concentrations. The accuracy of the method, however, is not as high as in laser-SNMS and XPS.

3.1.4
Spectral Information

A SSIMS spectrum, like any other mass spectrum, consists of a series of peaks of different intensity (i. e. ion current) occurring at certain mass numbers. The masses can be allocated on the basis of atomic or molecular mass-to-charge ratio. Many of the more prominent secondary ions from metal and semiconductor surfaces are singly charged atomic ions, which makes allocation of mass numbers slightly easier. Masses can be identified as arising either from the substrate material itself, from deliberately introduced molecular or other species on the surface, or from contaminations and impurities on the surface. Complications in allocation often arise from isotopic effects. Although some elements have only one principal isotope, for many others the natural isotopic abundance can make identification difficult.

Figure 3.5 shows the positive SSIMS spectrum from a silicon wafer, illustrating both the allocation of peaks and potential isobaric problems. SSIMS reveals many impurities on the surface, particularly hydrocarbons, for which it is especially sensitive. The spectrum also demonstrates reduction of isobaric interference by high-mass resolution. For reasons discussed in Sect. 3.1.3, the peak heights cannot be taken to be directly proportional to the concentrations on the surface, and standards must be used to quantify trace elements.

The relationship between what is recorded in a SSIMS spectrum and the chemical state of the surface is not as straightforward as in XPS and AES (Chap. 2). Because of the large number of molecular ions that occur in any SSIMS spectrum from a multicomponent surface (e. g. during the study of a surface reaction), much chemical information is obviously available in SSIMS, potentially more than in XPS. The problem in using the information from a molecular ion lies in the uncertainty of knowing whether or not the molecule represents the surface composition. For some materials,

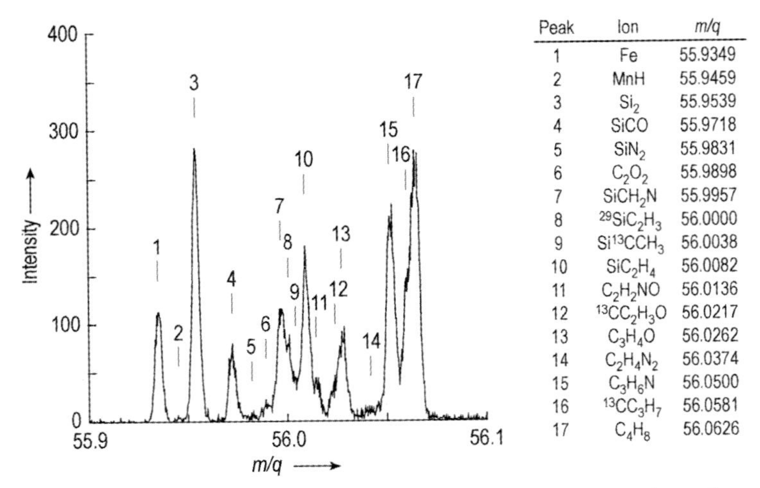

Peak	Ion	m/q
1	Fe	55.9349
2	MnH	55.9459
3	Si_2	55.9539
4	SiCO	55.9718
5	SiN_2	55.9831
6	C_2O_2	55.9898
7	$SiCH_2N$	55.9957
8	$^{29}SiC_2H_3$	56.0000
9	$Si^{13}CCH_3$	56.0038
10	SiC_2H_4	56.0082
11	C_2H_2NO	56.0136
12	$^{13}CC_2H_3O$	56.0217
13	C_3H_4O	56.0262
14	$C_2H_4N_2$	56.0374
15	C_3H_6N	56.0500
16	$^{13}CC_3H_7$	56.0581
17	C_4H_8	56.0626

Fig. 3.5. High-mass-resolution TOF SIMS spectrum of a contaminated Si wafer.

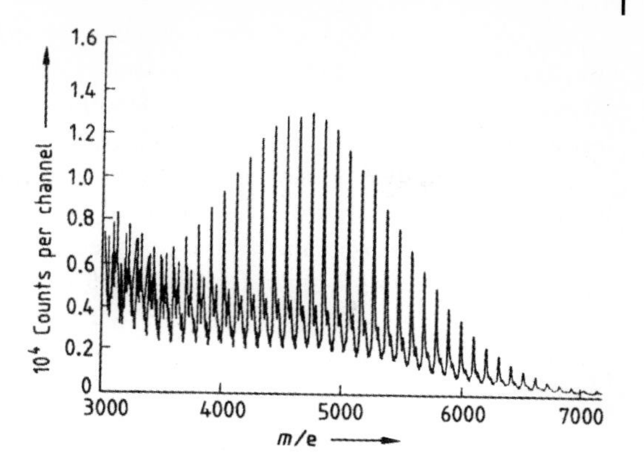

Fig. 3.6. Positive SSIMS spectrum of polystyrene ($M_n = 5100$, $M_{n,\text{calc}} = 4500$, $M_{n,\text{corr}} = 5600$) from a TOFMS [3.15].

e.g. polymers, the clusters observed are definitely characteristic of the material, as seen in Fig. 3.6, from Bletsos et al. [3.15], in which the SSIMS spectrum from polystyrene measured in a TOFMS contains peaks spaced at regular 104-mass-unit intervals, corresponding to the polymeric repeat unit. SSIMS measurements have also proved to be a very powerful tool in the characterization of additives in plastic materials. On the other hand, for example during oxidation studies, various metal–oxygen clusters are usually found in both positive and negative secondary ion spectra, the relationship of which to the actual chemical composition is problematic. Figure 3.7, from McBreen et al. [3.16], shows SSIMS spectra recorded after interaction of oxygen with half a monolayer of potassium adsorbed on a silver substrate. The positive and negative spectra both show evidence of K_iO_j clusters which the authors were unable to relate to the surface condition, although quite clearly much chemical information is available. Sometimes, the use of pure and well-characterized standards can be of help in the interpretation of complex spectra.

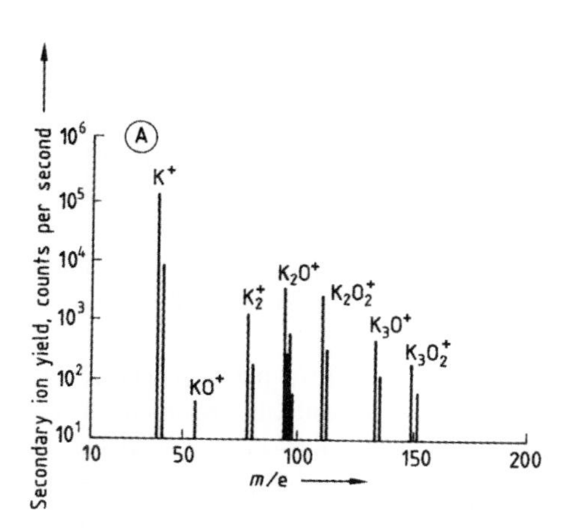

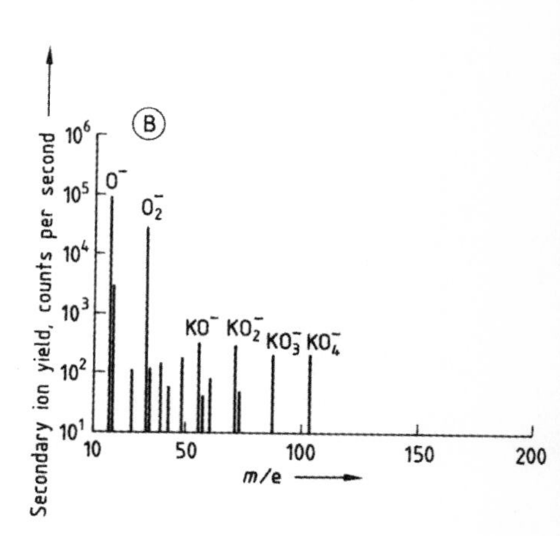

Fig. 3.7. SSIMS spectra after interaction of oxygen with half a monolayer of potassium adsorbed on silver, at 400 K [3.16]; (A) positive SSIMS spectrum; (B) negative SSIMS spectrum.

In conclusion, SSIMS spectra provide not only evidence of all the elements present, but also detailed insight into molecular composition. Quasimolecular ions can be desorbed intact up to 15 000 amu, depending on the particular molecule [3.17] and on whether an effective mechanism of ionization is present. Larger molecules can be identified from fragment peak patterns which are characteristic of the particular molecules. If the identity of the material being analyzed is completely unknown, spectral interpretation can be accomplished by comparing the major peaks in the spectrum with those in a library of standard spectra.

3.1.5 Applications

In general, SSIMS tends not to be used for the quantification of surface composition because of all the uncertainties described above. Its application has been more qualitative in nature, with emphasis on its advantages of high surface specificity, very high sensitivity for certain elements and molecules, multiplicity of chemical information, and high spatial resolution in the TOF imaging mode. It can be used to determine the surface composition, and the surface contaminants present, as a function of lateral position for all kinds of material, e. g. metals, alloys, semiconductors, oxides, carbides, ceramics, glasses, organic molecular species, organometallic compounds, biomolecules, and polymers, etc. It can also be used to study surface phenomena such as adhesion, chemi- and physisorption, wettability, wear and tear, lubrication, chemical reactivity, corrosion, surface diffusion, and segregation, etc. As a common analytical tool it is currently used for many applications, e. g. coatings, composite materials, micro- and nanoelectronics, heterogeneous catalysis, biosensors, combinatorial chemistry, pharmaceuticals, medical implants, chromatographies, sensors, painting, environment control and protection, and failure analysis, etc.

3.1.5.1 Oxide Films

Compared with XPS and AES, the higher surface specificity of SSIMS (1–2 monolayers compared with 2–8 monolayers) can be useful for more precise determination of the chemistry of an outer surface. Although from details of the O1s spectrum, XPS could give the information that OH and oxide were present on a surface, and from the C1s spectrum that hydrocarbons and carbides were present, only SSIMS could be used to identify the particular hydroxide or hydrocarbons. In the growth of oxide films for different purposes (e. g. passivation or anodization), such information is valuable, because it provides a guide to the quality of the film and the nature of the growth process.

One important area in which the properties of very thin films play a crucial role is the adhesion of aluminum to aluminum in the aircraft industry. For bonds of the required strength to form, the aluminum surfaces must be pre-treated by anodization, which creates not only the right chemical conditions for adhesion but also the right structure in the resulting oxide film. Films formed by anodizing aluminum in phosphoric acid under ac conditions have been studied by Treverton et al. [3.18] using both SSIMS and XPS. Knowledge both of the chemistry of the outer monolayers in terms of aluminum oxides and hydroxides, and of the nature and extent of any contamination introduced by the anodizing process, was important. Fig. 3.8 compares the positive and negative SSIMS spectra obtained after anodizing for 3 and 5 s at 20 V and 50 Hz. Both sets of spectra show that the longer anodizing treatment resulted in significant reduction in hydrocarbon contamination (peaks at 39, 41, 43,

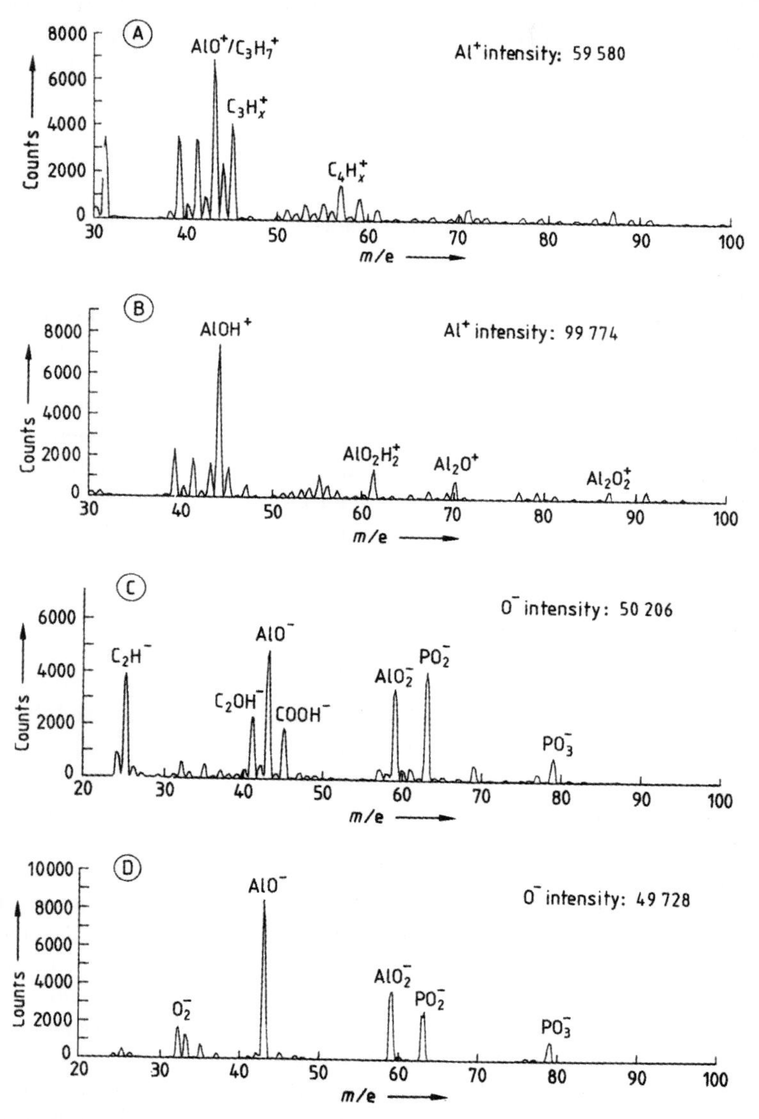

Fig. 3.8. SSIMS spectra from oxide films on aluminum after anodization in phosphoric acid for 3 or 5 s [3.18]. (A) positive SSIMS spectrum, 3 s; (B) positive SSIMS spectrum, 5 s; (C) negative SSIMS spectrum, 3 s; (D) negative SSIMS spectrum, 5 s.

and 45 amu in the positive spectrum, and 25, 41, and 45 amu in the negative), and marked increases in oxide- and hydroxide-associated cluster ions at 44, 61, 70, and 87 amu (positive) and 43 and 59 amu (negative), respectively. At higher masses, peaks allocated to $Al_2O_4H^-$, $Al_2O_3^-$, and $Al_3O_5^-$ were also identified. Residual PO_2^- and PO_3^- from the anodizing bath were reduced as a result of the longer anodizing time. The low level of contaminants is important, because the adhesive coating can then wet the surface more effectively, leading to improved adhesion. In addition, the change from AlO^+ (mass 43) to $AlOH^+$ (mass 44) in aluminum chemistry as a result of changing from 3- to 5-s anodizing is a significant effect that could not have been established by XPS.

3.1.5.2 **Interfaces**

The protection of steel surfaces by paint depends significantly on the chemistry of the paint–metal interface. The system has many variables, because the metal surface is usually pretreated in a variety of ways, including galvanizing and phosphating. Indeed, there are probably several interfaces of importance, and corrosion protection might be a function of the conditions at all of them.

As one of the final step in car production, visual checks are performed to determine the quality of the paint layers. If defects are observed, the car will not be processed further, and paint repair is necessary. Figs 3.9 and 3.10 describe the analysis of such a defect (crater in a red paint coat) [3.19]. The upper sequence of Fig. 3.9 depicts a series of optical images of the defect for different fields of view (the right image is approximately $150 \times 150\ \mu m^2$). Paint remnants can be observed in the center of the crater. The second row of Fig. 3.9 shows ion-induced secondary electron images taken from the crater with a highly focused liquid metal ion beam. The images clearly depict the crater topography.

Figure 3.10 shows the secondary ion spectra recorded from within the defect and its surroundings. The spectrum from the defective area clearly shows the presence of a polluting perfluorinated polyether structure (the $C_xF_y^+$ peaks and the peak at mass 47 amu (CFO^+) are diagnostic of the polyether structure).

The third sequence in Fig. 3.9 shows a mass-resolved secondary ion image recorded from the crater. On the far left, the lateral distribution of the uncharacteristic hydrocarbon $C_3H_5^+$ can be seen. The signal is enhanced in the crater because of increased sputtering yield. Topographical information can be obtained from hydrocarbon images. The next image shows the lateral distribution of the silicone oil, which is incorporated into the paint as a smoothing agent. It can be seen that silicone oil is detected only from the paint itself and not in the crater bottom. The silicone oil can, therefore, be excluded as the cause of the defect. The last three images show the lateral distributions of several $C_xF_y^+$ fragments, indicating unambiguously that the perfluorinated polyether is found exclusively in the crater.

Later it was found that the polluting lubricant droplets originating from the transport belts used in the production; they had fallen into the paint bath and prevented adhesion of the paint to the metal. It can be concluded that the high sensitivity of SSIMS in the detection of submonolayer coverage of organic species makes it an extremely powerful tool for solving such interface problems.

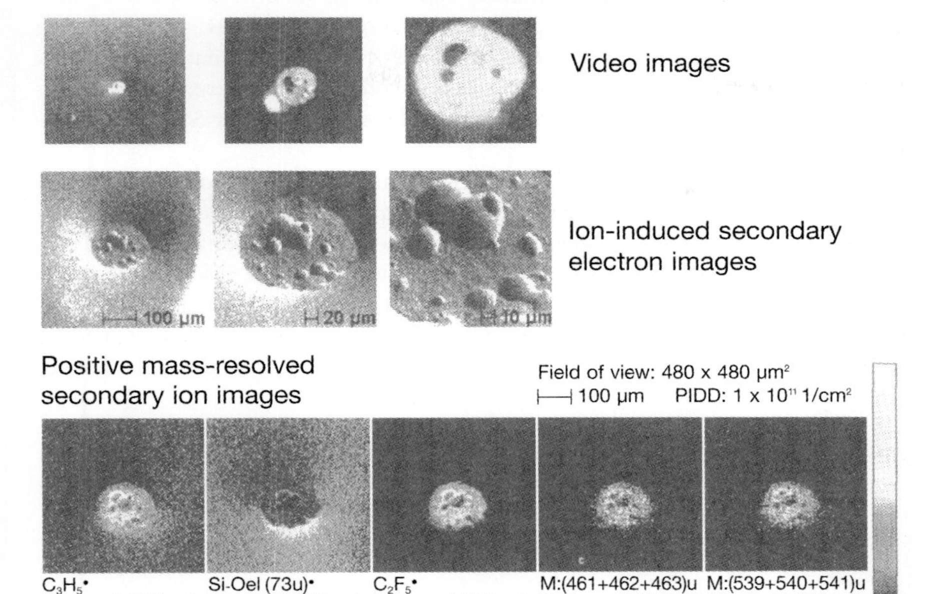

Fig. 3.9. Video images, ion-induced secondary electron images, and mass-resolved ion images of a defect in car paint; white signal corresponds to high intensities, black to low intensities [3.19].

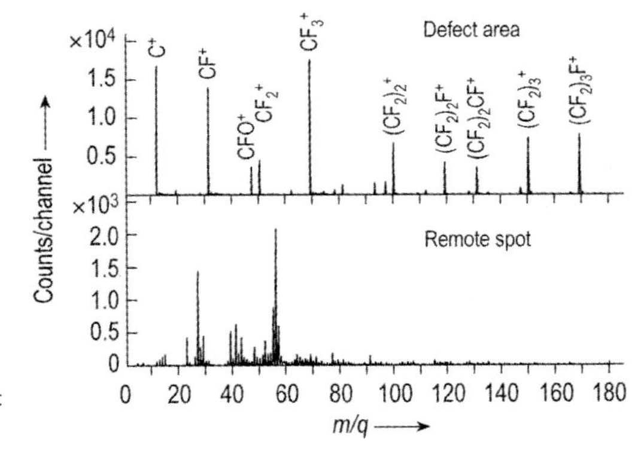

Fig. 3.10. Positive TOF SIMS spectra from the defective area and from a spot remote from the defect [3.19].

3.1.5.3 Polymers

Treatment of polymer surfaces to improve their wetting, water repulsion, and adhesive properties is now a standard procedure. The treatment is designed to change the chemistry of the outermost groups in the polymer chain without affecting bulk polymer properties. Any study of the effects of treatment therefore requires a technique that is specific mostly to the outer atomic layers; this is why SSIMS is extensively used in this area.

Plasma etching is a favored form of surface treatment. Depending on the conditions of the plasma discharge (i. e., nature of the discharge gas, gas pressure, rate of gas flow, discharge power and frequency, and substrate polarity), plasma etching can alter the chemical characteristics of a surface over a wide range. A plasma discharge is by nature highly controllable and reproducible. The effects of plasma treatment on the surface of polytetrafluoroethylene (PTFE) have been studied by Morra et al. [3.20] using combined SSIMS and XPS. They found, as is apparent from Fig. 3.11, that not

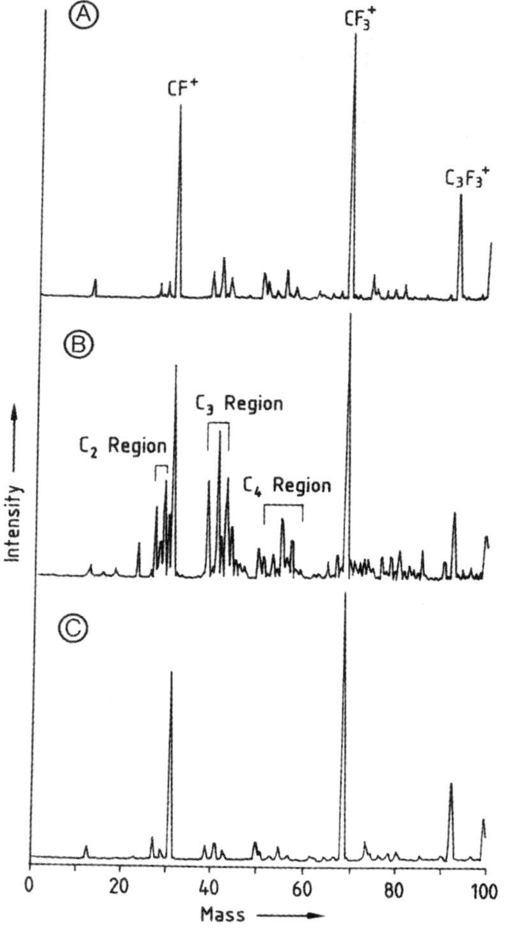

Fig. 3.11. Positive SSIMS spectra from PTFE [3.20]: (A) untreated; (B) 0.5 min oxygen plasma treatment; (C) 15 min oxygen plasma treatment.

only did the plasma change the surface chemistry, but the surface emission fragments themselves led to a temporary change of the plasma discharge conditions. Thus the untreated (Fig. 3.11A) and 15 min oxygen plasma-treated (Fig. 3.11C) positive-ion spectra are virtually identical, whereas the 0.5 min oxygen plasma-treated spectrum (Fig. 3.11B) is quite different. The initial breakup of the PTFE surface produced hydrocarbon fragments which reacted with the oxygen to produce water vapor, thus changing the local nature of the discharge gas. The important deduction to be made is that one of the vital conditions in plasma treatment is the length of contact time; in this instance the surface was unchanged if the contact time was too long.

Other SSIMS studies of polymer surfaces have included perfluorinated polyether [3.21], low-density polyethylene [3.22], poly(ethylene terephthalate) [3.23], and the oxidation of polyetheretherketone [3.24].

3.1.5.4 **Biosensors**

In recent years, biosensor chip technology has been a subject of growing interest for a variety of applications. In particular, DNA sequencing chips based on the method of sequencing by hybridization are becoming increasingly important. Unknown DNA sequences, which are typically labeled with radioactive or fluorescent markers, are hybridized to known complementary short DNA sequences called oligodeoxynucleotides (ODN), which have been immobilized on a solid surface [3.25–3.27]. This method has become valuable for clinical diagnostics, the sequencing of cDNAs or the partial sequencing of clones, DNA and RNA sequencing, gene polymorphism studies, and identification of expressed genes. Use of peptide nucleic acid (PNA) hybridization biosensor chips [3.26, 3.27] would even enable the sequencing of unlabeled DNA. PNA is a synthesized DNA analog in which both the phosphate and the deoxyribose of the DNA backbone are replaced by polyamides, resulting in a phosphate-free PNA backbone. This enables detection of the presence of DNA from a phosphate signal.

Figure 3.12 depicts TOF SIMS spectra obtained from ODN and PNA immobilized on silanized silicon wafers. The spectra clearly demonstrate that the masses corresponding to PO_2^- and PO_3^- provide the best correlation of the presence of ODN, enabling their use for precise distinction between ODN and PNA. The CF_3^- and $C_2O_2F_3^-$ peaks seen in the PNA spectra represent trifluoroacetic acid, which was part of the PNA solution. Deprotonated $(Cyt\text{-}H)^-$ and $(Thy\text{-}H)^-$ signals of the bases cytosine and thymine are observed for both immobilized PNA and ODN sequences and can be used to detect the presence of these bases.

Figure 3.13 shows the thermal stability of immobilized ODN and PNA. The signal for the Thy- and Cyt-bases obtained with temperature-programmed (TP) SIMS starts to decrease at approximately 150 °C for ODN and 200 °C for PNA. This variance is caused by the different strengths of binding between the bases and the sugar–phosphate and peptide backbones, respectively.

The data show that SSIMS can be used as a tool for characterizing the different steps in the production of biosensors, or even for sequencing. Similarly, SSIMS can be used to solve a variety of problems in bioanalytical chemistry, e.g. screening of combinatorial libraries, characterizing Langmuir–Blodgett layers, etc.

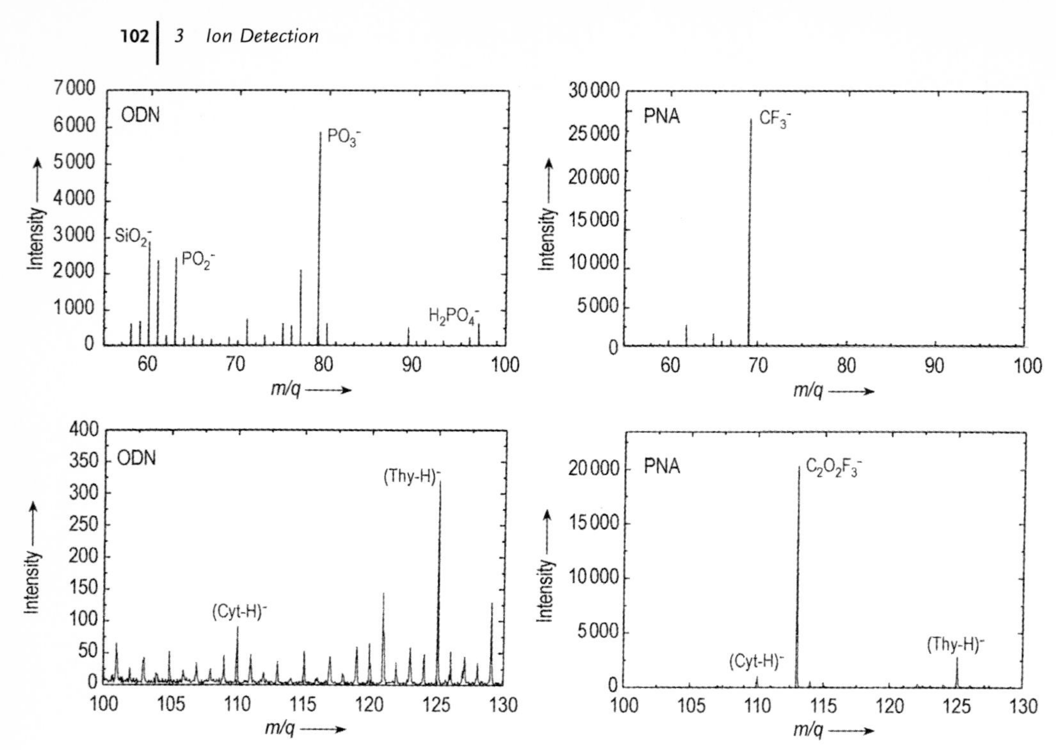

Fig. 3.12. Negative TOF SIMS spectra obtained from oligodeoxynucleotides (ODN) (left) and PNA (right) [3.26].

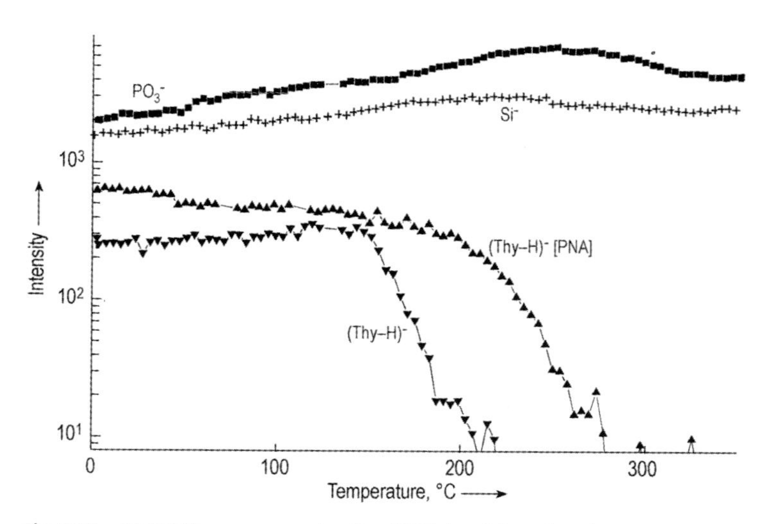

Fig. 3.13. TP SIMS measurements of an ODN ($c = 0.1$ mm) and PNA ($c = 0.1$ mm, curve labeled as PNA) TTTTCCCTCTCTC-sequence [3.26].

3.1.5.5 Surface Reactions

Because of the inherently destructive nature of ion bombardment, the use of SSIMS alone in the study of the reactions of surfaces with gases and vapor must be viewed with caution, but in combination with other surface techniques it can provide valuable additional information. The parallel techniques are most often XPS, TDS, and LEED, and the complementary information required from SSIMS normally refers to the nature of molecules on surfaces and with which other atoms, if any, they are combined.

A typical SSIMS spectrum of an organic molecule adsorbed on a surface is that of thiophene on ruthenium at 95 K, shown in Fig. 3.14 (from the study of Cocco and Tatarchuk [3.28]). Exposure was 0.5 Langmuir only (i.e. 5×10^{-7} torr s = 37 Pa s), and the principal positive ion peaks are those from ruthenium, consisting of a series of seven isotopic peaks around 102 amu. Ruthenium–thiophene complex fragments are, however, found at ca. 186 and 160 amu; each has the same complicated isotopic pattern, indicating that interaction between the metal and the thiophene occurred even at 95 K. In addition, thiophene and protonated thiophene peaks are observed at 84 and 85 amu, respectively, with the implication that no dissociation of the thiophene had occurred. The smaller masses are those of hydrocarbon fragments of different chain length.

SSIMS has also been used to study the adsorption of propene on ruthenium [3.29], the decomposition of ammonia on silicon [3.30], and the decomposition of methane thiol on nickel [3.31].

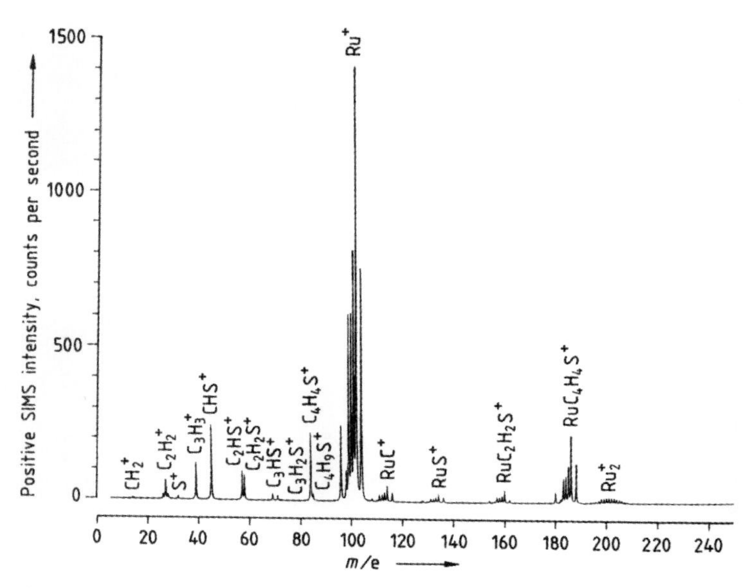

Fig. 3.14. Positive SSIMS spectrum from Ru (001) after exposure to 0.5 L thiophene at 95 K [3.28].

3.1.5.6 Imaging

SSIMS has been used in the TOF SSIMS imaging mode to study very thin layers of organic materials [3.32–3.36], polymeric insulating materials [3.37], and carbon fiber and composite fracture surfaces [3.38]. In these studies a spatial resolution of ca. 80 nm in mass-resolved images was achieved.

Figure 3.15 shows several mass-resolved images of polymer blends (Nylon 6 (Ny6) and bisphenol A polycarbonate (PC)), mixed for 20 and 40 min, respectively [3.39]. Reactive blending of polymers and exchange reactions that can occur during the melt-mixing process have attracted considerable interest from many researchers. Reactive mixing of immiscible polymers can lead to partial compatibilization of the two components with properties advantageous for technological applications. It is evident that the 20-min sample is quite inhomogeneous, with large distinct domains of different composition, whereas for the 40-min sample the distribution is more homogeneous. Detailed spectral analysis of the different imaging regions supported the picture that reactive mixing and compatibilization of Ny6/PC blends occurs via the incorporation of PC in the nylon phase.

TOF SIMS imaging can also be used to detect extremely small particles on a variety of substrates, e.g. silicon wafers or photographic films [3.40]. Figure 3.16 depicts

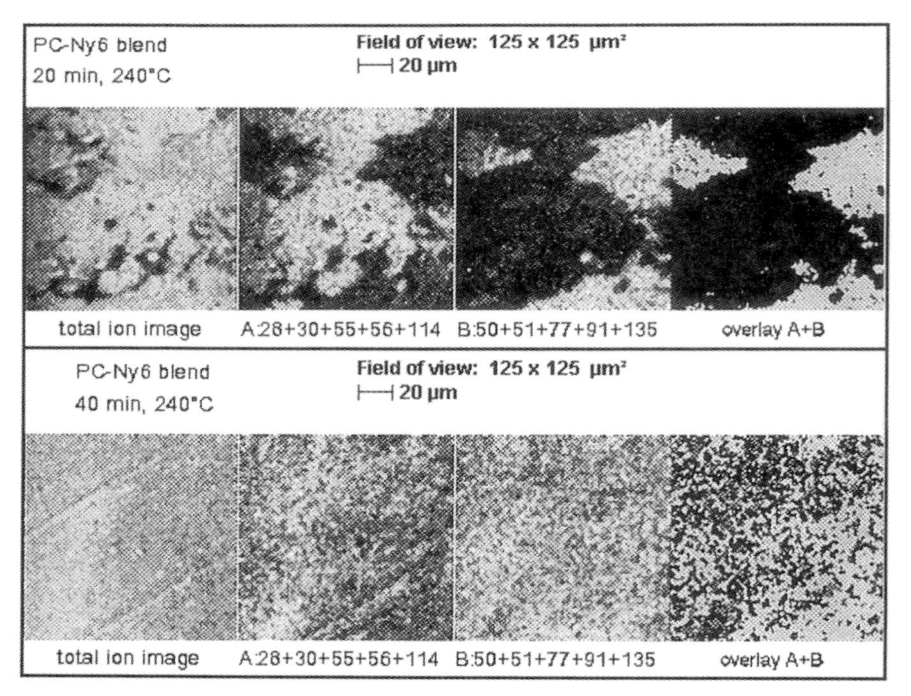

Fig. 3.15. Secondary ion images of blends melt-mixed for 20 min (top) and 40 min (bottom). Images A and B were obtained by adding the mass-resolved images pertaining to the masses listed below each image. Also shown are the overlays of A (black) and B (white) [3.39].

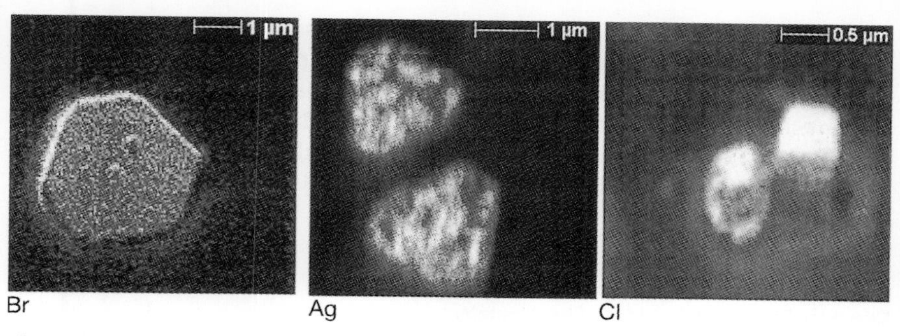

Fig. 3.16. High-resolution TOF SIMS images of silver bromide and silver chloride crystals.

images of silver bromide and silver chloride crystals, demonstrating the high spatial resolution (50 nm) possible with TOF SIMS. Current development is aiming to reach a spatial resolution for chemical mapping of approximately 20 nm.

3.1.5.7 **Ultra-shallow Depth Profiling**

In recent years TOF SIMS has also proved to be a very powerful tool for ultra-shallow depth profiling, having the advantage of simultaneously detecting all elements of interest. The dual beam mode [3.41], in particular, (see Sect. 3.2.2.1) enables optimized depth resolution, because sputtering conditions can be independently optimized.

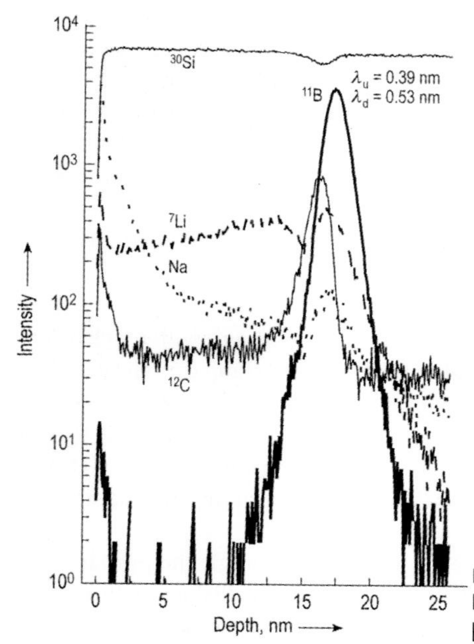

Fig. 3.17. Multi-element depth profile of a B layer in Si performed with 600-eV SF_5^+ sputtering [3.41].

Figure 3.17 depicts an ultra-shallow TOF SIMS depth profile of a 100-eV B-implant in Si, capped with 17.3 nm Si. The measurement was performed with 600-eV SF_5^+-sputtering and with O_2-flooding. The original wafer surface, into which the B was implanted, is indicated by the maxima of the alkali- and C-signals. Because of these contaminants, a minimum is observed in the ^{30}Si-signal. The dynamic range of the B-profile is more than 3.5 decades and the depth resolution is ≤ 0.5 nm.

3.2
Dynamic Secondary Ion Mass Spectrometry

Herbert Hutter

Today dynamic SIMS is a standard technique for measurement of trace elements in semiconductors, high performance materials, coatings, and minerals. The main advantages of the method are excellent sensitivity (detection limit below 1 µmol mol^{-1}) for all elements, the isotopic sensitivity, the inherent possibility of measuring depth profiles, and the capability of fast direct imaging and 3D species distribution.

The first SIMS instrument was built by Herzog and Viehboeck [3.42] in Vienna in the nineteen-forties. In the early nineteen-sixties, Herzog and Liebl [3.43] built the first sophisticated SIMS instrument at about the same time as Castaing and Slodzian in Paris [3.44]. In 1970, Benninghoven was the first to use the acronym SIMS [3.45].

3.2.1
Principles

The bombardment of a sample with a dose of high energetic primary ions (1 to 20 keV) results in the destruction of the initial surface and near-surface regions (Sect. 3.1.1). If the primary ion dose is higher than 10^{11} ions mm^{-2} the assumption of an initial, intact surface is no longer true. A sputter equilibrium is reached at a depth greater than the implantation depth of the primary ions. The permanent bombardment of the sample with primary ions leads to several *sputter effects* more or less present on any sputtered surface, irrespective of the instrumental method (AES, SIMS, GDOES ...).

Compensation of Preferential Sputtering. The species with the lower sputter yield is enriched at the surface. This effect is called preferential sputtering and complicates, e.g., Auger measurements. The enrichment compensates for the different sputter yields of the compound or alloy elements; thus in dynamic SIMS (and other dynamic techniques in which the signal is derived from the sputtered particles, e.g. SNMS, GD–MS, and GD–OES), the flux of sputtered atoms has the same composition as the sample.

Atomic Mixing. Depending on their mass, energy and impact angle the primary ions reach a mean depth until they are finally stopped by many collisions with sample atoms. Sample atoms are moved from their initial locations (see Sect. 3.1.1). This results in atomic mixing of the near-surface region and limits the depth resolution. Small primary ion energies, the use of molecular primary ions within which total energy is distributed to the split-off atoms according to their mass (see Sect. 3.1.5.7), and oblique bombardment angles reduce this effect.

Implantation of Primary Ions. The primary ions are implanted in the sample and thereby influence the chemical constitution. For energies in the 20 keV range the implantation depth is approximately 30 nm. The sputter yield, i.e. the ratio of secondary to primary particles (not only ions), is energy-dependent and has a maximum in the 10 keV range.

Crater Bottom Roughening. Depth resolution is also limited by roughening of the crater bottom under the action of ion bombardment. On polycrystalline samples this can be because of different sputter yields of different crystal orientations, because the sputter yields of single crystals can vary by a factor of two depending on their orientation. Because of this type of roughening, depth resolution deteriorates with increasing sputter depth.

Sputter-induced Roughness. Sputtering single crystal semiconductors, in particular II–VI semiconductors, with Cs^+ ions results in surface roughening. There is currently no satisfactory explanation of this effect. This effect, and the corresponding reduction of the depth resolution, can be avoided by rotating the sample during the measurement.

Charging Effects. The electrostatic potential of non-conducting samples and layers is changed due to the bombardment of charged atoms and the emission of secondary electrons and ions. Usually this results in a positive charging of the sample and a corresponding drift of the secondary ion energy. On good insulators this charging reaches such high values that the primary ions are repelled by the surface making analysis impossible. In order to compensate for this effect, low-energy electrons can be directed to the surface ("electron shower").

Because of the complex situation on the surface, satisfactory theoretical description of the ionization process leading to *secondary ion formation* has not yet been possible. Different ionization mechanisms have been proposed:

Electron-tunneling Model. Several models based on quantum mechanics have been introduced. One describes how an electron of the conducting band tunnels to the leaving atom, or vice versa. The probability of tunneling depends on the ionization potential of the sputtered element, the velocity of the atom (time available for the tunneling process) and on the work function of the metal (adiabatic surface ionization, Schroeer model [3.46]).

Broken Bond Model were developed to describe the process of ionization of ionic compounds, especially under primary oxygen-ion bombardment – or gas admission ("O_2

shower"). The model requires the presence of an oxide layer on the surface. The binding electrons remain at the oxygen atom and, therefore, the emission of positively ionized atoms results [3.47].

Local Thermodynamic Equilibrium (LTE). This LTE model is of historical importance only. The idea was that under ion bombardment a near-surface plasma is generated, in which the sputtered atoms are ionized [3.48]. The plasma should be under local equilibrium, so that the Saha–Eggert equation for determination of the ionization probability can be used. The important condition was the plasma temperature, and this could be determined from a knowledge of the concentration of one of the elements present. The theoretical background of the model is not applicable. The reason why it gives semi-quantitative results is that the exponential term of the Saha–Eggert equation also fits quantum-mechanical expressions.

3.2.2 Instrumentation

The basic instrumental set-up for dynamic SIMS is the same as for SSIMS (Sect. 3.1.2). Depending on the intensity, beam diameter, and ion species needed, different ion sources are used. Several mass analyzers with different characteristics enable a broad field of applications.

3.2.2.1 Ion Sources

Bombardment with reactive species increases the ionization yield and, therefore, the SIMS sensitivity. Electropositive elements are more sensitively detected by use of oxygen bombardment, electronegative elements with Cesium bombardment (Sect. 3.2.3). In dynamic SIMS, therefore, Cs^+ or O_2^+ ions are usually used for sample sputtering. Because of the lower melting point of gallium and, as a consequence, the higher achievable brightness of a Ga^+ source, this type is also used for fine-focus imaging applications, despite its lower sensitivity. A liquid metal ion source (LMIS) described in Sect. 3.1.2.1 can also be run with Cs^+. O_2^+ beams are produced in electron impact/extraction or duoplasmatron sources.

Electron impact (EI) ion sources are the simplest type. O_2, Ar, or another (most often noble) gas flows through an ionization region similar to that depicted in Fig. 3.30. Electrons from an incandescent filament are accelerated to several tens of eV by means of a grid anode. A 20–100 eV electron impact on a gas atom or molecule typically effects its ionization. An extraction cathode accelerates the ions towards electrostatic focusing lenses and scanning electrodes.

The Duoplasmatron (Fig. 3.18). In the Duoplasmatron, gas-discharge ion sources are used for bombardment with oxygen or argon. In dynamic SIMS, especially, the use of O_2^+ ions is common because of the chemical enhancement effect. With a duoplasmatron ion beam currents of several microamps can be generated. The diameter of the beam can be focused down to 0.5 μm.

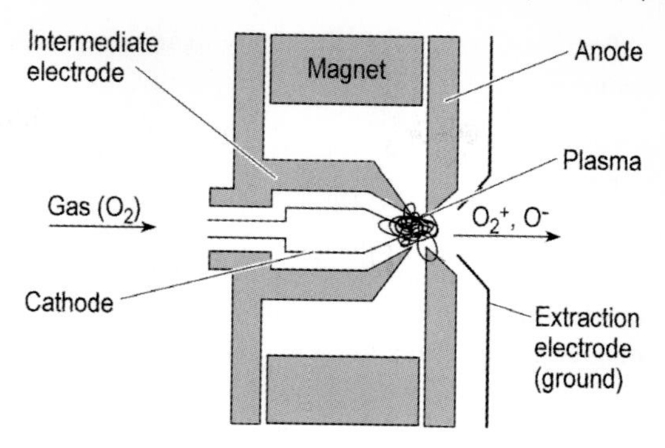

Fig. 3.18. Schematic diagram of a Duoplasmatron ion source.

The attainable particle current density per solid angle of the beam (ions $\mu m^{-2}\,s^{-1}\,sr^{-1}$) is an inherent property of the ion source, the so-called brightness. Because of this, reduction of the beam diameter is effected by reducing the beam current.

Dual-beam Technique for TOF Instruments. Because of their low duty cycle, TOF instruments are less suitable for common depth profiles. If only one fine-focus pulsed ion gun is used, the sample is bombarded by primary ions for only 0.1% of the time; it would, therefore, take a long time to reach deeper regions. But by the use of a second ion gun working anti-cyclically to the "measuring" beam, without an applied accelerating potential, substantial amounts of sample material can be removed. One advantage of this dual-beam technique is that sputtering for measurement and sputtering for material removal are not coupled; beam diameter, energy, mass, and angle of the "removal" gun can, therefore, be optimized independently of the "measuring" gun. One disadvantage is that the atoms sputtered during the material-removal phase are not analyzed. This reduces the detection power. By simultaneously detecting all masses one can only partially reduce this disadvantage, especially in implantation profiles where only one mass is of interest.

3.2.2.2 Mass Analyzers

Apart from the quadrupole and TOF analyzers described in Sect. 3.2.2, the most important types of mass analyzer used in common dynamic SIMS instruments employ a magnetic-sector field.

Magnetic Sector Field. In a magnetic field **B** an ion with the velocity **v** and the charge q experiences a centripetal force, the Lorentz force **F**:

$$\mathbf{F} = q(\mathbf{v} \times \mathbf{B}) \tag{3.10}$$

This force must be balanced by the centrifugal force of the ion mv^2/r, where r is the radius of curvature of the trajectory. Hence,

$$\frac{mv^2}{r} = q \cdot (\mathbf{v} \times \mathbf{B}) \tag{3.11}$$

Because of this force particles with an energy $E = mv^2/2$ are held on circular paths of radius r:

$$r = \sqrt{\frac{2E(m/q)}{|\mathbf{B}|}} \tag{3.12}$$

Secondary ions are accelerated from the surface to an energy E. After passing an entrance slit different masses are forced to different radial paths in the homogeneous magnetic field. Only mass m ($\pm\Delta m$) defined by:

$$\frac{m}{q} = \frac{|\mathbf{B}|^2\, r^2}{2\,E} \tag{3.13}$$

can pass the exit slit and reach the detector. Because the probability of formation of single ionized atoms is considerably greater than that of more highly charged particles, q in Eq. (3.4) can normally be set to unity. By scanning the magnetic field all masses are detected sequentially.

Two types of commercial magnetic sector field instrument will be described in the following:

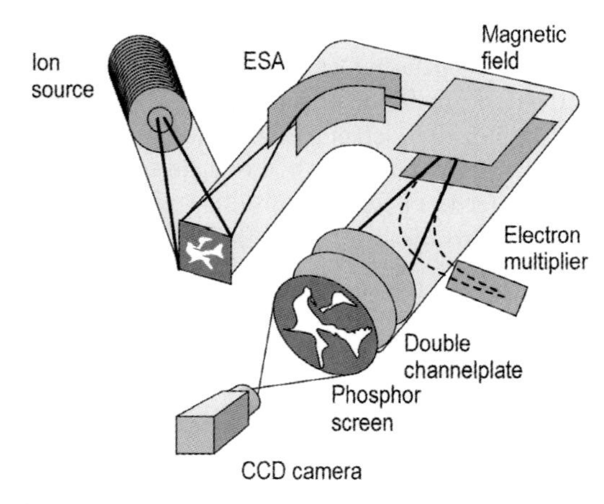

Fig. 3.19. Basic set-up of a direct imaging magnetic sector instrument. The stigmatic secondary ion optics consists of an electrostatic analyzer (ESA) and a magnet sector field.

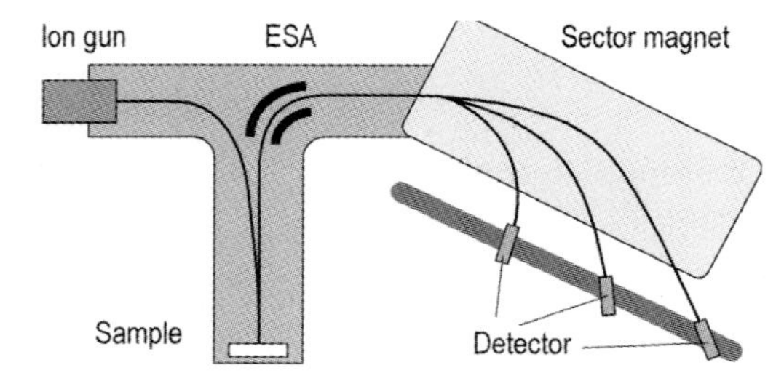

Fig. 3.20. Schematic drawing of a Mattauch–Herzog magnetic-sector instrument with simultaneous mass detection.

The *direct imaging magnetic sector mass analyzer* (Fig. 3.19) has the unique property that all parts (lenses, electrostatic analyzer and magnetic sector field) of the secondary ion optics are stigmatic (comparable with light microscopes). This means that all points of the surface are simultaneously projected into the analyzer.

The *Mattauch–Herzog geometry* (Fig. 3.20) enables detection of several masses simultaneously and is, therefore, ideal for scanning instruments [3.49]. Up to five detectors are adjusted mechanically to locations in the detection plane, and thus to masses of interest. Because of this it is possible to detect, e. g., all isotopes of one element simultaneously in a certain mass range. Also fast, sensitive, and precise measurements of the distributions of different isotopes are feasible. This enables calculation of isotope ratios of small particles visible in the image. The only commercial instrument of this type (Cameca Nanosims 50) uses an ion gun of coaxial optical design, and secondary ion extraction; the lateral resolution is 50 nm.

3.2.2.3 Detector

Ion intensities up to a count rate of 2×10^6 are measured using a secondary electron multiplier (SEM) . When it becomes saturated above that value, it is necessary to switch to a Faraday cup. Its ion-current amplification must be adjusted to fit to the electron multiplier response.

For measurement of local ion intensities in the direct imaging mode (see Fig. 3.19), amplification ensuring laterally uniform-single ion detection is necessary. Depending on the sensitivity of the detector a single or double channel plate is used. Two imaging devices are in use:

CCD Camera. For standard CCD cameras a double-channel plate (amplification $>10^6$) is necessary. For high-sensitivity cameras (sensitivity $>10^{-3}$ lux, cooled or with internal amplification) a single channel plate suffices. By controlling the channel plate high-voltage, i. e. amplification, a high dynamic range can be achieved with this system [3.50].

Resistive Anode Encoder (RAE). This detector has the advantage that the single-ion events are detected digitally. It therefore it delivers quantitative results, irrespective of local differences in the amplification of the channel plate. One disadvantage is that the count rate is limited to 200 000.

3.2.3 Spectral Information

The element sensitivity is determined by the ionization probability of the sputtered atoms. This probability is influenced by the chemical state of the surface. As mentioned above, Cs^+ or O_2^+ ions are used for sample bombardment in dynamic SIMS, because they the increase ionization probability. This is the so-called chemical enhancement effect.

The near-surface region is partially oxidized during O_2^+ bombardment. During the sputter process the chemical bonding of the oxides is broken. Because the binding electrons have greater affinity for oxygen, the (metallic) binding partner leaves the surface positively ionized.

Bombardment with Cs^+ results in covering of the surface with Cs. This reduces the work function and results in a higher probability that one electron from the conducting band tunnels from the surface to the leaving atom during the sputter process. This effect results in a higher rate of negatively charged atoms, especially of electronegative elements.

Both models have been described theoretically and verified experimentally. Because of the complex surface structure of real samples, no quantification algorithm based on physical models is yet available.

3.2.4 Quantification

Because the probability of ionization is very dependent on the matrix, it is necessary to use standards.

3.2.4.1 Relative Sensitivity Factors

The most accurate – and most popular – method of quantifying matrix effects is to analyze the unknown sample with a similar sample of known composition. The relationship between measured intensity and the content of each sample is, usually, defined by the *relative sensitivity factor* (RSF):

$$\frac{I_{\mathrm{el}}}{I_{\mathrm{ref}}} = RSF \cdot \frac{c_{\mathrm{el}}}{c_{\mathrm{ref}}} \tag{3.14}$$

Conveniently, a matrix element is chosen as the reference element (index ref); (el) indicates the element to be quantified. By use of a standard sample with known concentrations c_{el} and c_{ref}, RSF is adjusted to the specific matrix. Relative precision up to 1% is possible by use of standards.

3.2.4.2 Implantation Standards

Production of homogeneous solid-state standards is costly. Dynamic SIMS has the advantage that non-homogeneous ion implantation standards can also be used. Knowing the implantation dose of element (el), its RSF can be calculated by use of the integrated (summed) intensities of a depth profile according to Eq. (3.15):

$$RSF = \frac{f_{\mathrm{ref}} \cdot c_{\mathrm{ref}} \cdot \rho \cdot N_{\mathrm{A}} \cdot d}{f_{\mathrm{el}} \cdot Q_{\mathrm{el}} \cdot M \cdot Cyc} \cdot \sum_{i=1}^{Cyc} \frac{I_{\mathrm{el}}(i)}{I_{\mathrm{ref}}(i)} \tag{3.15}$$

where f_{ref} is the isotope abundance of the reference (e.g. matrix) element, c_{ref} is the concentration of a reference (e.g. matrix) element [mol/mol], ρ is the matrix density, N_{A} is Avogadro's number (= 6.022×10^{23} mol^{-1}), d is the crater depth, f_{el} is the isotope abundance of the element the RSF of which is required, Q_{el} is the dose of the implanted element the RSF of which is required [atoms/unit area], M is the relative molar mass of the matrix, *Cyc* is the number of measured sputter cycles, I_{el} (i) is the signal intensity of the element the *RSF* of which is required in cycle i, and I_{ref} (i) is the signal intensity of the reference element in cycle i.

For quantification of multilayers data for each layer must be calculated separately. When some of the layers are oxides, especially, quantification is not possible, because of large differences between the probability of ionization in the different layers.

Modern materials have a complex three-dimensional internal structure with many different phases. Although for these samples quantification is not possible, technologists are often interested in relative differences between several samples, or they already know the bulk concentrations and are only interested in the element distribution.

3.2.4.3 MCs^+ Ions

Under Cs bombardment the matrix effect can be significantly reduced by using the MCs^+ ion signals for quantification of species M. The detection limit is increased, i. e. the detection power deteriorates, by two or more orders of magnitude, but sometimes even standard-free quantification has been reported [3.51]. MCs^+ ions have high masses; this is a disadvantage because many mass interferences occur in this mass range.

3.2.5 Mass Spectra

Magnetic sector field instruments have mass resolutions up to $m/\Delta m = 20\,000$; quadrupole instruments are limited to a mass resolution of approximately $m/\Delta m = 500$. For both types of instrument the a mass range extends to 500.

High-mass resolution is needed to separate mass interferences of molecular and atom ions. Because of the mass defect of the binding energy of the nucleus, atomic ions have a slightly smaller mass than the corresponding molecular ions. To observe this typical mass resolutions between 5000 and 10 000 are necessary.

There is a second means of avoiding mass interferences. Molecular ions have a steeper energy distribution than atomic ions, i. e. abundance less is at higher energies, because of the decreasing probability of escape (or formation) from (or in) the collision cascade. Low-energy molecular ions can be discriminated by energy filtering. Because atomic ions are also suppressed to some extent, detection power might occasionally be reduced. Figure 3.21, however, shows an example in which energy filtering hardly reduces atom ion intensities but strongly suppresses molecular ions. The ^{63}Cu signal is not visible without offset because of the many interferences, e. g. $^{51}V^{12}C$.

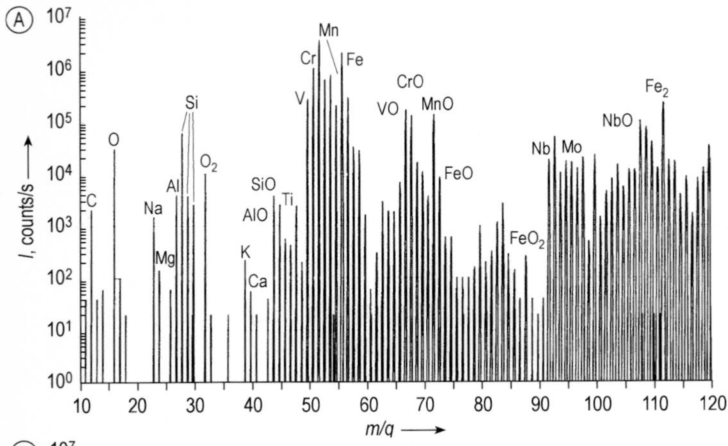

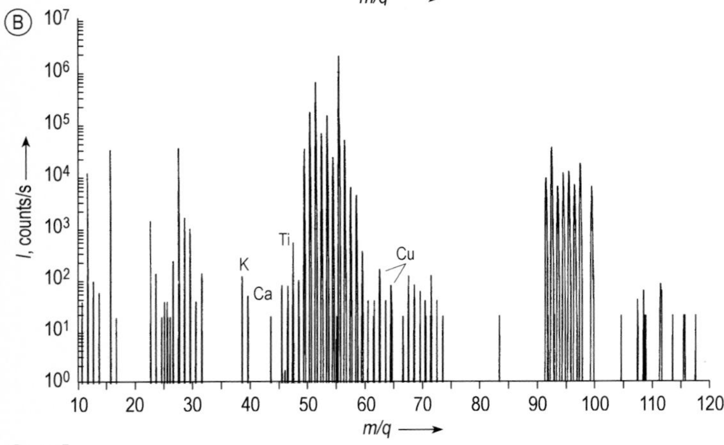

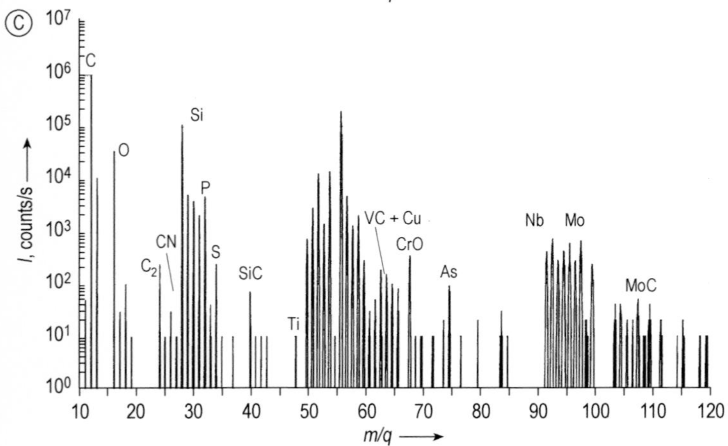

Fig. 3.21. SIMS spectra obtained from a high-speed steel. (A) primary ions O_2^+; no secondary ion energy limitation; electropositive elements are sensitive; many molecule ions are visible. (B) same conditions but 300 V offset was used; the molecule ion intensities are reduced significantly. (C) Primary ions Cs^+; 300 V offset was used; therefore electronegative elements are detected more sensitively.

3.2.6
Depth Profiles

Measurement of depth profiles is based on detection of the masses of interest during sputter removal of the sample material. Such experiments have several limitations:

Depth resolution is limited by atomic mixing, the roughness of the initial sample surface, and roughening processes during sputtering. Calibration of the depth scale by measuring the crater depth can be hampered, because multilayer systems consist of layers with different sputter yields. The different chemical constitution of the layers can, furthermore, result in different *RSF* values for a given element in different layers. These effects might necessitate determination of depth scale and *RSF* for each individual layer.

Although SIMS spectra and depth profiles are not affected by significant background or noise, there is a background of the elements H, C, and O, because of adsorbed residual gas. This background depends solely on the pressure in the sample chamber, not on the primary ion beam current. To test the influence of the residual gas background, the ion beam current can be reduced for several cycles. All signals should be reduced in proportion; if this reduction is not observed it indicates that the signal originates from residual gas.

To illustrate the use of the isotope sensitivity of SIMS, Fig. 3.22 shows the depth profile of a passivation layer on high-purity chromium. The layer was produced by a two-step experiment. First the sample was oxidized for 30 min in air with natural isotope concentrations. In the second step the gas was changed to a mixture of $^{18}O_2$ and $^{15}N_2$. The profile shows that the ^{18}O layer is located on the top. This indicates that chromium diffuses to the surface and reacts with the oxygen there. The high dynamic range of SIMS necessitates the use of logarithmic scales over 6 or 7 orders of magnitude. This must be kept in mind when interpreting depth profiles, e.g. diffusion profiles appear as straight lines. The diffusion of oxygen through the oxide layer to the metal interface is slower by a factor of 50. The enrichment of carbon at the interfaces is because of the gas change during sample production.

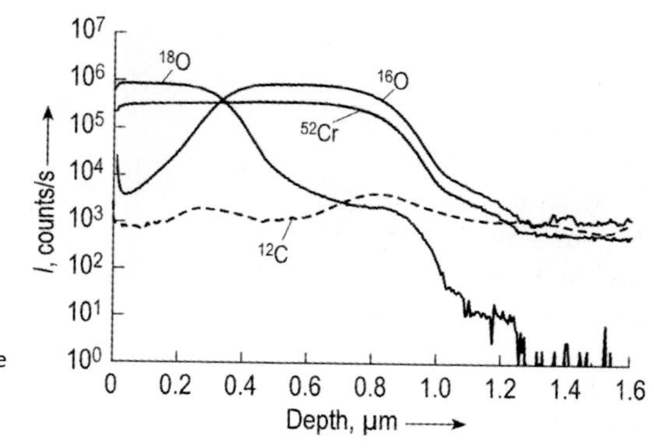

Fig. 3.22. Depth profile of a passivation layer on high-purity chromium. The ^{18}O layer is on the top, the ^{16}O layer at the interface with the metal.

3.2.7
Imaging

3.2.7.1 Scanning SIMS

In the scanning (or microprobe) mode the image is measured sequentially point-by-point. Because the lateral resolution of the element mapping in scanning SIMS is dependent solely on the primary beam diameter, LMISs are usually used. Beam diameters down to 50 nm with high currents of 1 nA can be reached.

The oxygen ion beam diameter is limited to 0.5 μm by the duoplasmatron source used. For mapping electropositive elements this drawback must be tolerated because of the chemical enhancement effect.

The counting time for one pixel, and the number of pixels, determine the measurement time for one image. For low concentrations of the elements of interest the Pois-

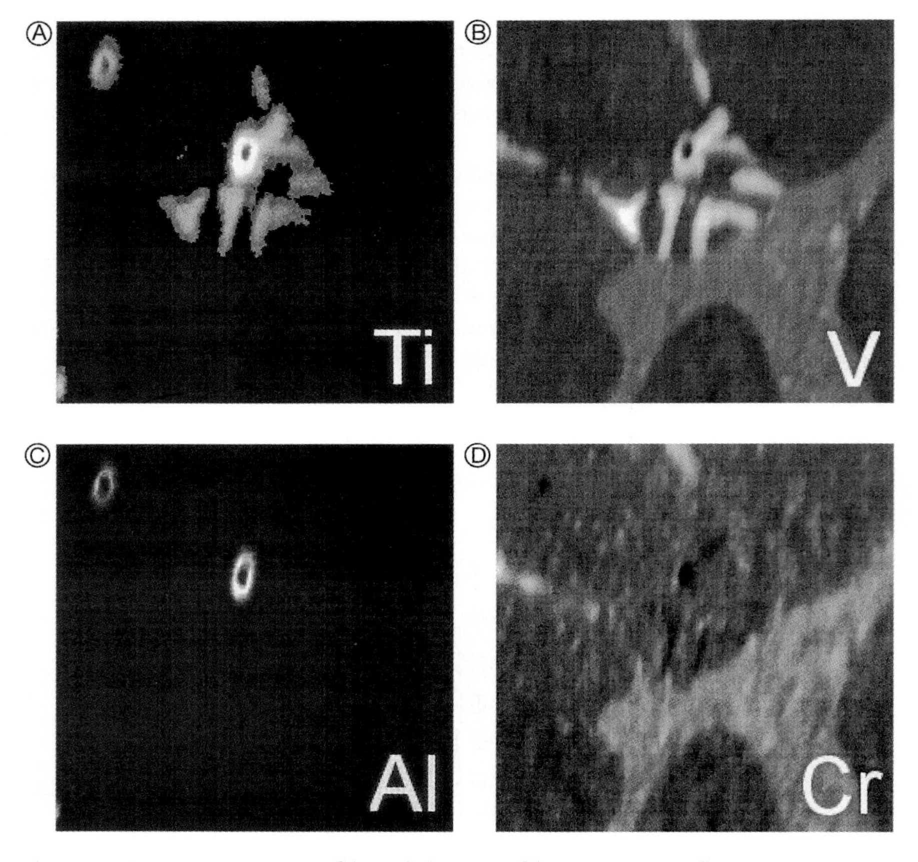

Fig. 3.23. Scanning SIMS image of the carbide structure of a high-speed steel [3.52]. In the V distribution different phases (MC and M_2C) are visible. The Al distribution shows the shell structure of the NMI (non-metallic impurities) acting as condensation nucleus. Measurement time 10 min; image size 64 × 64 μm; primary ions O_2^+; primary energy 5.5 keV.

son statistics of the small integer values of the secondary ions determine the counting time (Fig. 3.23).

Reduction of the measurement time for element distributions is possible by simultaneous detection of several masses. This can be achieved only by use of a magnetic sector field spectrometer with Mattauch–Herzog geometry [3.49] (Fig. 3.20) and parallel detection of up to five masses by mechanically adjusted electron multipliers.

3.2.7.2 **Direct Imaging Mode**

In the direct imaging (or microscope) mode (see, e.g., Fig. 3.24) all pixels of the element distribution are detected simultaneously by use of stigmatic secondary ion optics. The lateral resolution of the image is not affected by the diameter of the primary ion beam. The resolution is determined by the energy distribution of the secondary ions. The energy and angle distributions result in a chromatic (i.e. ion energy) aberration of the first electrostatic lens, the emission lens [3.6]. The resolution can be increased only by limiting the energy (or angle) acceptance of the instrument, which inherently reduces transmission and therefore detection limit. Chromatic aberration effects limit lateral resolution to approximately 1 μm. For the main components 0.5 μm is possible.

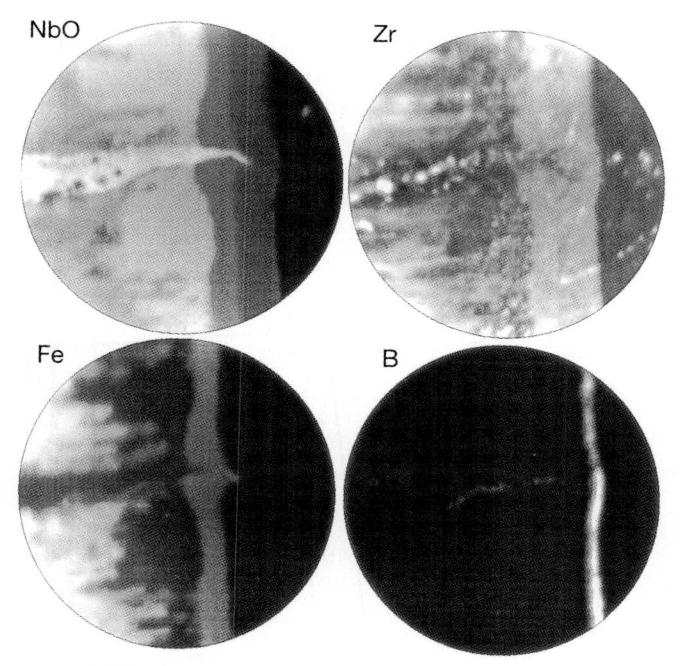

Fig. 3.24. Direct-imaging mode SIMS image of a passivation layer on a niobium alloy [3.54]. Boron enrichment at the interface is not visible with EPMA. Measurement time 10 s; image diameter 150 μm; primary ions O_2^+; primary energy 5.5 keV.

In practice image quality is also reduced by use of high mass resolution and energy offset. Often, therefore, mass interference cannot be avoided. Determination of element distributions is possible by use of image processing tools for classification of mappings of different masses [3.53].

For detection of secondary ions a laterally resolving detector is necessary. In the first step a channel plate for amplification is used; secondary electrons from the output of this device are accelerated either to a fluorescent screen or to a resistive anode. If a fluorescent screen is used the image is picked up by a CCD camera and summed frame by frame by use of a computer. The principal advantage of this system is unlimited secondary ion intensities, but compared with the digital detection of the resistive anode encoder the lateral and intensity linearity is not as well-defined.

The advantage of the imaging mode is fast data acquisition. Because all pixels are projected and detected simultaneously the measurement time for one distribution is extremely low.

3.2.8 3D SIMS

3D-SIMS, an example of which is given in Fig. 3.25, is a further development combining in-depth and imaging analysis. In the imaging mode the acquisition time of single-ion distributions is low (1 to 5 s). This enables the possibility of cyclic image acquisition during a normal depth profile. Measurements of large and, therefore, representative volumes ($150 \times 150 \times 10\ \mu m^3$) in reasonable times (1 h) are feasible [3.55].

In scanning mode the sequential detection of single pixels (picture elements) and voxels (volume elements) results in long measurement times; in practice, therefore, only small volumes ($10 \times 10 \times 1\ \mu m^3$) can be measured [3.56].

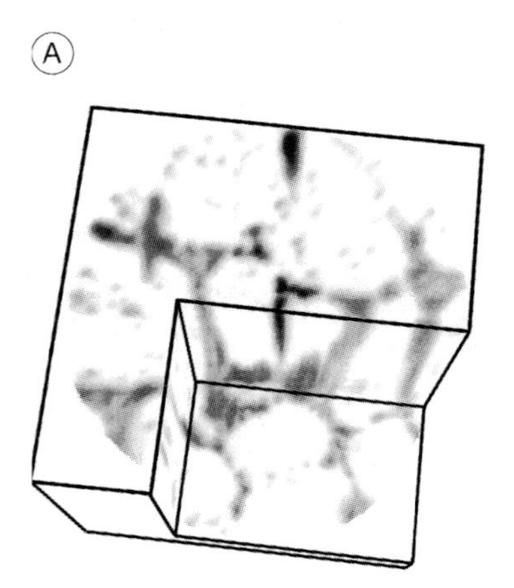

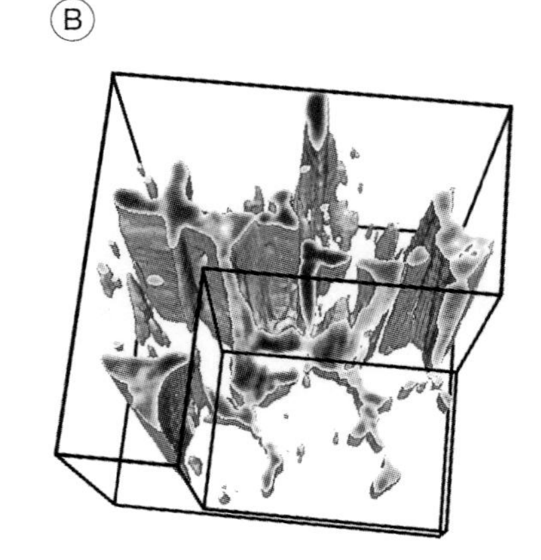

Fig. 3.25. 3D SIMS measurement of the V distribution in high-speed steel. Dimension $150 \times 150 \times 10\ \mu m^3$, top: block view, bottom: isosurface view. Primary ions oxygen; primary energy 5.5 keV; primary intensity 2 μA.

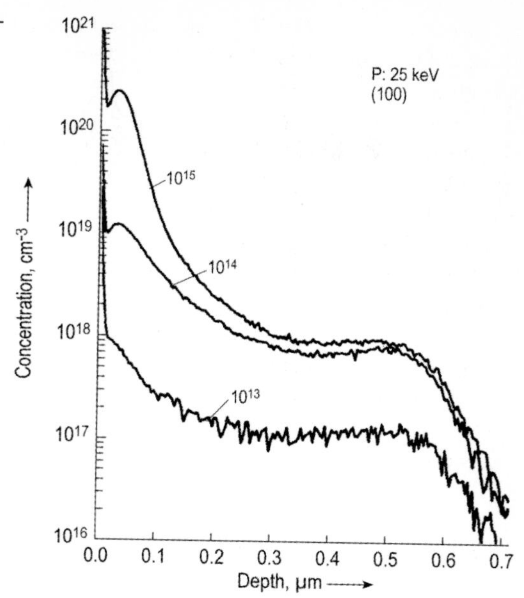

Fig. 3.26. Depth profiles of P implantations in Si. Primary ions Cs^+; primary energy 14.5 keV.

3.2.9 Applications

3.2.9.1 Implantation Profiles

Ion implantation is a method commonly used for doping semiconductors. Because the concentrations of the dopants (mostly B and P) are very low, a dynamic range of more than five orders of magnitude is often necessary. Measurement of ^{31}P is more difficult than that of B, because of the mass interference of $^{30}Si^{1}H$. High mass resolution of $m/\Delta m = 5000$, or an energy offset of 300 V, is necessary.

To obtain high implant concentrations at great depths together with low damage, the channeling effect is employed. The wafer is bombarded in a high-indexed crystal direction so that the implant ions can penetrate the solid mostly along the "channels" between the occupied lattice positions. Small but non-zero defect production, however, limits the channeling effect to a dopant and target-specific implantation concentration.

The depth profiles in Fig. 3.26 show that the typical flat channeling implantation profile is generated with low doses only. Increasing the dose superimposes the normal implantation profile shape. Undertaking such experiments with homogeneous wafers enables the production of calibrating models for semiconductor production.

3.2.9.2 Layer Analysis

Coating techniques have the common goal of increasing productivity and reducing costs [3.57]. CrN provides excellent material properties to meet these requirements. It is characterized by high hardness, superior oxidation stability, high corrosion resistance, low friction coefficient, and good adhesion. Because of its outstanding proper-

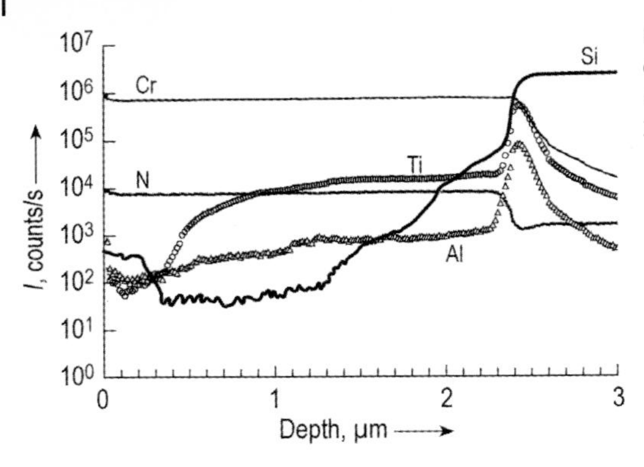

Fig. 3.27. Depth profile of CrN layer deposited on Si. Primary ions O_2^+; primary energy 5.5 keV; primary intensity 2 µA; measuring time 1 h.

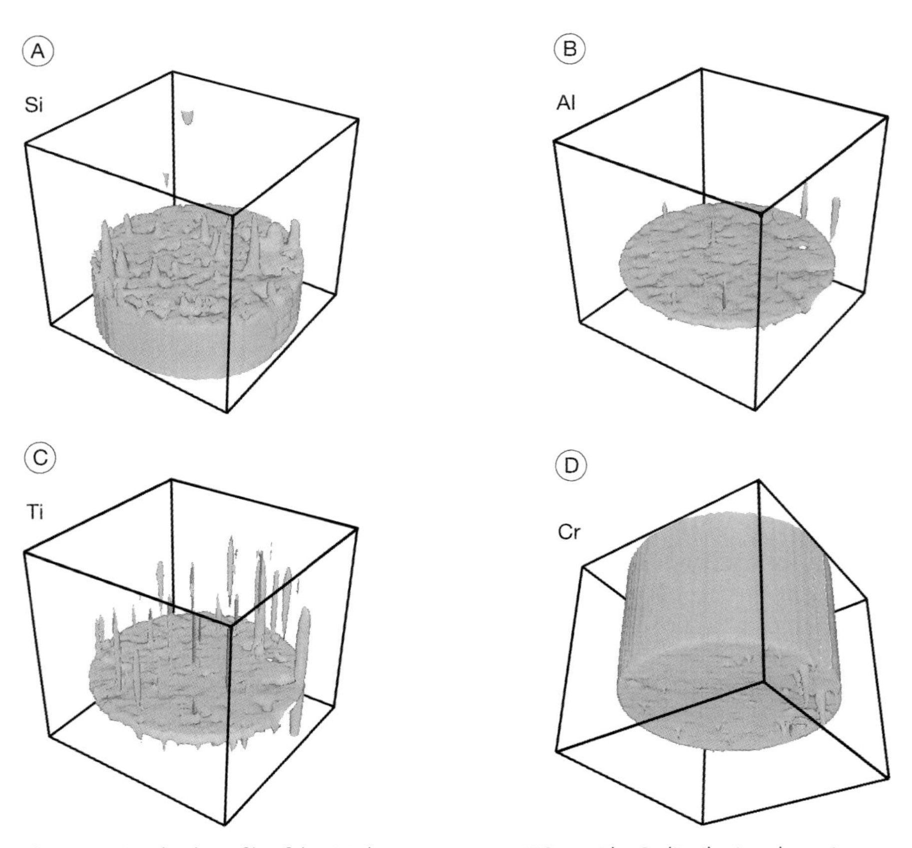

Fig. 3.28. 3D depth profile of the CrN layer on Si substrate, same measurement as Fig. 3.27. The thickness of the CrN layer is 2.3 µm, diameter 150 µm. The Cr distribution shown is seen from the bottom.

ties, CrN will be often used to machine iron-based materials and wear parts. Although CrN still has extraordinary material properties, further improvement is essential. For this reason it is very important to have not only exact but also complete information about the presence of desired or undesired trace elements within these layers.

Figure 3.27 shows the depth profile of such a layer. Enrichment of Ti and Al at the layer–substrate interface is visible. The Si signal in the layer increases with depth. It will subsequently be shown that it is not possible to determine from the depth profile alone whether there is diffusion of Si into the CrN layer.

Figure 3.28 shows 3D-SIMS distributions of the elements Si, Al, Ti, and Cr. The Cr distribution is shown from the bottom, to illustrate the rough interface. It is apparent that the interfaces are not smooth. This is the reason for the slowly decreasing Cr signal in the depth profile. As is apparent in the 3D-distribution, the different depth profiles of Si, Ti, and Al in the layer are a result of respective particulate inclusions.

From the depth profile alone it is not possible to distinguish whether the profile shape is a result of diffusion of elements or of geometric effects. This example demonstrates the capacity of 3D SIMS to improve the information content of depth profiles.

3.2.9.3 3D Trace Element Distribution

Steel alloyed with ~1% Al is more resistant to oxidation at high temperatures. The macro- and microscopic homogeneity of the distribution of all elements is better for powder metallurgical, hot isostatically pressed steels. Combining these advantages should result in better performance of high-speed steels.

The steel particles are covered with an Al_2O_3 layer before hot isostatic pressing [3.58]. 3D SIMS measurements (Fig. 3.29) show that this layer is still present in the compacted material. Carbonitrides are precipitated in small particles with a diameter of 30 nm inside the particles. The distribution of Ca^+ ions is shown as an example of a trace element with a concentration in the ppm range.

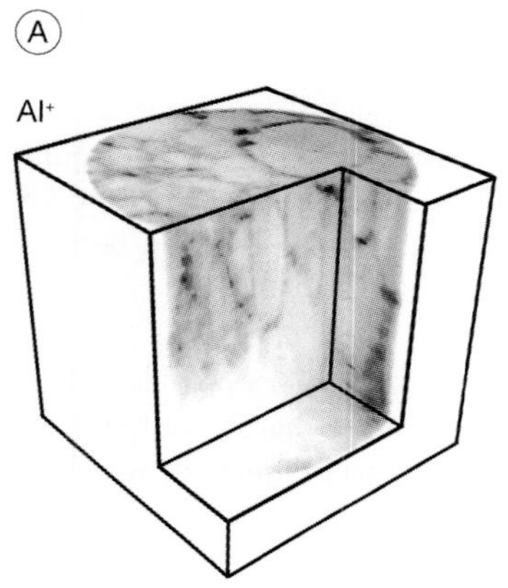

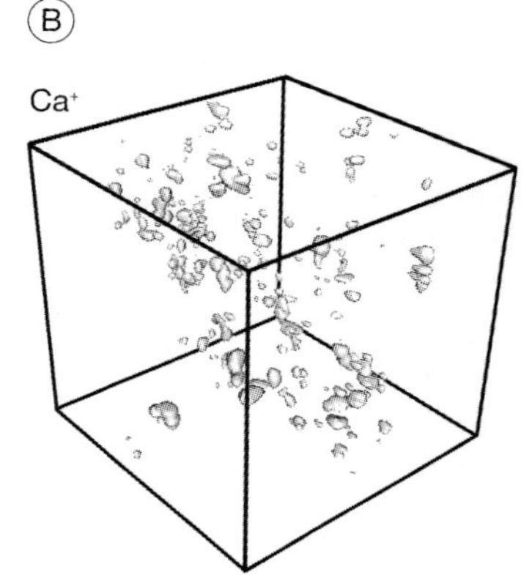

Fig. 3.29. 3D element distribution of powder metallurgically produced steel. Diameter of the cube 150 μm. Primary ions top 0.1 μA Cs^+, bottom 2 μA O_2^+; depth top 1 μm, bottom 11 μm.

3.3
Electron-impact (EI) Secondary Neutral Mass Spectrometry (SNMS)

Holger Jenett

3.3.1
General Principles of SNMS

The limitations of SIMS – some inherent in secondary ion formation, some because of the physics of ion beams, and some because of the nature of sputtering – have been mentioned in Sect. 3.1. Sputtering produces predominantly neutral atoms; for most of the elements in the periodic table the typical secondary ion yield is between 10^{-2} and 10^{-5}. This leads to a serious sensitivity limitation when extremely small volumes must be probed, or when high lateral and depth resolution analyses are needed. Another problem arises because the secondary ion yield can vary by many orders of magnitude as a function of surface contamination and matrix composition; this hampers quantification. Quantification can also be hampered by interferences from molecules, molecular fragments, and isotopes of other elements with the same mass as the analyte. Very high mass-resolution can reject such interferences but only at the expense of detection sensitivity.

To improve sensitivity and quantification, several post-ionization methods have been developed. These techniques decouple the sputtering and ionization processes by ionizing the *sputtered neutrals* (SN) after emission from the sample surface. Although the *secondary ion* (SI) yield can vary widely with surface composition, this is not true for the SN yield. For example, variation of the SI yield over three orders of magnitude (10^{-2} to 10^{-5}) produces less than 1% change in the SN yield. As a result, matrix effects are much smaller for SNMS than for SIMS, although changes in sputtering yield (as a function of matrix) can still cause quantification errors. These can be corrected by using standards or by determining the different sputter yields. Recombination effects and incomplete atomization are other possible sources of matrix effects in SNMS.

Compared with the other methods of surface and thin film analysis, the main advantages of SNMS are:

Compared with XPS and AES sputter depth profiling: After achieving sputter equilibrium, and until a layer with different sputtering behavior is reached [3.59], the SN flux represents stoichiometry and not altered layer concentrations evolving because of preferential sputtering effects.

Compared with GD-OES (and -MS, if used for depth profiling): SNMS provides somewhat better depth resolution (1 nm range); HF-plasma SNMS hardly suffers from molecule formation in the plasma gas (Ar), as do the GD techniques, in which argides are formed because of the comparatively high pressure.

Depending on the matrix and the post-ionization technique, SN^+ spectra can be dominated by atomic or molecular signals. In particular, compounds with high mass differences between the constituting elements and strong chemical bonds lead to significant amounts of molecules (10–100% of the metal monomer), mostly dimers, even if post-ionized by electrons. In such circumstances isobaric interferences from molecular signals must be considered.

SNMS sensitivity depends on the efficiency of the ionization process. SNs are post-ionized (to SN^+) either by electron impact (EI) with electrons from a broad electron (*e*-)beam or a high-frequency (HF-) plasma (i. e. an *e*-gas), or, most efficiently, by photons from a laser. In particular, the photoionization process enables adjustment of the fragmentation rate of sputtered molecules by varying the laser intensity, pulse width, and/or wavelength.

SNMS is suitable for quantitative element depth profiling of metallic and electrically insulating samples. Laser-SNMS enables the additional acquisition of 2D element distributions; with HF-plasma SNMS bulk analysis is also feasible.

3.3.2
Principles of Electron-beam and HF-plasma SNMS

The basic principle of *e*-beam SNMS as introduced by Lipinsky et al. in 1985 [3.60] is simple (Fig. 3.30) – as in SIMS, the sample is sputtered with a focused keV ion beam. SN post-ionization is accomplished by use of an *e*-beam accelerated between a filament and an anode. The applied electron energy $E_e \approx 50 \pm 20$ eV is higher than the range of first ionization potentials (IP) of the elements (4–24 eV, see Fig. 3.31). Typical probabilities of ionization are in the 0.01% range. SI^+ and residual gas suppression is achieved with electrostatic lenses before SN post-ionization and energy filtering, respectively.

Oechsner's basic set-up of HF-plasma (*e*-gas) SNMS (Fig. 3.32) [3.63] was first published in 1966 [3.64]. An inductively coupled, low-pressure, HF noble gas plasma (Ar, density $n_{pl} \approx 10^{10}$ cm^{-3} was usually used) serves for sputtering using the ion component, and for electron impact post-ionization using the Maxwellian electron gas. Its temperature $T_e \approx 10^4$–10^5 (i. e. electron energy 1–10 eV) fits the given IP range well; depending mostly on IP and on the plasma (= electron) density n_e, probabilities of ionization are mostly in the 0.1% range. Conducting samples introduced into the plasma are set to a negative potential U_{DBM} (DBM = direct bombardment

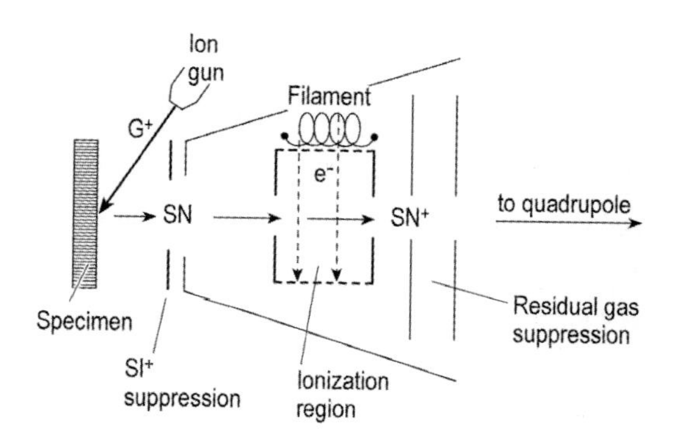

Fig. 3.30. Schematic diagram of an electron beam SNMS set-up [3.61].

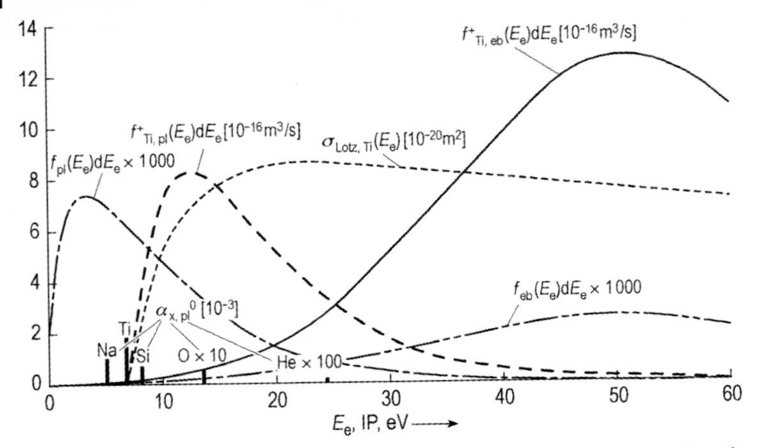

Fig. 3.31. Distributions (i) $f(E_e)\,dE_e$ of electron energy (E_e) for a low-pressure HF-plasma (suffix pl, Maxwellian with temperature $T_e = 80\,000$ K) and an electron beam (suffix eb, simplified to Gaussian shape with 40 eV half-width); (ii) $\sigma_X(E_e)$ of the E_e dependent electron impact ionization cross-section for X=Ti calculated according to Lotz [3.62]; and (iii) of the differential ionization rate constant (Sect. 3.2.2.4); (iv) respective calculated post-ionization probabilities for X=Na, Ti, Si, O, and He, in the order of their first ionization potentials IP, for the HF-plasma parameters given in Sect. 3.2.2.4.

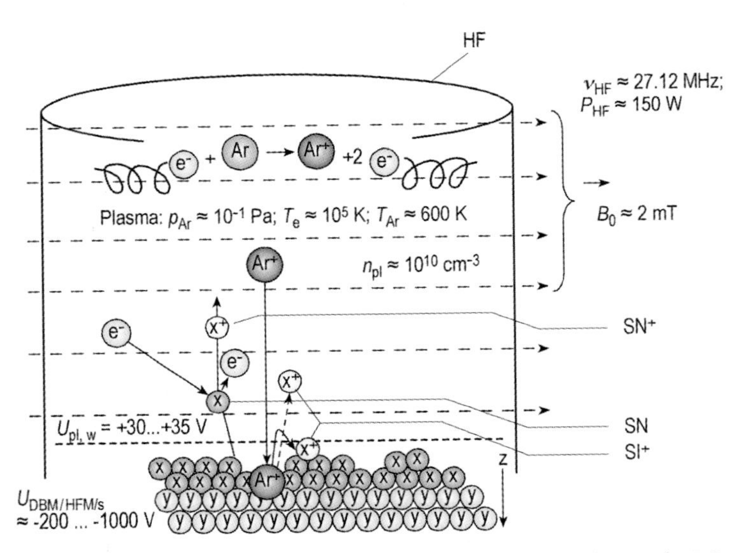

Fig. 3.32. Basic principles of HF-plasma SNMS. ν_{HF}, P_{HF} are the HF generator frequency and power, respectively, p_{Ar} the plasma gas (Ar) pressure; T_e and T_{Ar} the electron and plasma gas temperatures, respectively. $n_{pl} = n_e$ is the plasma (e^-, Ar^+) density, B_0 the magnetic flux, U_{DBM} and U_{HFM} the DC or HFM bombardment voltages, respectively (see text), U_s the surface potential relative to ground (HFM), and $U_{pl,w}$ the plasma sheath edge potential. $SN^{(+)}$ are the (post-ionized) secondary neutrals and SI^+ the positive secondary ions.

mode) of a few hundred volts. Accelerated by the difference between U_{DBM} and the plasma potential ($\approx +30 \pm 5$ V), Ar^+ ions can induce comparatively shallow sub-surface collision cascades with sputter yields, *Y*, of the order of 0.1–1.

Positive secondary ions (SI^+) are repelled by an electrode in *e*-beam SNMS or completely (in DBM) or widely (in HFM, see below) re-attracted by the sample surface in HF-plasma SNMS. Ionized plasma species (Ar^+, contamination) are suppressed by energy filtering.

3.3.3
Instrumentation

The *e*-beam SNMS set-up has been shown schematically in Fig. 3.30. The SN flux through the few centimeters long ionization region is crossed by an electron current in the lower mA range. Instead of the simple lens depicted in Fig. 3.30, an electrostatic analyzer can serve for the energy discrimination of SN^+ against ionized plasma gas (1–10 eV compared with 10 meV). In contrast with the other SNMS techniques, *e*-beam SNMS modules can easily be added to many existing, UHV-based surface analysis instruments. Compared with a plasma, the lack of positive charge carriers in an electron beam gives rise to space charge limitations.

A schematic diagram of the commercially available INA-3 type spectrometer (SPECS, Berlin, Germany) is given in Fig. 3.33. A constant Ar pressure of 0.3 ± 0.1 Pa is maintained by means of a Piezo valve between the high-pressure supply and the cylindrical plasma chamber (volume 1.26 L; typically applied values are given). The HF power of 150 ± 30 W is supplied from an HF generator by a single turn coil. Electron cyclotron wave resonance [3.65] is effected, and skin effects of the electron gas are suppressed, by the static magnetic field in the mT range maintained by a constant current of 5 ± 2 A flowing through two parallel, rectangular Helmholtz coils with 184 turns each and 25 cm distance. Delivering primary ion current densities of 0.1–1 mA cm^{-2}, the plasma can effect sputter rates in the 0.1–1 nm s^{-1} range. If U_{DBM} is fitted appropriately ($\rightarrow U_{DBM}$*) to n_e and T_e with a given distance between sample surface and plasma sheath edge, sputter erosion happens with perfect lateral

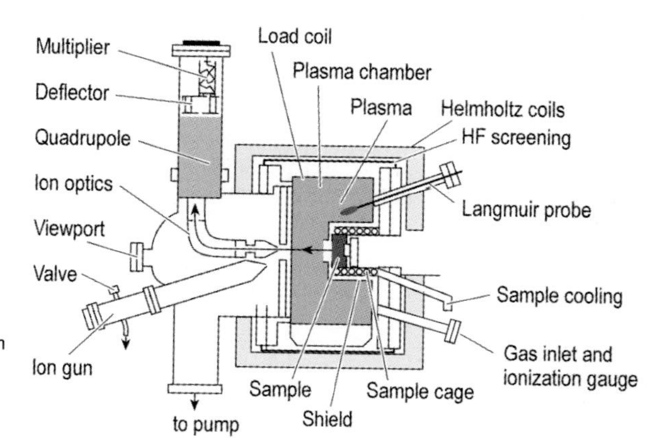

Fig. 3.33. Schematic diagram of the commercial INA-3 HF-plasma SNMS apparatus (SPECS, Berlin, Germany).

homogeneity, i.e. nanometer depth resolution, on a spot 2–5 mm in diameter [3.66]. This compares with removing sand grains from a soccer field layer by layer; rough surfaces yield less ideal depth resolutions. The large area sputtered with high current density is one reason why typical signal intensities are an order of magnitude higher than in *e*-beam SNMS.

The gas temperature of the Ar plasma in the range of 600 ± 100 K effects temperatures of water or LN_2 cooled samples of 350 ± 30 or 200 ± 30 K, respectively. In addition, C, N, and O species are desorbed from the chamber walls and introduced samples, effecting a plasma contamination level in the 0.01–0.1% range.

After post-ionization in the 3 cm long cylindrical plasma space between sample surface and the opposite wall, SN^+ enter a 90° electrostatic ion energy analyzer (ion optics) suppressing ionized plasma gas particles to a degree of 10^8–10^9; noise levels are correspondingly low (1 cps). The transmission of the electrostatic ion optics is in the range of a few per cent.

In both electron post-ionization techniques mass analysis is performed by means of a quadrupole mass analyzer (Sect. 3.1.2.2), and pulse counting by means of a dynode multiplier. In contrast with a magnetic sector field, a quadrupole enables swift switching between mass settings, thus enabling continuous data acquisition for many elements even at high sputter rates within thin layers.

With useful yields typically in the 10^{-10} range – a value also valid for *e*-beam SNMS – typical measured intensities are in the 10^4–10^6 range for HF-plasma SNMS, depending on the material and U_{DBM}. With a typical plasma and low U_{DBM} near −300 V one effects ultimate depth resolution and low intensities, whereas $U_{DBM} \geq 800$ V enables bulk analysis in the ppm range (apart from C, N, and O being implanted from contamination) but no longer with good depth resolution.

Electrically insulating materials can be analyzed in HF-plasma SNMS by applying a square-wave HF in the 100 kHz range to the sample (Fig. 3.34). Dielectric charge transfer at the start of a period shifts the surface potential to the amplitude U_{HFM} applied. Ar^+ ions are attracted from the plasma and sputter the surface until the end of Δt^-. The potential increase ΔU_s = 1–100 V caused by their charge is then converted to a positive absolute ΔU_s which is reduced to less than 1 V within <0.1 μs by the

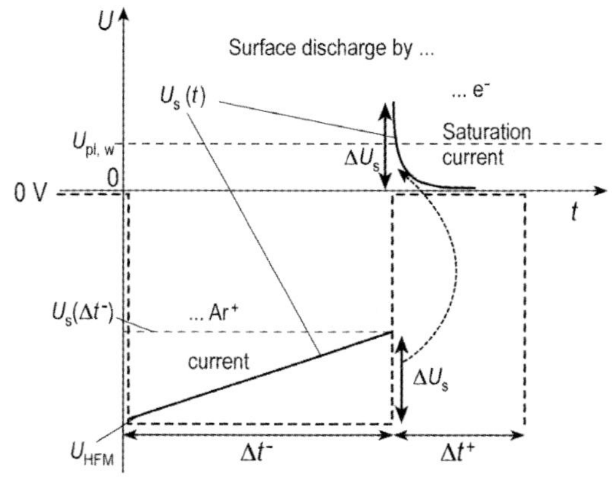

Fig. 3.34. The principles of HFM. – – –, applied voltage; ——, variation of surface potential U_s with time *t*.

plasma electron current. The missing sputtering time of Δt^+ and, more importantly, the often low sputter yield of dielectrics, especially oxides, give rise to lower intensities than given above.

A novel HF-plasma SNMS instrument which can be combined with XPS has recently been developed [3.67]. Detection limits in the nmol/mol range have been achieved with a dedicated HF-plasma instrument attached to a double-focussing mass spectrometer [3.68].

3.3.4
Spectral Information

Spectral information from *e*-beam and HF-plasma SNMS, both of which are dynamic sputtering methods, is similar to that from dynamic SIMS. Because molecules from a heavily sputtered surface survive a collision cascade *and* a 1 eV-electron impact to a comparatively low extent, emphasis lies on elemental analysis based on mostly atomic spectra. Binary molecules formed from the main constituents might add restricted speciation information. Multiple SN ionization occurs rarely, usually because of insufficient electron energy and the low probability of a second electron impact. As a consequence, *e*-beam and HF-plasma SNMS spectra a characterized by a small amount of interference and are easy to interpret. Figure 3.35 shows a typical HF-plasma SN^+ mass spectrum as obtained from a TiAl alloy. The quadrupole is the reason for ΔM being unity throughout the spectrum. As discussed above, molecule intensities are in the percentage range of the parent peaks. Doubly ionized species appear in the per mil range, i.e. for the main constituents only, and adsorbed or implanted plasma gas contamination gives rise to a few other low-intensity peaks.

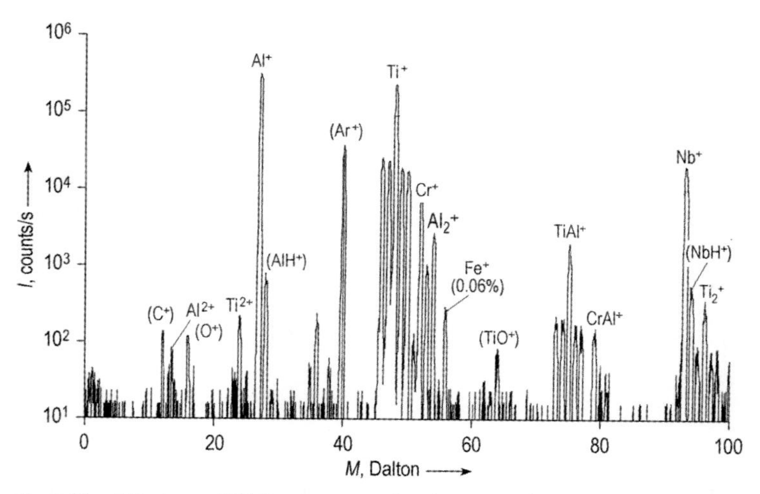

Fig. 3.35. HF-plasma SNMS survey spectrum of a powder metallurgically-produced TiAl alloy sample. In brackets: signals coming from adsorbed and re-sputtered plasma gas contamination.

3.3.5 Quantification

The measured SN$^+$ intensity $I_X(E_{X,0})$ of element X, its atomic SN being sputter emitted with energy E_0, is given by:

$$I_X(E_{X,0}) = \frac{I_{G^+}}{e_0} \Delta Y_X(E_{X,0}) W_X n_e t_X(E_{X,0}) F_X^+ T_{X^+} \quad (3.16)$$

where I_{G+}/e_0 is the primary ion particle current (e_0 is the elementary charge); $\Delta Y_X(E_{X,0})$ is the sputtering yield of X valid for the emission energy range $E_{X,0} \pm \Delta E_f/2$, (ΔE_f is the width of the energy window of the ion optics); W_X is the probability of an X emission into the ionization volume (e-beam, plasma) within the acceptance angle of the detection system (10^{-3} range); n_e is the electron density; $t_X(E_{X,0})$ is the dwell time of an X particle with $E_{X,0}$ in the ionization region; and T_{X+} is the combined spectrometer transmission and detector response probability for X$^+$ ions.

F_X^+ is the ionization rate constant [m^3 s^{-1}] [3.69] (see also Fig. 3.31), given by:

$$F_X^+ = \int_{E_{I,X}}^{\infty} \sigma_X(E_e) \sqrt{\frac{2E_e}{m_e}} f(E_e) \, dE_e = \int_{E_{I,X}}^{\infty} f_X^+(E_e) \, dE_e \quad (3.17)$$

where $E_{I,X} = IP_X$, the ionization potential of X; $\sigma_X(E_e)$ is the ionization cross-section for an X particle hit by an electron with energy E_e; $f(E_e)$ is the electron energy (E_e) distribution; and m_e is the rest mass of the electron.

The product $n_e t_X(E_{X,0}) F_X^+$ is equal to the post-ionization probability α_X^0:

$$\alpha_X^0 = n_e t_X(E_{X,0}) F_X^+ \quad (3.18)$$

The values $\alpha_{X,pl}^0$ in Fig. 3.31 have been calculated with $n_{pl} = n_e = 6 \times 10^{15}$ cm^3, $E_{X,0}$ = 30 eV (optics setting), and an l_{pl} = 3 cm long path through the plasma. With a mean electron density $n_{eb} = 8 \times 10^{14}$ m^{-3}, $E_{X,0}$ = 10 eV, and l_{eb} = 3 mm, the corresponding $\alpha_{X,eb}^0$ amount to 6, 9, 13, 45, and 180% of the given $\alpha_{X,pl}^0$ values (X = Na, Ti, Si, O, He).

The absolute detection (or sensitivity) factor under stationary sputtering conditions is

$$S_X = W_X \alpha_X^0 T_{X^+} \quad (3.19)$$

and this can be related to a reference value S_{ref} according to:

$$S_{(ref)X} \equiv \frac{S_X}{S_{ref}} = \frac{I_X}{I_{ref}} \frac{c_{ref}}{c_X} \quad (3.20)$$

Equation (3.19) is valid for any species X. If, however, a multi-component material emits only atomic SN after attaining sputter equilibrium, X stands for elements and atoms only, and the total sputter yield Y can be written as:

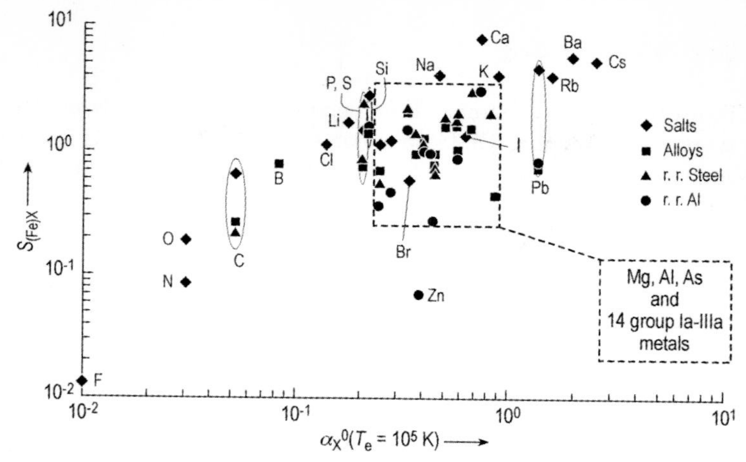

Fig. 3.36. Experimental, Fe-related HF-plasma SNMS sensitivity factors $S_{(Fe)X}$ with elements X ordered according to their post-ionization probabilities calculated according to [3.74]; from Ref. [3.71] (salts); [3.72] alloys, [3.73] round robins (r.r.).

$$Y_X = c_X Y \tag{3.21 a}$$

and

$$Y = \sum_{i=1}^{n} Y_i = \frac{e_0}{I_{G^+}} \sum_{i=1}^{n} \frac{I_i}{S_i} \tag{3.21 b}$$

where $i = 1; 2; \ldots; X; \ldots; n$ sample elements.

The different post-ionization probabilities, α_M^0, for M = Ti and Al on one hand and Cr and Nb on the other, both pairs being present at the same concentrations, are the main reason the respective signal intensities in Fig. 3.35 are similar, but not equal. Fig. 3.36 [3.70] shows that experimental relative sensitivity factors vary approximately with theoretical α_X^0. This variation by three orders of magnitude is mostly because of F, O, N, C, and Zn (yet unexplained), with more than one order of magnitude, and some alkali and alkaline-earth metals with another half order of magnitude. The $S_{(Fe)X}$ of all other elements represented in Fig. 3.36 vary between 0.3 and 3, irrespective of matrix and measurement conditions; depending on the latter, identical $S_{(Fe)X}$ vary by factors of 2–5. It is this, compared with SIMS, narrow range of $S_{(ref)X}$ variation which has made SNMS famous for its (relative) absence of matrix effects and ease of quantification.

3.3.6
Element Depth Profiling

Figure 3.37 gives an example of the depth resolution routinely achieved with both *e*-beam and HF-plasma SNMS on suitable samples. If Eq. (3.21) is applicable, i.e. all sputtered material has been recorded with ΣI_i, then Ye_0/I_{Ar^+} equals $\Sigma I_i/S_i$ [3.75], and depths z can be calculated according to:

$$z = \frac{I_{G^+}\, t\, Y}{e_0\, A_s\, n_u} \tag{3.22 a}$$

$$n_s = \frac{\rho\, N_A\, N_u}{M_u} \tag{3.22 b}$$

where t is the sputter time; A_s is the sputtered sample area; n_u is the number density of formular units u (= Ta_2O_5, $Ti_{0.48}Al_{0.5}Cr_{0.02}$, ...); ρ_s is the sample density; N_A is Avogadro's number; N_u is the number of atoms per formula unit; and M_u is the relative mass (in Daltons) of u.

The absolute sensitivity factors S_X must be determined for this procedure by integrating intensities over time while sputtering suitable pure element samples and determining the crater volume; for HF-plasma SNMS the weight loss can also be measured.

Figure 3.38 shows that good depth resolution can also be maintained sputtering through an insulating μm-thick layer.

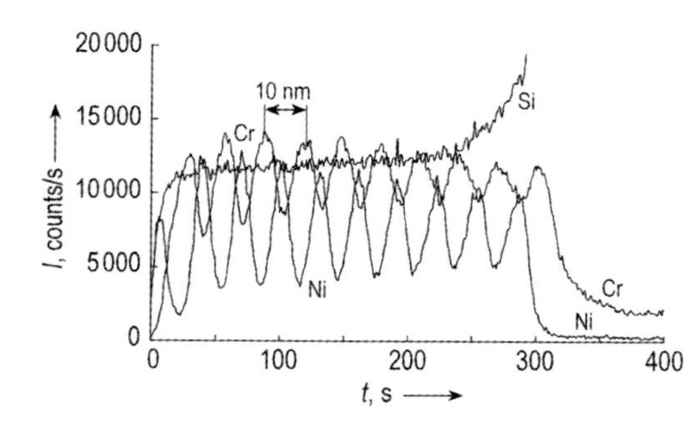

Fig. 3.37. HF-plasma SNMS sputter time profile (DBM) of a home-made multilayer system consisting of ten double layers, 5 nm Cr + 5 nm Ni each, on a Si wafer.

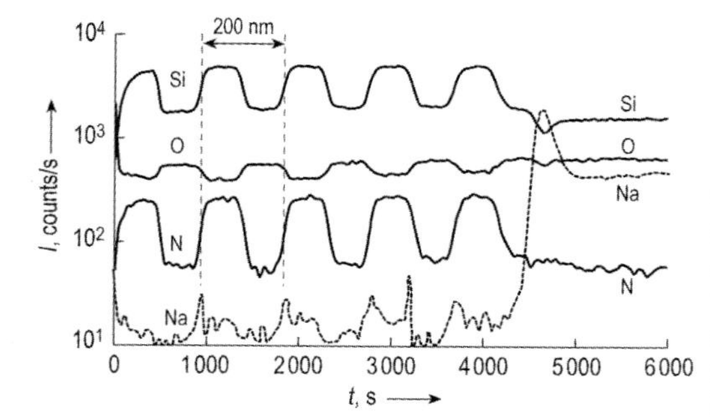

Fig. 3.38. HF-plasma SNMS sputter time profile (HFM) of a multilayer system consisting of five double layers, 100 nm SiO_2 + 100 nm Si_3N_4, each, on a glass substrate (courtesy: V.-D. Hodoroaba, BAM Berlin, and Schott Glas, Mainz, Germany).

3.3.7 Applications

Figure 3.39 exemplifies the use of HF-plasma SNMS in materials research [3.76]: The intermetallic compound γ-TiAl to which 2 mol-% Cr have been added in order to increase ductility has been heated to 1,070 K a) for less than 4 s only and cooled down again, and b) kept at this high temperature for 8 h. In the first case, an about 150 nm thick oxide layer is formed with Al_2O_3 enriched topmost. Fig. 3.39 b) shows that after this initial oxide layer formation, nitrogen has started to penetrate to the interface between oxide and alloy oxidizing the latter by forming nitrides. The N enrichment at the interface and tracer isotope experiments described in [3.76] furthermore suggest that nitride has continuously been converted to oxide between the oxide layer and the nitride enriched interface layer. Fig. 3.39 b) also shows that Ti has diffused outward forming a topmost TiO_2-rich layer.

Recent applications of *e*-beam and HF-plasma SNMS have been published in the following areas: aerosol particles [3.77], X-ray mirrors [3.78, 3.79], ceramics and hard coatings [3.80–3.84], glasses [3.85], interface reactions [3.86], ion implantations [3.87], molecular beam epitaxy (MBE) layers [3.88], multilayer systems [3.89], ohmic contacts [3.90], organic additives [3.91], perovskite-type and superconducting layers [3.92], steel [3.93, 3.94], surface deposition [3.95], sub-surface diffusion [3.96], sensors [3.97–3.99], soil [3.100], and thermal barrier coatings [3.101].

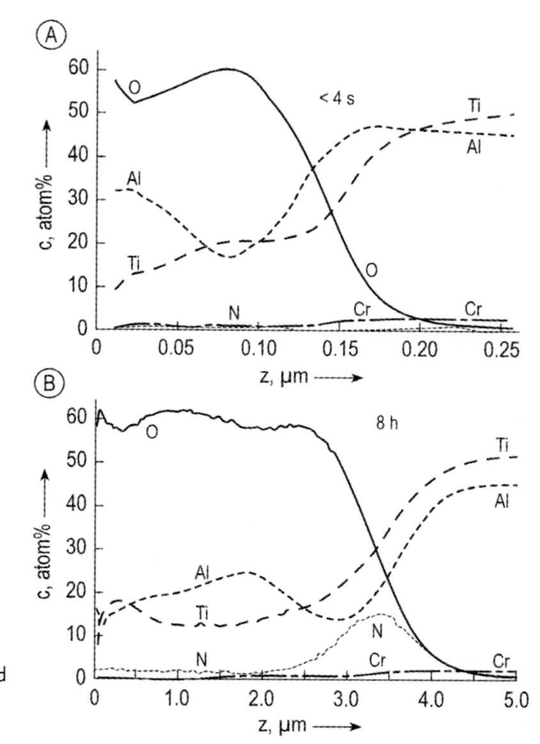

Fig. 3.39. HF-plasma SNMS sputter depth profiles (DBM) of Ti–48Al–2Cr (atom%) samples having been oxidized at 800 °C in air for (a) less than 4 s, (b) 8 h.

3.4 Laser-SNMS

Heinrich F. Arlinghaus

3.4.1 Principles

3.4.1.1 Non-resonant Laser-SNMS

Surface analysis by non-resonant (NR-) laser-SNMS [3.102–3.106] has been used to improve ionization efficiency while retaining the advantages of probing the neutral component. In NR-laser-SNMS, an intense laser beam is used to ionize, non-selectively, all atoms and molecules within the volume intersected by the laser beam (Fig. 3.40 b). With sufficient laser power density it is possible to saturate the ionization process. For NR-laser-SNMS adequate power densities are typically achieved in a small volume only at the focus of the laser beam. This limits sensitivity and leads to problems with quantification, because of the differences between the effective ionization volumes of different elements. The non-resonant post-ionization technique provides rapid, multi-element, and molecular survey measurements with significantly improved ionization efficiency over SIMS, although it still suffers from isobaric interferences.

3.4.1.2 Resonant Laser-SNMS

Resonant (R-) laser-SNMS [3.107–3.112] has almost all the advantages of SIMS, *e*-SNMS, and NR-laser-SNMS, with the additional advantage of using a resonance laser ionization process which selectively and efficiently ionizes the desired elemental species over a relatively large volume (Fig. 3.40 C). For over 80% of the elements in the periodic table, R-laser-SNMS has almost unity ionization efficiency over a large volume, so the overall efficiency is greater than that of NR-laser-SNMS. Quantification is also simpler because the unsaturated volume (where ionization is incom-

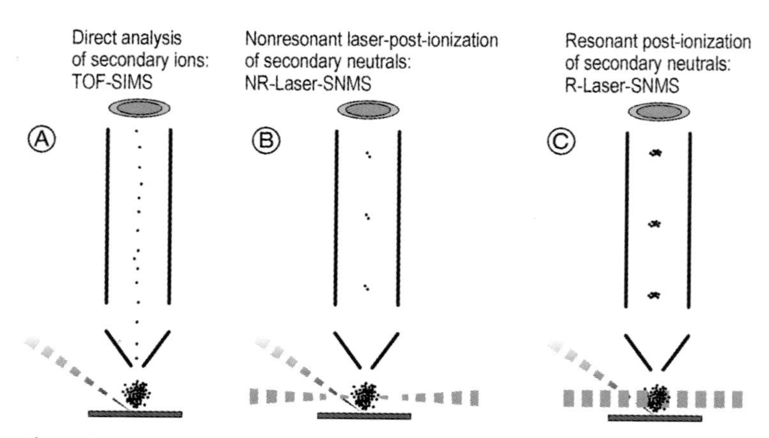

Fig. 3.40. Comparison of TOF SIMS and Laser-SNMS.

plete) is insignificant compared with the saturated volume. This leads to consistently high sensitivity and great accuracy in R-laser-SNMS measurements. In particular, the extremely high selectivity prevents almost all isobaric and molecular interference. This improves the accuracy of the data and simplifies its interpretation without requiring very high-resolution mass spectrometry. The main drawbacks accompanying this immunity to interference are that only one element is analyzed at a time and that low-repetition rate pulsed lasers impose a low duty cycle. The use of multiple lasers enables simultaneous ionization of multiple elements but this is practically limited to two to three elements. Advances in diode-pumped Nd:YAG lasers and optical parametric oscillators (OPOs) will enable increased repetition rates of over 1000 Hz from the currently available 30–100 Hz.

3.4.1.3 Experimental Set-up

In Laser-SNMS experiments, the sample is bombarded with a finely focused pulsed primary ion beam (Sect. 3.1.2.1). During the time the primary ion pulse is striking the target, voltages on the extraction electrodes are set so that electric fields retard positive secondary ions. After the primary ion pulse, the voltages are switched so that ions formed by laser post-ionization of the neutral particles in the sputtered cloud have energy significantly different from that of the secondary ions formed in the sputtering process. Ions are analyzed by means of a time-of-flight mass spectrometer with an electrostatic energy analyzer or a reflectron which discriminates against the secondary ions. The TOFMS enables simultaneous measurement of all photoions in each laser shot.

Laser-SNMS imaging is performed either by scanning the ion beam over the sample or by translating the sample under a fixed ion beam. If an LMIS is used for sputtering, nanoanalysis down to 50 nm can be achieved. Charge compensation for insulator analysis is possible using pulsed low-energy electrons, which are introduced during the time interval between sputtering pulses. Laser-SNMS depth profiles are obtained by sputtering the sample with a continuous ion beam and taking data with a pulsed ion beam in the center of the crater (dual beam mode, see Sect. 3.2.2.1). In laser-SNMS, as in other microprobe techniques (Sect. 3.2.8), the combination of 2D imaging with depth-profile capability enables visualization of the three-dimensional distribution of elements. However, the extremely high overall efficiency of R-laser-SNMS (typically 3 to 8% useful yield for most elements) enables measurement of low concentrations of analytes in much smaller volumes, thus enabling higher lateral and depth resolution in these 3D visualizations. This capability could contribute greatly to the study of both dopant and contaminant distributions in semiconductor devices and trace elements in biological samples.

3.4.1.4 Ionization Schemes

As illustrated in Fig. 3.41, several laser schemes can be used to ionize elements and molecules. Scheme (a) in this figure stands for non-resonant ionization. Because the ionization cross-section is very low, a very high laser intensity is required to saturate the ionization process. Scheme (b) shows a simple single-resonance scheme. This is the simplest but not necessarily the most desirable scheme for resonant post-ionization. Cross-

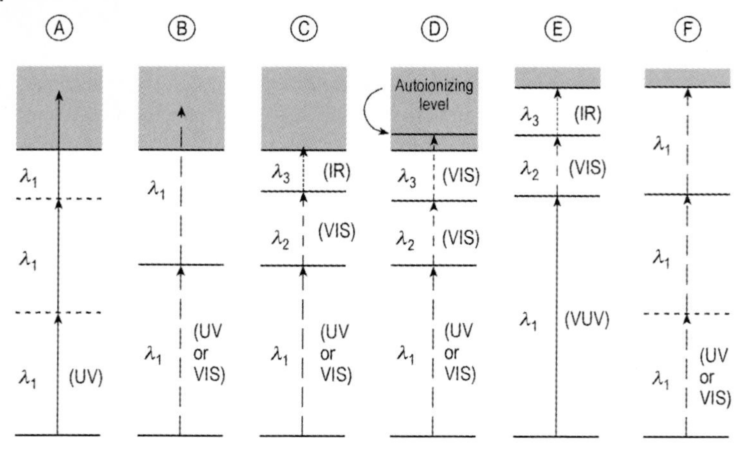

Fig. 3.41. Schemes of laser post-ionization steps.

sections for (photo-)ionization are much lower than those for resonant excitation steps; the photoionization laser intensity must, therefore, be proportionately higher to ionize the atoms efficiently. High-intensity visible or UV laser beams can cause non-resonant ionization of interfering species; this is rarely a problem with high-intensity IR beams. Whenever selectivity is an issue, therefore, IR wavelengths for the photoionization step are preferable. Scheme (c) is the most versatile of the schemes shown, and gives the best performance for metals. H, He, C, N, O, F, Ne, P, S, Cl, Ar, Br, Kr, I, Xe, and Rn cannot be analyzed by use of scheme (c) with commercial laser systems and frequency up-conversion in non-linear crystals. Resonance ionization scheme (d) can be used when an autoionizing structure occurs in the ionization continuum. Because the cross-section for autoionization is significantly larger than for normal photoionization, the intensity of the ionizing laser can be reduced. This enables the use of visible wavelengths without extensive non-resonant ionization of interfering species. Scheme (e) is used when even the lowest-energy-excited states of the analyte element require vacuum UV wavelengths. It differs from scheme (c) only in the experimental method used to generate the wavelength for the first transition. For wavelengths below approximately 189 nm, either four-wave mixing or anti-Stokes Raman must be used. This makes scheme (e) the most experimentally complex of the resonance ionization schemes. With this technique it is possible to measure the amount of ^{81}Kr (a radioactive isotope with approximately 1 part in 10^{18} natural abundance in the atmosphere) in a water sample after several isotope enrichment steps [3.113]. Scheme (f) is generally used when the energy of the excited state structure is too high for schemes (b), (c), or (d) and the researcher does not wish to go to the experimental complexity of scheme (e). Examples are C, N, O, P, S, Cl, Br, Kr, I, and Xe. The two-photon transition used in scheme (f) is a second-order process, which is why its cross-section is quite low compared with the one-photon transitions described in the other schemes. This means that high-intensity light is needed, and focusing of the laser beam is generally required. In fact, the intensity of the light needed for efficient two-photon excitation is usually sufficient to ionize the excited state, and so scheme (f) often requires only a single wavelength. This experimental simplicity is counterbalanced by the fact that the high-intensity light can cause non-resonant ionization processes, which could lead to isobaric interferences when a mass spectrometer with low mass resolution is used.

Isotope shifts for most elements are small in comparison with the bandwidth of the pulsed lasers used in resonance ionization experiments, and thus all the isotopes of the analyte will be essentially resonant with the laser. In this case, isotopic analysis is achieved with a mass spectrometer. Time-of-flight mass spectrometers are especially well-suited for isotopic analysis of ions produced by pulsed resonance ionization lasers, because all the ions are detected on each pulse.

3.4.2 Instrumentation

A versatile Laser-SNMS instrument consists of a versatile microfocus ion gun, a sputtering ion gun, a liquid metal ion gun, a pulsed flood electron gun, a resonant laser system consisting of a pulsed Nd:YAG laser pumping two dye lasers, a non-resonant laser system consisting of a high-power excimer or Nd:YAG laser, a computer-controlled high-resolution sample manipulator on which samples can be cooled or heated, a video and electron imaging system, a vacuum lock for sample introduction, and a TOF mass spectrometer.

3.4.3 Spectral Information

Compared with electron impact on molecules, laser photoionization can induce fragmentation of molecules to a much lesser extent by optimizing the laser settings intensity, pulse width and wavelength. Especially, the use of a fs-laser for post-ionization can significantly reduce molecular fragmentation. More chemical information is, therefore, available from laser-SNMS spectra. On the other hand, significant multiple ionization of atoms typically occurs when high laser power is used to achieve high post-ionization rates.

3.4.4 Quantification

Only a portion of the neutral particles present in the ionization volume (*IV*) of the laser beam is available for ionization. For this description the geometrical yield $Y_{IV}(\mathrm{X_i^0})$ is introduced (for notation, see Sect. 3.1.3):

$$Y_{IV}\,(\mathrm{X_i^0}) = \frac{\text{No. of } \mathrm{X_i^0} \text{ in the ionization volume}}{\text{No. of } \mathrm{X_i^0}} \tag{3.23}$$

Post-ionization occurs with the probability $\alpha\,(\mathrm{X_i^0} \rightarrow \mathrm{X_j^{\oplus}})$, (the sign $\oplus$ is used to denote a photoion):

$$\alpha\,(X_i^0 \rightarrow X_j^{\oplus}) = \frac{\text{No. of generated photoions } X_j^{\oplus}}{\text{No. of } X_i^0 \text{ in the interaction volume}} \tag{3.24}$$

Molecular particles can dissociate during the ionization process. This is taken into account by the different indices i and j.

For post-ionized particles, it is possible to define a generalized transformation probability $P^{\oplus}\,(A \rightarrow X_j^{\oplus})$:

$$P^{\oplus}\,(A \rightarrow X_j^{\oplus}) = \frac{\text{No. of generated photoions } X_j^{\oplus}}{\text{No. of sputtered A}} \tag{3.25}$$

$P^{\oplus}\,(A \rightarrow X_j^{\oplus})$ is given by Eq. (3.26):

$$P^{\oplus}\,(A \rightarrow X_j^{\oplus}) = \sum_i P\,(A \rightarrow X_i^0) \cdot Y_{IV}(X_i^0) \cdot \alpha\,(X_i^0 \rightarrow X_j^{\oplus}) \tag{3.26}$$

The summation over i is necessary, because different precursors can be transformed into the same photoion $X_j^{\oplus}$ by fragmentation.

For detection of the generated ionic particles, the transmission of the mass spectrometer and the detection probability are considered in the same way as discussed in Sect. 3.1.3.

An important quantity is, again, the useful yield $Y_u\,(X_i^{\oplus}(A))$:

$$Y_u\,(X_i^{\oplus}(A)) = \frac{\text{No. of detected } X_i^{\oplus}}{\text{No. of sputtered A}} \tag{3.27}$$

Values of $Y_u\,(X_i^{\oplus}(A))$ for elements typically range from 10^{-4} to 10^{-2} in NR-laser-SNMS, and from 10^{-2} to 10^{-1} in R-laser-SNMS. If the experimental conditions are not well known, the concentration of A can also be quantified by using the relative sensitivity factor (*RSF*) method (Eqs (3.8) and (3.9) in Sect. 3.1.3).

3.4.5 Applications

3.4.5.1 Non-resonant Laser-SNMS

Figure 3.42 [3.105] shows a typical NR-laser-SNMS spectrum of a contaminated Si wafer under low dose sputtering conditions (primary ion fluence $< 10^{14}$ cm^{-2}) . For post-ionization, the radiation of an excimer laser operating at a wavelength of 248 nm was focused into the sputtered cloud of particles (intensity 10^{10} W cm^{-2}). Because at 248 nm the ionization cross-section of Si is relatively low, the intensity of the matrix signals (Si, SiO) is lower than expected. Because of the use of a high-resolution TOF mass spectrometer (mass resolution $m/\Delta m$ = 3000), the metal contaminants (Mn 20 ppm, Ni 130 ppm, Cu 500 ppm, and Zn 400 ppm) are clearly separated from the molecular signals at the same integral mass. The ultimate detection limits (DL) of NR-laser-SNMS are given by the amount of sample material available and by the useful yield. If the outer monolayer is completely consumed within an area of 10^{-4} cm^2

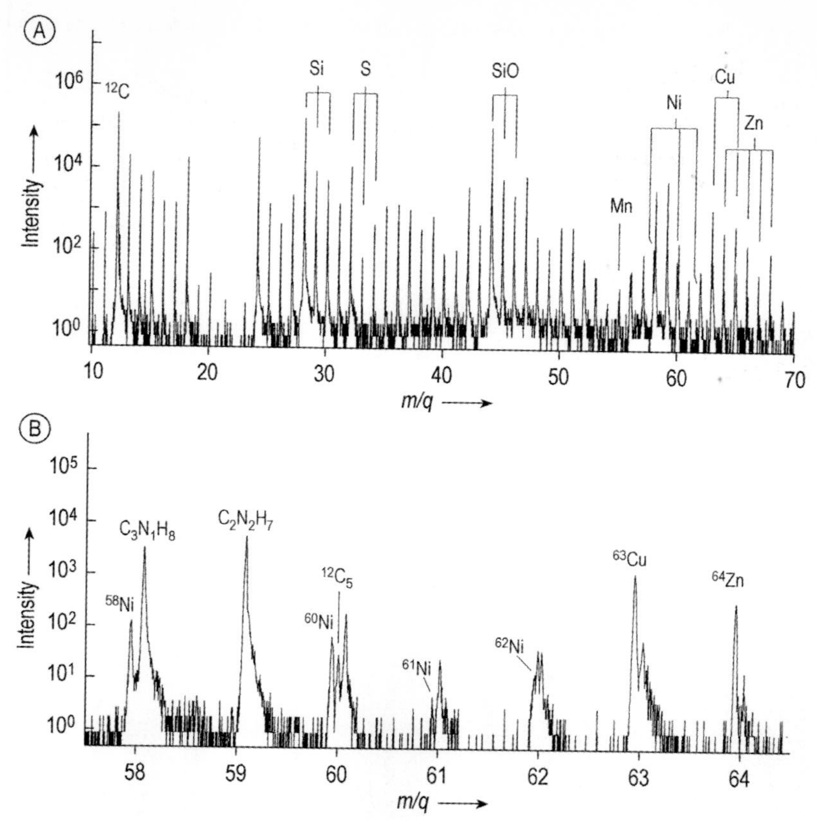

Fig. 3.42. Non-resonant laser-SNMS spectrum of a Si wafer contaminated with a variety of transition metals and hydrocarbons [3.105].

and if the useful yield is 10^{-2}–10^{-3}, a surface coverage of approximately 10^7 atoms cm^{-2} can be detected by at least 10 ions. This limit can only be reached if the corresponding ion signal is free from background and from mass interference. In Table 3.1 are listed the relative sensitivity factors S (M^+, Si^+ + SiO^+) in respect to the sum of the matrix signals Si^+ and SiO^+ and the DL determined for several metals on Si wafers by non-resonant post-ionization with 248 nm and 193 nm light [3.105]. Because of the low matrix-dependence of laser-SNMS, the same sensitivity can be achieved for the detection of trace metals in matrices such as polymers or large biomolecules, as for semiconductor matrices.

Element mapping with non-resonant laser-SNMS can be used to investigate the structure of electronic devices and to locate defects and microcontaminants [3.114]. Typical SNMS maps for a GaAs test pattern are shown in Fig. 3.43. In the subscript of each map the maximum number of counts obtained in one pixel is given. The images were acquired by use of a 25-keV Ga^+ liquid metal ion source with a spot size of approximately 150–200 nm. For the given images only 1.5 % of a monolayer was consumed – "static SNMS".

Tab. 3.1. NR-laser-SNMS: Relative sensitivity factors S (Me, ΣSi) and detection limits DL for metals on Si wafer surfaces.

Element	*S (M, ΣSi)*		*DL (atoms cm^{-2})*	
	193 nm	*248 nm*	*193 nm*	*248 nm*
Al	0.9	0.9	8×10^9	3×10^{10}
Ti	1.9	9	5×10^9	5×10^9
V	2.1	42	3×10^8	3×10^8
Cr	2.9	28	3×10^9	3×10^9
Fe	1.6	31	5×10^9	5×10^9
Co	1.1	6.9	5×10^8	5×10^8
Ni	0.9	14	1×10^9	1×10^9
Cu	1.2	24	5×10^9	5×10^9
Ga	4.4	0.4	5×10^8	8×10^9
As	1.9	10	3×10^8	3×10^8
Mo	1.8	32	3×10^8	3×10^8
W	5.1	39	5×10^8	5×10^8

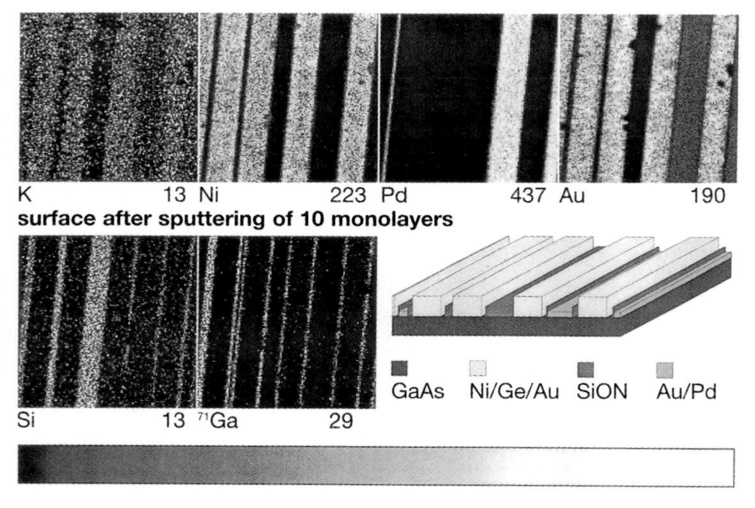

Fig. 3.43. Non-resonant laser-SNMS mapping of a contact test structure on GaAs. Field of view 40 × 40 μm^2 [3.114].

3.4.5.2 Resonant Laser-SNMS

Figure 3.44 shows Cu depth profiles obtained from silicon and 0.8 μm silicon oxide on silicon samples. For this experiment, half a silicon wafer without an SiO_2 layer and half an SiO_2-covered Si wafer were simultaneously implanted with 5×10^{14} ^{63}Cu atoms cm^{-2} at 150 keV. For analysis, both samples were cooled to 107 K to reduce migration effects, and depth profiles were taken under the same experimental conditions. The integrated Cu signal for both samples, i.e. the Cu sensitivity factor for Cu in silicon and for Cu in silicon oxide, was almost the same (less than 3% difference, which is because of the differences in sputter yield), demonstrating the matrix independence and quantification accuracy possible with R-laser-SNMS. The dynamic range for ^{63}Cu is better than six orders of magnitude. Because of the limited mass discrimination of the implanter, 2% ^{65}Cu is present. At the SiO_2–Si interface a higher concentration of Cu was observed. The isotopic ratio at this pile-up near the interface and below is closer to that of natural copper, indicating that the sample was contaminated with Cu when the oxide was grown. The Cu profile in Si was broadened because of channeling effects which occur in crystalline silicon but not in amorphous silicon oxide [3.115]. The lowest measured signal corresponds to a concentration of 400 ppt which is still not the detection limit.

Figure 3.45 shows R-laser-SNMS images of copper concentration in the region of Te and Cd inclusions in a CdZnTe (CZT) film. It is clear from the images that the copper concentration is approximately ten times higher in the Te inclusion (peak concentration approximately 50 ppb) than in the surrounding CZT matrix and approximately 500 times higher in the Cd inclusion. These data support the theory that the copper migrates preferentially to Te/Cd second phase regions inside the CZT matrix [3.116]. Further experiments must be performed to verify whether the solubi-

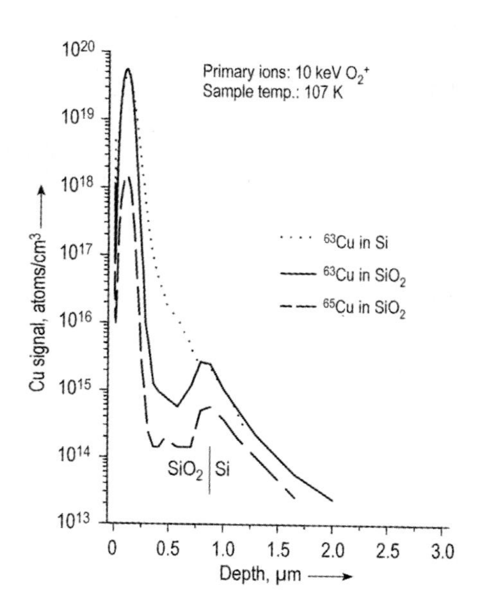

Fig. 3.44. Depth profile of ^{63}Cu in Si and of ^{63}Cu and ^{65}Cu in SiO_2 [3.115].

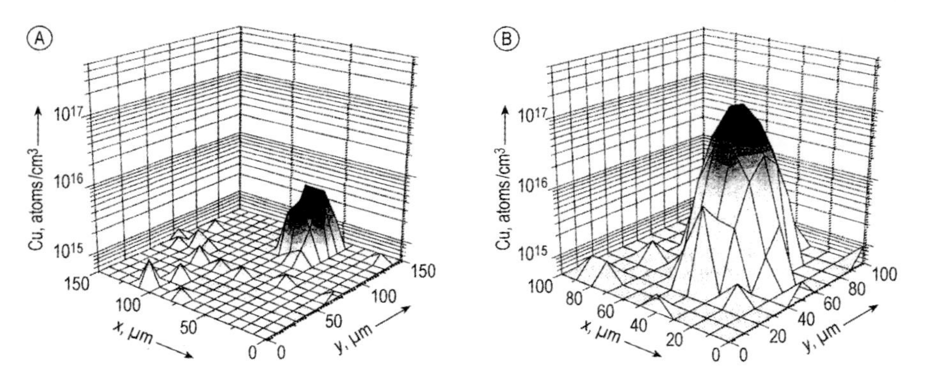

Fig. 3.45. R-laser-SNMS images of copper around (a) tellurium and (b) cadmium inclusions [3.116].

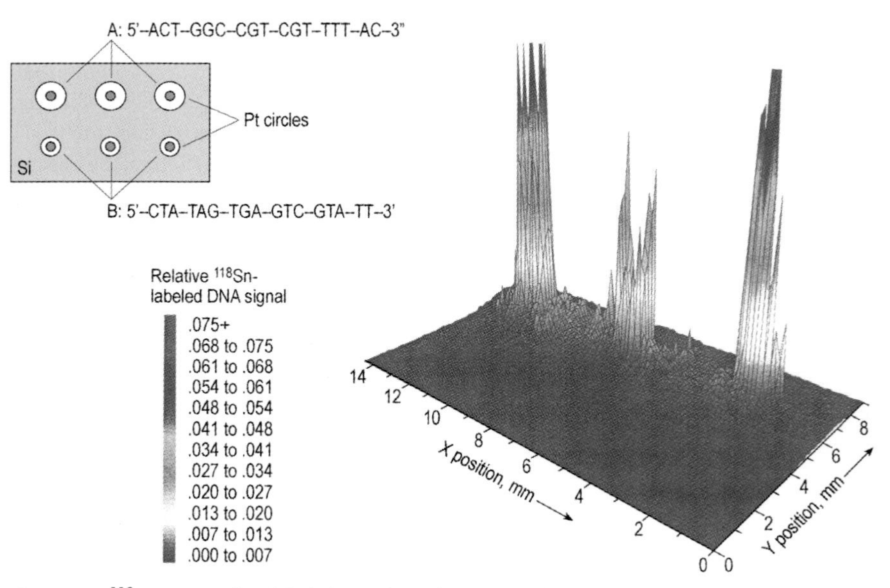

Fig. 3.46. ^{118}Sn image of Sn-labeled DNA hybridized to six oligonucleotide spots; three complimentary and three non-complementary.

lity of Cu is much higher in Cd than Te. The same imaging technique can be used to locate and quantify pharmaceutical products in tissue, with sub-cellular resolution and sub-ppm detection limits [3.117].

R-laser-SNMS can also be applied to detect Sn-labeled DNA at positively hybridized and unhybridized sites on a DNA biosensor chip which can be used, e.g., for genetic disease and cancer diagnostics [3.118, 3.119]. DNA diagnostics by sequencing by hybridization (SBH) involve binding a small (typically 18–20-mer) oligonucleotide to a chip. The chip might be glass, silicon, gold, or platinum. By polymerase chain reaction fragments of the genomic DNA are produced with an attached label. The size of the fragment can vary from a few dozen to several hundred or several thousand nucleotides.

Figure 3.46 shows an R-laser-SNMS image of ^{118}Sn obtained from a DNA biosensor chip. For this experiment, two different 17-mer oligonucleotides were bound to platinum circles on a silicon wafer (three locations each). Enriched ^{118}Sn-labeled DNA, which was completely complementary to one of these oligonucleotides and non-complementary to the other, was hybridized to this chip. After the hybridization, the sample was washed to remove unhybridized probes. The image shows three peaks at the complementary DNA sites. The discrimination between hybridized and unhybridized sites is better than 100. A single point on the top center spot was analyzed twice before the full image was taken, leading to a slight reduction of the signal. The same technique can be readily applied to a variety of problems in bioanalytical chemistry, e.g. screening of combinatorial libraries, etc.

3.5 Rutherford Back-scattering Spectroscopy (RBS)

Leopold Palmetshofer

Rutherford back-scattering spectroscopy (RBS) is one of the most frequently used techniques for quantitative analysis of composition, thickness, and depth profiles of thin solid films or solid samples near the surface region. It has been in use since the nineteen-sixties and has since evolved into a major materials-characterization technique. The number and range of applications are enormous. Because of its quantitative feature, RBS often serves as a standard for other techniques.

3.5.1 Principles

In RBS, a beam of monoenergetic ions, usually H^+ or He^+ of typical energy 0.5 to 2.5 MeV, is directed at a target, and the energies of the ions which are scattered backwards are analyzed. In the back-scattering collision, energy is transferred from the impinging particle to the stationary target atom. The energy ratio between the projectile energy E_1 after collision and the energy E_0 before collision, derived from binary collision theory, is [3.120, 3.121]:

$$\frac{E_1}{E_0} = K = \left[\frac{(M_2{}^2 - M_1{}^2 \sin^2\theta)^{1/2} + M_1 \cos\theta}{M_2 + M_1}\right]^2 \tag{3.28}$$

The energy ratio E_1/E_0, called the kinematic factor K, shows that the energy after scattering depends only on the mass M_1 of the projectile, the mass M_2 of the target atom, and the scattering angle θ (i.e. the angle between incident and scattered beams). If M_1, E_0, and θ are known, M_2 may be determined and the target element identified.

For direct back-scattering through 180°, the lowest value of the energy ratio is given by:

$$\frac{E_1}{E_0} = \left(\frac{M_2 - M_1}{M_2 + M_1}\right)^2 \tag{3.29}$$

If $M_1 = M_2$, the incident particle is at rest after a central collision and all the energy is transferred to the target atom. For target atoms with $M_2 < M_1$ no back-scattering occurs.

The energy of the back-scattered ion is given by Eq. (3.28) only for scattering by an atom at the surface of the target. In RBS, however, the ion beam penetrates the target and an ion might be back-scattered by target atoms at any point along its path. In the energy region used for RBS the ion trajectory is a straight line (apart from the back-scattering collision) along which the ions lose energy primarily through excitation and ionization of atomic electrons (electronic energy loss). The energy loss per unit path length, dE/dx, is called the stopping power. These additional energy losses broaden the peak to be observed in an RBS spectrum of a thin film.

The situation is illustrated in Fig. 3.47. The upper part shows a thin film of Ni deposited on a Si substrate. Only particles scattered from the front surface of the Ni film have an energy given by the kinematic equation, Eq. (3.28), $E_1 = K_{Ni}E_0$. As particles traverse the solid, they lose energy along the incident path. Particles scattered from a Ni atom at the Si–Ni interface therefore have an energy smaller than $K_{Ni}E_0$. On the outward path the particles again lose energy. On emerging from the surface, the particles scattered at the interface have a total energy difference ΔE from particles scattered at the surface; this results in a broad peak in the back-scattering spectrum. The peak width ΔE is related to the thickness of the Ni film (see below).

For Si atoms the kinematic factor is smaller, $K_{Si} < K_{Ni}$, because of their lower mass. Back-scattered particles from Si atoms appear at lower energies in the spectrum. Because no Si atoms are on the surface, the energy spectrum produced by scattering from Si starts at an energy lower than $K_{Si}E_0$ and then extends to zero energy, because Si atoms form the substrate with effectively infinite depth.

For a surface layer consisting of atoms of lower mass than the substrate atoms, the peak arising from the surface layer merges with the broad continuum and appears as a small feature on top of it. For surface analysis by RBS, conditions must be such that the mass of surface atoms is considerably higher than the mass of substrate atoms if the peak from surface atoms is to be completely resolved. Only in these circumstances is the sensitivity of RBS comparable with that of AES and XPS, i.e., up to ~10^{-3} atomic layers. No chemical information is possible.

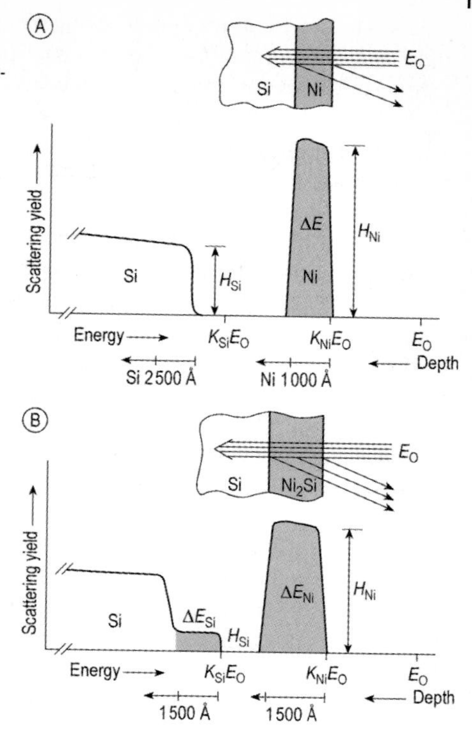

Fig. 3.47. Schematic back-scattering spectra for MeV He^+ ions incident on a 100-nm Ni film on Si (top) and after reaction to form Ni_2Si (bottom). Depth scales are indicated below the energy axes [3.120].

So far we have tacitly assumed that all the target atoms are equally visible to the projectiles. In a single crystal, atomic rows and planes can guide energetic ions along the channels between rows and planes so that the ion beam penetrates deeply into the crystal, an effect known as channeling [3.120, 3.122]. Channeling occurs when the ion beam is carefully aligned with a major symmetry direction of the single crystal. Fig. 3.48 shows a side view of this process in which most of the ion beam is steered through the channels formed by the strings of atoms. Channeled particles cannot get close enough to the atomic nuclei to undergo large-angle Rutherford scattering, hence scattering is drastically reduced by a factor of approximately 100. Atoms at the surface produce a surface peak in the back-scattering spectrum because the ion beam is scattered from the surface atoms with the same intensity as from a random array of

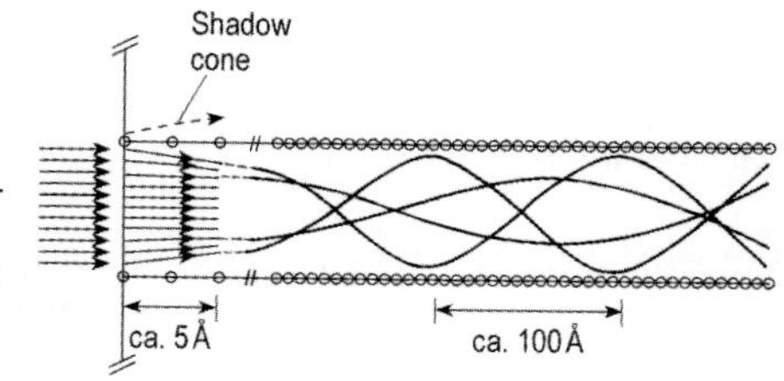

Fig. 3.48. Schematic diagram of particle trajectories undergoing scattering at the surface and channeling within the crystal. The depth scale is compressed relative to the width of the channel, to display the trajectories [3.120].

atoms. The second atom in each string of atoms is completely shadowed in a perfect rigid lattice, because of the "shadow cone" formed by the ion trajectories. Thus, channeling enormously improves the sensitivity of RBS to atoms at the surface.

3.5.2 Instrumentation

Ion beams suitable for RBS are produced in particle accelerators. By far the most widely used device for the production of MeV ion beams is the Van de Graaff electrostatic accelerator, either single ended or double ended (tandem Van de Graaff). The reader is referred elsewhere [3.123] for detailed information about accelerator technology. Accelerators normally produce ions in several charge states and even a multiplicity of species. To extract a beam suitable for materials analysis, the beam is passed into the field of an analyzing magnet. A mass- and charge-selected beam then enters an UHV environment via differentially pumped apertures and is steered to the target by electrostatic or magnetic lenses. The beam size at the target is, typically, 1 mm^2. Small beam spots, if desired, can be obtained by suitable lens systems. Selected area analysis using small spots is usually achieved by specimen manipulation. For general analysis the ions strike the specimen at normal incidence. When structural information is sought (RBS + channeling), the specimen is mounted on a multiple axis manipulator and can be rotated about the point of ion impact to vary the angle of incidence about channeling directions.

The scattering angle for optimum mass resolution would be 180° (a certain ΔM_2 gives the largest change in K when $\theta = 180°$, Eq. 3.28). Because of detector size, in practice $\theta \approx 170°$ is chosen. The detection of back-scattered ions is usually performed with a solid state detector, either a silicon surface barrier detector or a passivated implanted planar silicon (PIPS) detector. For conventional RBS with He^+ or H^+ ions at 1–2 MeV the PIPS detector has the better energy resolution (about 10 keV compared with 15 keV for surface barrier detectors [3.124]). The detector signals, which are highly proportional to the energy of the incident particle, are amplified and assorted in energy in a multichannel analyzer.

Recent years have seen increasing use of medium-energy ions and of heavier ions to optimize certain features of back-scattering analyses. Two examples are given. *Medium-energy ion scattering* (MEIS) employs ions with energies of 100–500 keV together with electrostatic analyzers or *time-of-flight* (TOF) detectors. MEIS enables very shallow analysis depth with resolution up to monolayers [3.125]. *Heavy-ion back-scattering spectroscopy* (HIBS) uses ions such as ^{12}C, ^{16}O, or ^{35}Cl together with TOF detectors to obtain extremely high sensitivities (~10^9 at cm^{-2}) for trace analysis [3.126].

3.5.3 Spectral Information

An RBS spectrum contains information about the mass of the scattering atoms, the composition of the surface layer, the depth of scattering atoms, and the thickness of a surface layer.

Evaluation of the mass of the target atoms from the energy of back-scattered particles has been described above. An example of the nature of the compositional information obtainable from a RBS spectrum is based on Fig. 3.47. The lower part of the figure shows, schematically, a Ni film on Si reacted to form Ni_2Si. After reaction, the Ni signal ΔE_{Ni} has spread slightly, because of the presence of Si atoms contributing to the energy loss. The Si signal has a step in the energy $K_{Si}E_0$ corresponding to Si in the Ni_2Si. The hatched peak areas in the figure are a measure of the number of particles scattered by Si or Ni atoms. Because the probability of a collision between the projectile and a target atom is easily obtained (Eq. 3.32), the number of Si and Ni atoms in the target and hence the composition N_{Ni}/N_{Si} can be calculated.

Knowing the composition of a layer, it is possible to establish a depth scale for the distribution of an element or to measure the layer thickness from the energy of the scattered particles. This depends on the energy loss of the projectile on its inward and outward paths, as described in Sect. 3.5.1. The energy difference, ΔE, for a particle scattered at the surface and a particle scattered at a depth x is given by:

$$\Delta E = [S]x \tag{3.30}$$

where $[S]$, called the energy loss factor, depends on the stopping power on the inward and outward paths, the kinematic factor, K, and the orientation of the sample both to the incident beam and to the detector direction [3.121]. In Fig. 3.47 the depth scales are indicated.

RBS with channeling can be used to detect and measure the thickness of an amorphous layer on an otherwise crystalline substrate. The method is commonly used to obtain structural information about the damaged surface layer of ion-implanted semiconductors and to study removal of the damage on annealing. An example is shown in Fig. 3.49 for GaAs [3.127]. If the incident ion beam is directed on to the GaAs crystal in a $\langle 110 \rangle$ channeling direction, the back-scattering yield is drastically reduced (d) compared with the yield when the beam is incident in a non-channeling or "random" direction (a). Note the small surface peak in the channeled spectrum. Ion implantation with 120 keV Si^+ ions to a dose of 5×10^{15} cm^{-2} produced an amorphous surface layer in GaAs. Because no channeling is possible in this layer, the intensity of the back-scattered signal under channeling conditions increases and is as high as for the random direction. Ions passing through the amorphous layer experience the crystalline nature of GaAs, which results in a reduced back-scattering yield (b). From the width of the broad peak in curve b the thickness of the amorphous layer is found to be ~140 nm. After annealing at 950 °C the intensity of the back-scattering yield under channeling conditions dropped substantially (c). The spectrum closely resembles that of crystalline GaAs, which indicates that the implantation damage is almost completely annealed out.

One of the most fascinating applications of channeling RBS is the study of lattice locations of impurity atoms. By measuring the angular dependence of the back-scattering yield of the impurity and host atoms around three independent channeling axes it is possible to calculate the position of the impurity. Details can be found elsewhere [3.122].

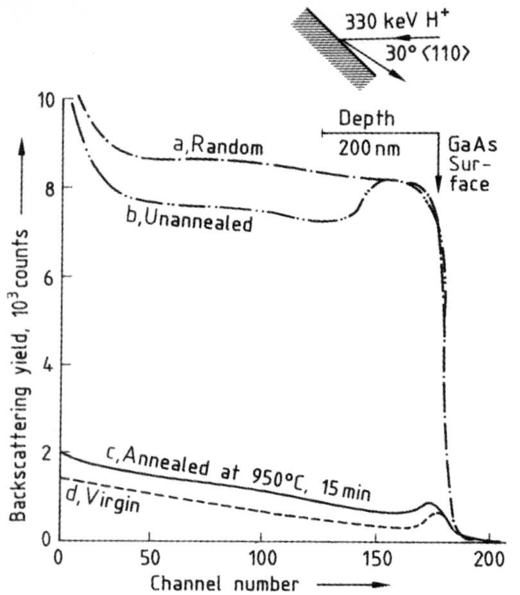

Fig. 3.49. RBS spectra from GaAs implanted with Si (120 keV, 5×10^{15} cm^{-2}) before and after annealing at 950 °C. The uppermost spectrum is taken in a random direction, the others are in the channeling direction [3.127].

3.5.4 Quantification

For an ion beam with the total number Q of ions impinging on a thin film, the number, Q_A, of particles back-scattered from atoms of type A and registered in the detector (also called yield, Y_A), is given by:

$$Q_A = Y_A = QN_A \sigma_A \Delta\Omega \tag{3.31}$$

where N_A is the areal density of atoms A in the film (atoms cm^{-2}), σ_A the differential scattering cross-section (cm^2 sr^{-1}), and $\Delta\Omega$ the solid angle of the detector. The differential scattering cross-section describes the probability of a projectile being scattered by a target atom through an angle, θ, into a solid angle, dΩ, centered about θ. If the interaction potential between the particle (M_1, Z_1) and the target atom (M_2, Z_2) during scattering is given by the Coulomb potential, the cross-section is given by the Rutherford formula:

$$\sigma(E, \theta) = \left(\frac{Z_1 Z_2 e^2}{4E}\right)^2 \frac{4\left[(M_2{}^2 - M_1{}^2 \sin^2\theta)^{1/2} + M_2 \cos\theta\right]^2}{M_2 \sin^4\theta \left(M_2{}^2 - M_1{}^2 \sin^2\theta\right)^{1/2}} . \tag{3.32}$$

Equation (3.32) is given in cgs units. For practical calculations, the number $e^2 \approx 1.44 \times 10^{-13}$ MeV cm is useful. For standard RBS with 1–2 MeV He$^+$ ions, the use of the Rutherford cross-section is justified (giving the technique its name). Deviations occur at both higher and lower energies [3.121].

For a particular primary ion and fixed experimental conditions, the scattering cross-section is a function only of the mass and atomic number of the scattering atom, and can be calculated. Thus in principle, all the terms in Eq. (3.31) except the required N_A are known. RBS is, therefore, an absolute method that does not require the use of standards.

For a compound film A_mB_n, the composition can be calculated from Eq. (3.31) to be:

$$\frac{n}{m} = \frac{N_B}{N_A} = \frac{Q_B}{Q_A} \frac{\sigma_A(E,\theta)}{\sigma_B(E,\theta)} \tag{3.33}$$

Note that this ratio depends only on the ratio of measured yields, Q_A/Q_B, and knowledge of the cross-section ratio, σ_A/σ_B. The hard-to-measure quantities Q and $\Delta\Omega$ have cancelled.

For targets containing several elements which might produce overlapping peaks, RBS spectra are analyzed by use of computer simulations. The energy spectrum of the back-scattered particles is calculated for the actual experimental conditions and an assumed target composition. The target composition is then altered (either manually or by using a least-square fitting procedure) until the calculated and measured spectra are closely matched. A widely used RBS analysis program is the RUMP code [3.128]. Commercial software packages are also available [3.121].

The accuracy in RBS results is ~3% for areal densities and better than 1% for stoichiometric ratios. This high accuracy is obtained only when all relevant quantities are measured or evaluated carefully. Pitfalls which often prevent RBS from achieving its full accuracy are described elsewhere [3.129]. Calibration can be achieved by measuring standards obtained by either implanting into or depositing on a light element (silicon) a known amount of a much heavier element (e.g. Ta or Sb).

3.5.5 Applications

Because RBS is a major technique for the analysis of thin solid films and surface layers, the number of published applications in these areas is enormous. The number of publications reporting the use of RBS either alone or with other analytical techniques, is approximately 1000 per year. Most applications are in the field of semiconductor technology. Ion-implantation damage and damage annealing, implantation profiles, multilayer systems grown by molecular beam epitaxy (MBE) or chemical vapor deposition (CVD), silicide formation, and diffusion barriers for contacts are among the topics routinely investigated by RBS.

A typical example is the formation of a buried β-$FeSi_2$ layer in Si [3.130]. The system is a favorable one for RBS, because Fe is much heavier than Si. The $FeSi_2$ was synthesized by implanting Fe to a high dose (2.7×10^{17} cm^{-2}) into Si and performing high temperature annealing. $FeSi_2$ occurs as different phases, which complicates the formation of the low-temperature semiconducting β-phase. After brief annealing

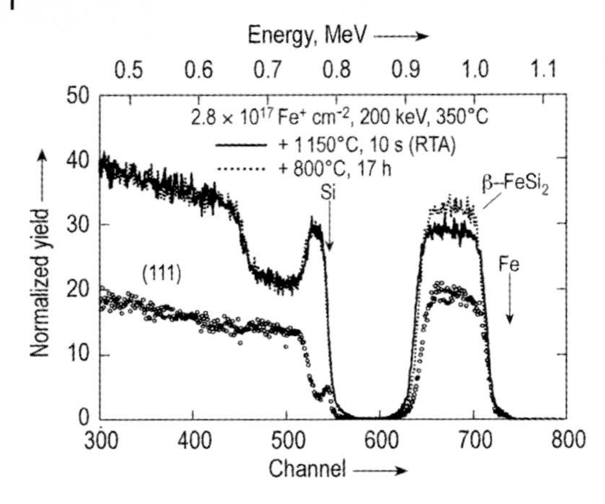

Fig. 3.50. RBS spectra of buried α-$FeSi_2$ (full line) and β-$FeSi_2$ (dotted line) layers taken in a random direction. The ⟨111⟩ channeling spectrum (circles) refers to the β-$FeSi_2$ layer [3.130].

(10 s) at 1150 °C α-$FeSi_2$ is formed. The transformation to β-$FeSi_2$ is performed by long-term annealing (17 h) at 800 °C, a temperature well below the α–β transition. Figure 3.50 shows the RBS spectra. The Si yield starts at the surface energy $K_{Si}E_0$ and has a large dip at lower energies, whereas the Fe peak starts at energies below $K_{Fe}E_0$, which indicates that the $FeSi_2$ layer is buried below the Si surface. The difference between Fe signal heights for the α and β phases indicates a change in density during phase transformation. It is estimated that the α phase contains structural vacancies of ~17%. The channeling spectrum of the β phase shows that the buried layer is crystallographically aligned with the Si – (001) planes of β-$FeSi_2$ are parallel or slightly off-oriented to (111) Si.

RBS and channeling are extremely useful for characterization of epitaxial layers. An example is the analysis of a $Si_{1-x}Ge_x$/Si strained layer superlattice [3.131]. Four pairs of layers, each approximately 40 nm thick, were grown by MBE on a ⟨100⟩ Si substrate. Because of the lattice mismatch between $Si_{1-x}Ge_x$ ($x \approx 0.2$) and Si, the $Si_{1-x}Ge_x$ layers are strained. Figure 3.51 shows RBS spectra obtained in random and channeling directions. The four pairs of layers are clearly seen in both the Ge and Si

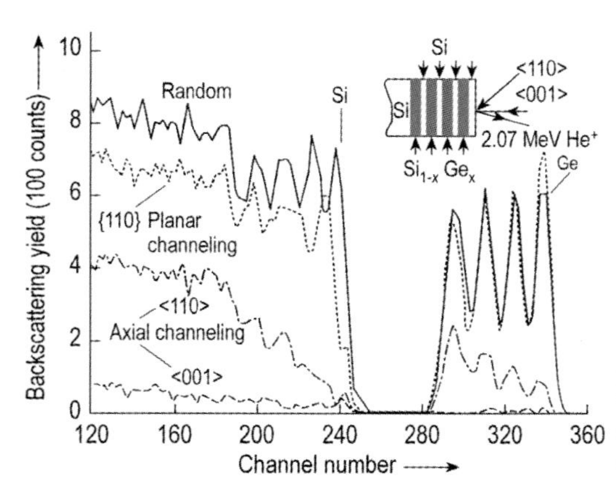

Fig. 3.51. RBS spectra of 2.07 MeV He^+ ions back-scattered from a $Si_{1-x}Ge_x$/Si strained layer superlattice. The full line refers to a random direction, the other spectra are taken along a variety of channeling directions [3.131].

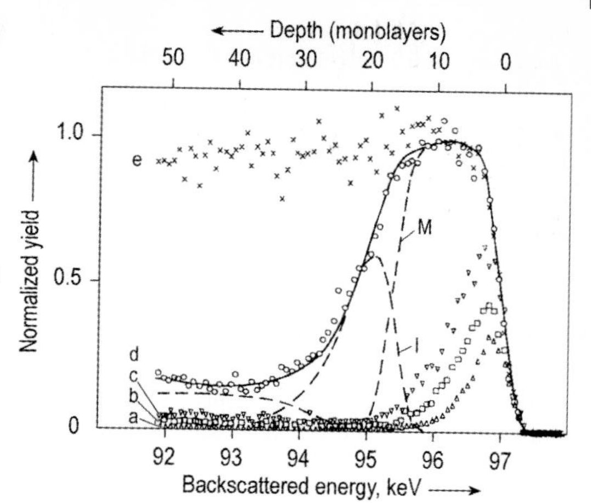

Fig. 3.52. Normalized back-scattering yields of H^+ ions from Pb near the melting point, with the incident beam and scattered beam directed along ⟨101⟩ crystal axes (double alignment): curve a, 295 K; curve b, 506 K; curve c, 561 K; curve d, 600.5 K; curve e, 600.8 K. Spectrum d is fitted by a sum of contributions M, from a liquid surface layer, and I, from a partially ordered transition layer [3.133].

yield. The spectrum for channeling in the ⟨100⟩ direction is typical, whereas large dechanneling is observed in the ⟨110⟩ direction, because of the lattice distortion at the interface $Si_{1-x}Ge_x/Si$. (The strained $Si_{1-x}Ge_x$ layer grows such that the in-plane lattice constant is the same as that of Si, whereas the perpendicular lattice constant is larger. ⟨100⟩ channels are, therefore, rather undistorted and diagonal ⟨110⟩ channels are distorted by a small angle.) The large dechanneling yield is a direct consequence of the lattice strain. The distortion can be measured as a small angular shift in angular scans at energies corresponding to specified Si and $Si_{1-x}Ge_x$ layers. By applying high-resolution RBS, even the intermixing of Si and Ge during epitaxial growth has been studied [3.132].

Other fields for RBS analysis include optical and dielectric materials, hard and protective coatings, superconductors, and magnetic materials.

RBS as an extremely useful method for applied materials science has also contributed to basic science. A famous example is surface melting [3.133]. Fig. 3.52 shows normalized back-scattering yields of H^+ from Pb near the melting point in the ⟨101⟩ direction (double alignment). The surface peak increases with increasing temperature and develops into a broad peak at 600.5 K. At 600.8 K, just above the melting temperature (600.7 K), the random spectrum is obtained, as is expected for a bulk liquid. The broad peak below the melting temperature (curve d) can be fitted by assuming a disordered liquid surface layer (M) plus a partially ordered transition layer (I) between the liquid and the crystalline substrate. These were the first direct measurements of surface melting. Later experiments showed that surface melting is very dependent on the orientation of the surface.

3.6
Low-energy Ion Scattering (LEIS)

Peter Bauer

3.6.1
Principles

In principle, a low energy ion scattering (LEIS) experiment is like RBS at low energies – an ion beam is sent to a target at a large angle relative to the surface plane, and back-scattered projectiles are detected at a large scattering angle [3.134, 3.135]. In contrast with RBS, in a typical LEIS experiment only positive ions are detected. Consequently, also the singly charged ion fraction P^+, i. e. the yield of singly charged ions divided by the total back-scattered yield, has to be included when calculating the ion yield, Q_A^+, back-scattered from an atom of species A:

$$Q_A^+ = Q_0 \, N_A \, \frac{d\sigma_A}{d\Omega} \, P^+ S \, T \Delta \Omega \tag{3.34}$$

where Q_0 is the number of incident projectiles, N_A the areal density of atoms A in the surface (atoms cm^{-2}), $d\sigma_A/d\Omega$ the differential scattering cross-section, S a steric factor describing the influence of multiple scattering, and T and $\Delta\Omega$ are the transmission function and the solid angle of the spectrometer, respectively.

The scattering cross-section is considerably different from the Rutherford cross-section, because the distance of closest approach, R_{min}, is rather large at low energies. Thus, electronic screening of the interaction between the nuclei is important. The screened scattering potential $V(r)$ reads:

$$V(r) = V_C(r) + V_e(r) = V_C(r) \, \Phi(r/a), \tag{3.35}$$

where V_C and V_e are the potentials of the target nucleus and the target electrons, respectively, and $\Phi(r/a)$ is the screening function, which describes how the Coulomb potential is weakened by electronic screening. Φ is a function of the reduced distance, r/a, where the screening length, a, is characteristic of the ion-target combination. Within the Thomas–Fermi-Molière model [3.120], Φ and a are given by Eqs (3.36):

$$\Phi(r/a) = \sum_{i=1}^{3} \delta_i \exp\left(-\frac{c_i \, r}{a}\right) \tag{3.36 a}$$

$$a = \frac{0.8852 \, a}{\sqrt[3]{\left(\sqrt{Z_1} + \sqrt{Z_2}\right)^2}} \tag{3.36 b}$$

where $a_0 = 0.529$ Å (the Bohr radius), and the constants $\delta_1 = 0.35$, $\delta_2 = 0.55$, $\delta_3 = 0.1$, $c_1 = 0.3$, $c_2 = 1.2$, and $c_3 = 6$. Thus, the influence of screening is negligible for small distance ($\Phi \to 1$ for $r \to 0$) and limits the interaction to distances comparable with the screening length ($\Phi \to 0$ for $r \gg a$).

From the screened potential the scattering angle, $\theta(b)$, that corresponds to a given impact parameter, b, is obtained by solving the scattering integral, i. e. the general relationship $\theta(b)$ for an arbitrary central potential [3.136]. From $\theta(b)$, one readily obtains the differential scattering cross-section:

$$\frac{\mathrm{d}\sigma}{\mathrm{d}\Omega} = \left|\frac{b}{\sin\theta}\frac{\mathrm{d}b}{\mathrm{d}\theta}\right| \tag{3.37}$$

In Eq. (3.34), the major unknown is the charge fraction, P^+, which is still a subject of theoretical investigation [3.137]. Current knowledge is that when noble gas ions are used as projectiles the detection of back-scattered positive ions leads to the extreme surface sensitivity typical of LEIS, because $P^+ > 0$ is observed only when the projectile is scattered by a surface atom (because of their large binding energy, the mean charge state of low energy noble gas ions in any matter is neutral).

When only two charge states (0, +1) are of relevance, there are two contributions to P^+:

(1) projectile ions that have escaped neutralization along their trajectory, and
(2) those which were neutralized on their incoming path, re-ionized in the surface collision and remained in the charged state when leaving the surface:

$$P^+ = P^+_{\mathrm{surv,in}}\, P^+_{\mathrm{surv,out}} + (1 - P^+_{\mathrm{surv,in}})\, P_{\mathrm{Rel}}\, P^+_{\mathrm{surv,out}} \tag{3.38}$$

where $P^+_{\mathrm{surv,in}}$ and $P^+_{\mathrm{surv,out}}$ are the probabilities of the ion surviving neutralization on the incoming and the outgoing paths, respectively, and P_{ReI} is the probability of re-ionization in a surface collision. P_{ReI} is highly dependent on the distance of closest approach in the surface collision and on the detailed electronic structures of projectile and target atom, and is more important at higher energies [3.138]. At low energies, therefore, survivals usually dominate the charge fraction. As long as only one neutralization process (Auger or resonance process) dominates in the energy range of interest, the probability that the ion escapes neutralization, i. e. P^+_{surv}, is obtained from the transition rate $R(t)$ (transitions/second):

$$P^+_{\mathrm{surv}} = \exp\left[-\int \mathrm{d}t\, R(t)\right] \tag{3.39}$$

Taking into account that neutralization means tunneling of a target conduction-band electron to the ion, the time integral can easily be replaced by integration over the distance from the surface, s, by use of the identity $\mathrm{d}t = \mathrm{d}s/v_\perp$, where $v_\perp$ is the component of the ion velocity perpendicular to the surface. From this, the velocity-dependence of the survival probability, P^+_{surv}, is obtained:

$$P^+_{\mathrm{surv},l} = \exp\left[-\frac{1}{v_{l,\perp}}\int \mathrm{d}s\, R(t)\right] = \exp\left[-\frac{v_{\mathrm{c}}}{v_{l,\perp}}\right] \tag{3.40}$$

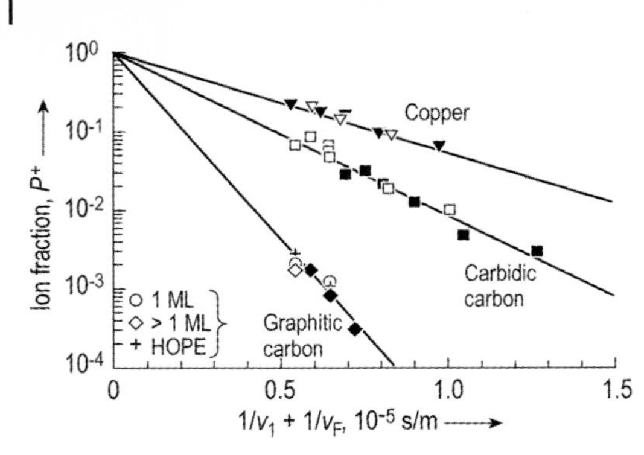

Fig. 3.53. Ion fraction of ^{3}He and ^{4}He ions (open and full symbols, respectively) in LEIS from Cu (▽,▼) and different types of C (carbidic carbon □, ■; graphitic carbon ○ ,◇,+,◆) as a function of the sum of the reciprocal velocities of the incoming and the scattered ion [3.139].

In Eq. (3.40), l can mean in and out, taking into account that the ion velocity after scattering is reduced, because of the recoil energy transferred to the target atom. The characteristic velocity v_c is obtained by integrating the transition rate along the trajectory and illustrates the neutralization efficiency: large values for v_c correspond to strong neutralization.

From Eq. (3.40) we may draw two conclusions:

(1) At high energies (small values of $1/v_\perp$) the ion fraction is high, because there is little time for neutralization.

(2) For a given geometry, the ion fraction decreases exponentially with the inverse of the velocity.

These theoretical predictions have been verified experimentally for numerous target materials (Fig. 3.53 [3.139]). Note that in Fig. 3.53 there is a pronounced difference between the neutralization of carbon atoms in a carbide and in graphite, respectively. This is one of the rare examples where matrix effects are observed.

3.6.2 Instrumentation

When the equipment used for RBS and LEIS is compared the following differences are apparent:

RBS	***LEIS***
accelerator	ion source (< 5 keV)
deflection magnet	Wien filter
PIPS-detector	Electrostatic spectrometer or time-of-flight spectrometer + micro-channel plate
H^+, He^+ ions	noble gas ions (He^+, Ne^+)

Because of its lower beam energy, a LEIS ion source is much more compact than an accelerator for MeV ions, but it has the same purpose – to provide a beam of ions with well defined energy and mass. In LEIS, the latter demand is fulfilled by a Wien filter that selects ions of one specific mass, as does the deflection magnet in RBS. In

LEIS the energies of the ions are too low for PIPS detectors and would lead to a signal-to-noise ratio less than unity. A stack of microchannel plates is, therefore, used to detect the ions in LEIS. The detection efficiency of the microchannel plates is included in the experimental transmission factor T in Eq. (3.34).

Energy selection is achieved either with an electrostatic analyzer or by use of a time-of-flight spectrometer. Both types of spectrometer have advantages and disadvantages. In time-of-flight measurement ions and neutrals are separated by post-acceleration of the ions. Without post-acceleration, neutrals also are detected, thereby increasing the intensity by orders of magnitude; usually, however, this leads to the loss of surface sensitivity. For analysis of surface structures, however, this is not a problem, because shadow cones can be used to regain surface sensitivity (see below). A conventional electrostatic spectrometer works with ions only and is, therefore, surface sensitive. Its efficiency is, however, low, because only ions in a narrow energy window around a well-defined pass energy are detected simultaneously; the pass energy must be scanned to obtain a spectrum. The detection efficiency can be increased by orders of magnitude by use of a double-toroidal energy analyzer in combination with a position-sensitive detector, because such a system measures simultaneously an energy spectrum of back-scattered ions in a larger energy window. Such an arrangement, called ERISS, has been developed at Eindhoven University [3.140] (Fig. 3.54). ERISS enables static LEIS measurements (without noticeable damage), even at polymer surfaces (see below) for which the sputter rate even under LEIS conditions is very high.

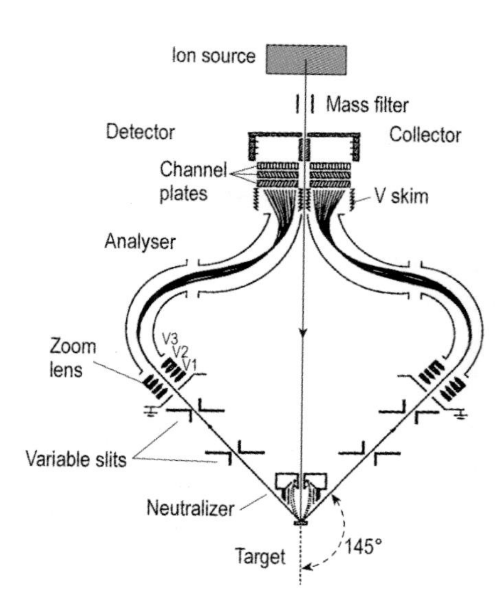

Fig. 3.54. Schematic diagram of the ERISS set-up [3.140].

3.6.3
Information

The energy spectrum in LEIS contains two types of information – energy and ion yield.

3.6.3.1 Energy Information

In the discussion so far we have considered the typical LEIS experiment only, i. e. large angles of incidence of exit relative to the surface plane. Under these conditions, in general, quantitative composition analysis is possible, because the ion-target interaction can be considered as a binary collision, because of the absence of matrix effects (see below).

Surface composition analysis by LEIS is based on the use of noble gas ions as projectiles, making use of the superb surface sensitivity of LEIS under these conditions. A consequence of this surface sensitivity is that the LEIS energy spectrum consists of lines, one per element, if the masses differ sufficiently. The lines are narrow, because inelastic energy losses play a minor role here. Thus, the information on the atomic species present is deduced from the energy of the back-scattered ions, which can be converted to the mass of the scattering center. (Fig. 3.55 [3.141]). In Fig. 3.55 it is shown that the mass range, where LEIS is sensitive, depends on the projectile mass.

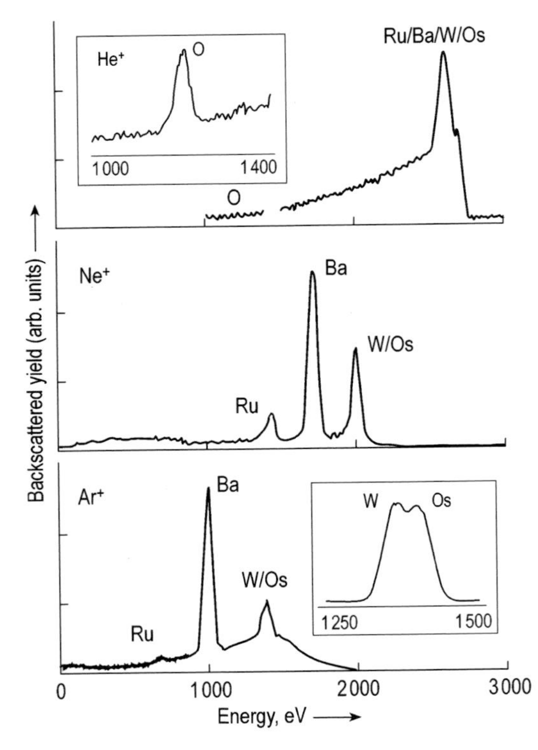

Fig. 3.55. LEIS spectra obtained from an Os/Ru top layer dispenser cathode with 3 keV He, Ne, and Ar projectiles, respectively. The He spectrum demonstrates the absence of O on an undamaged cathode surface, the insert in the He spectrum was obtained for a cathode exposed to 20 Langmuir oxygen at room temperature [3.141].

3.6.3.2 Yield Information

As is apparent from Eq. (3.34), the back-scattered ion-yield in LEIS is a measure of the concentration of an element in the surface, if the other quantities are known. This is generally so – the scattering cross-section can be calculated in accordance with Eqs (3.36) and (3.37) or found in tabulations; S is less problematic for large angles of incidence and exit (and can be considered constant); the transmission probability T is proportional to E for any electrostatic spectrometer that varies the pass energy (as long as the detection efficiency is energy-independent); the solid angle $\Delta\Omega$ of the spectrometer can be assumed constant. Thus, for fixed experimental conditions (constant primary energy, constant geometry, etc.), Eq. (3.34) can be simplified to:

$$Q_A^+ = \eta_A c_A \tag{3.41}$$

where η_A is the sensitivity factor which contains all element-specific quantities and c_A is the concentration of element A in the surface.

Figure 3.56 depicts LEIS spectra for two completely different types of Al_2O_3 sample, i.e. α-alumina (sapphire) and γ-alumina (a powder with high specific surface area) which show very similar results in both cases after thermal treatment at 400 °C [3.142]. Reduction of the Al signal in γ-alumina was ascribed to shielding by hydroxyl groups formed by water molecules, which are typical adsorbates on γ-alumina.

When measuring LEIS at small angles to the surface, for both incoming and for the outgoing particles, information on atomic structure can be obtained for a given surface (Fig. 3.57). Both, TOF-SARS (scattering and recoiling analysis) and ICISS (impact collision ion scattering spectroscopy) make use of the shadow cone and of focusing of the enhanced ion intensity at the shadow onto the neighboring atom – at glancing incidence all surface atoms are hidden in the shadow cone of the first atom in a plane or terrace and the scattered intensity is low; when the angle of incidence is increased (relative to the surface), at a critical angle the neighboring atoms leave the shadow cone and are hit by an enhanced flux of projectiles, which leads to a peak in the scattered intensity as a function of the angle of incidence (see below).

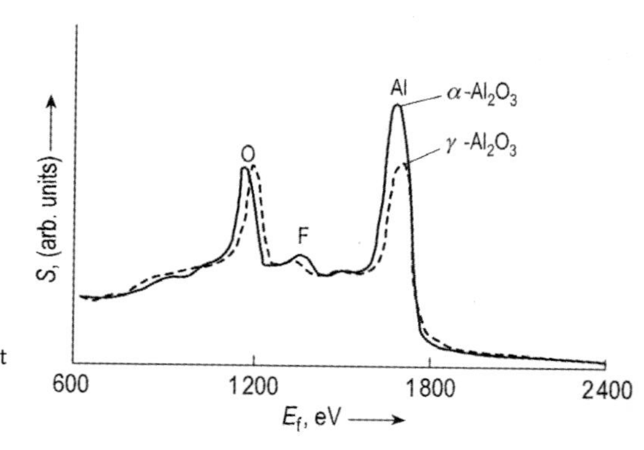

Fig. 3.56. LEIS obtained from different types of alumina with 3 keV He ions [3.142].

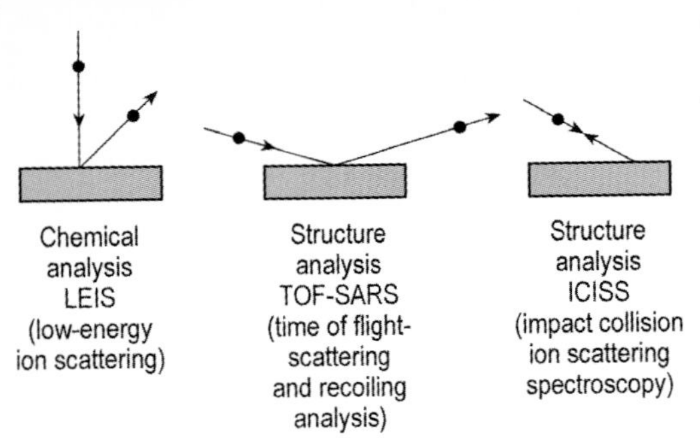

Fig. 3.57. Different experimental geometries for low energy ions spectroscopies.

3.6.4
Quantification

The usefulness of Eq. (3.41) depends crucially on whether or not the sensitivity factor η_A depends on the presence of other elements in the surface ('matrix effects'). It is an experimental finding that in general neutralization depends only on the atomic number of the scattering center, and matrix effects occur rarely. An instructive example is the neutralization of He by Al in the pure metal and in alumina. The slopes of the neutralization curves turn out to be the same for both materials, i.e. matrix effects are absent [3.143]. This is a strong indication that in the neutralization process not only the valence/conduction electrons, but also atomic levels below the valence/conduction band are involved.

Practically it is more convenient to measure intensity ratios instead of absolute intensities. Thus, e.g., Cu may serve as a reference material, relative to which the ion intensities back-scattered from the atoms of the surface under consideration are measured:

$$\frac{Q_A^+}{Q_{Cu}^+} = \frac{\eta_A \, c_A}{\eta_{Cu} \, c_{Cu}} \tag{3.42}$$

A further example where quantitative surface composition analysis is possible for a non-trivial surface is shown in Fig. 3.58, where for the systems Ta + O and Nb + O adsorption the ion signal from the metal is shown as a function of the ion signal from O. In this binary example Eq. (3.43) are valid for the concentration:

$$c_{Ta} + c_O = 1 \tag{3.43 a}$$
$$c_{Nb} + c_O = 1 \tag{3.43 b}$$

Inserting Eq. (3.43 a) into Eq. (3.42) yields:

$$Q_O^+ = \eta_O - Q_{Ta}^+ \frac{\eta_O}{\eta_{Ta}} \tag{3.44}$$

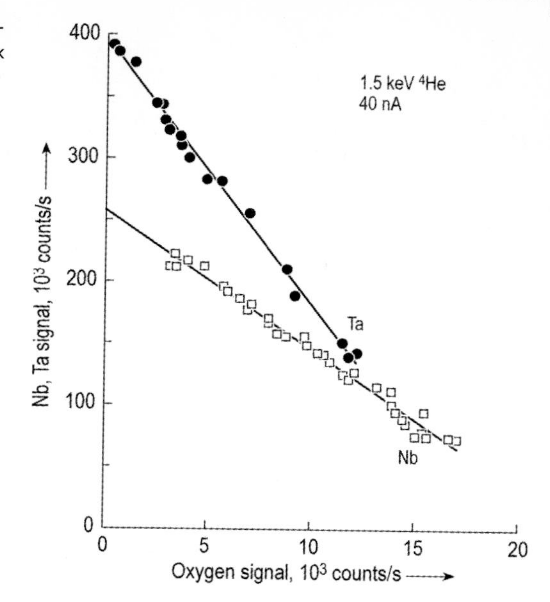

Fig. 3.58. Dependence of peak intensities of Nb and Ta on the oxygen peak intensity for adsorption of oxygen on the pure metals [3.144].

From Eq. (3.44), one expects a linear relationship between Q_O^+ and Q_{Ta}^+ and a corresponding relationship for Q_O^+ and Q_{Nb}^+. Indeed, this has been observed experimentally (Fig. 3.58) [3.144].

3.6.5 Applications

Applications of LEIS are widespread, going far beyond the needs of a typical surface science laboratory, because LEIS is capable of yielding information also on insulating samples and on very rough structures (catalysts!). The main problem with rough insulating surfaces is that they need charge compensation , i. e. flooding by thermal electrons which neutralize the charging of the surface by incoming ions and emitted energetic electrons.

An example is a LEIS study on a specific spinel, namely $ZnAl_2O_4$, for which cations (Zn) in tetrahedral sites are expected [3.145] to be less stable and therefore move to sites below the surface where they are better shielded, yielding a lower LEIS signal. This has been confirmed by Brongersma et al. [3.146] (Fig. 3.59). This figure shows that LEIS is very sensitive to Zn, as shown by LEIS from ZnO, but for the spinel no Zn is visible in the surface.

An even more ambitious goal is to characterize an unsupported catalyst, because the surface is extremely rough and the target rapidly deteriorates under bombardment. Energy deposition leads to enormous erosion, because the substrate cannot get rid of the energy deposited, owing to the low heat conductivity. As a consequence static LEIS conditions have to be used to obtain information on the surface alone. In Fig. 3.60 a we show a series of LEIS spectra obtained with 5 keV Ne^+ ions on a

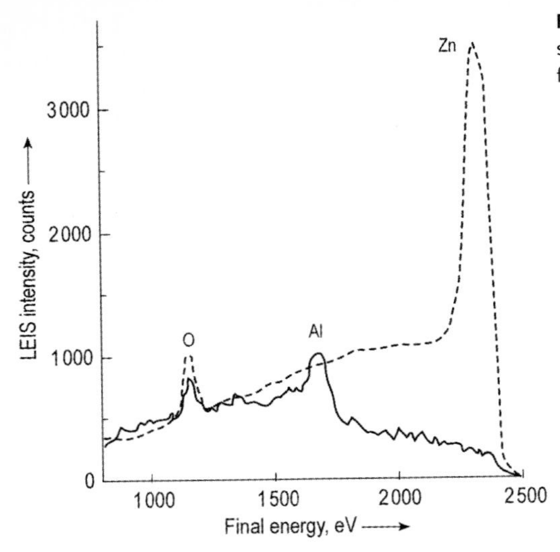

Fig. 3.59. LEIS spectra of 3 keV He^+ scattering from $ZnAl_2O_4$ (solid line) and from ZnO (dashed line) [3.146].

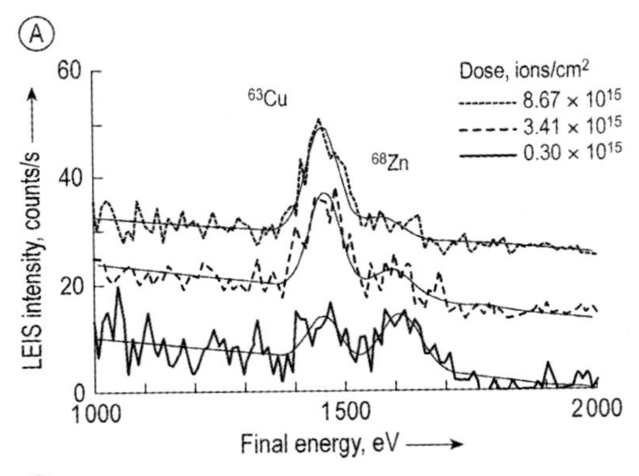

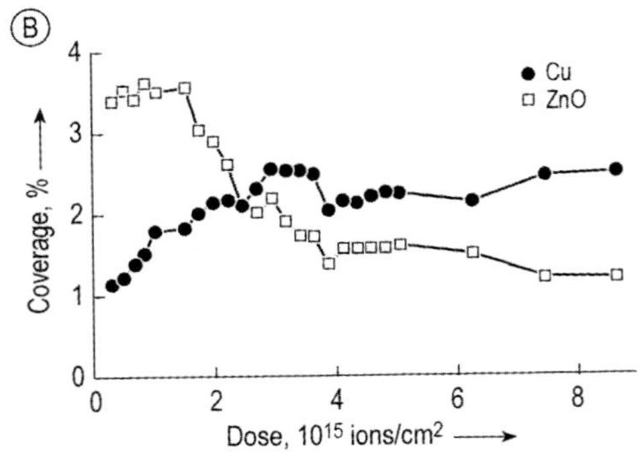

Fig. 3.60. (a) LEIS spectra of the $^{63}Cu/^{68}ZnO/SiO_2$ catalyst obtained with different doses of 5 keV Ne^+ ions (see insert, spectra are shifted vertically for clarity). Catalyst reduction temperature 700 K. Solid lines: fitted Gauss peaks [3.147]. (b) The relative coverage of Cu and ZnO on the silica-supported catalyst, reduced at 700 K, as a function of the ion dose [3.147].

$^{63}Cu/^{68}ZnO/SiO_2$ catalyst reduced at 700 K [3.147]. These LEIS spectra were obtained at three different ion doses – 3×10^{14}, 3.41×10^{15} and 8.67×10^{15} Ne^+ cm^{-2}. Because of the use of isotopically enriched Cu and Zn, and of Ne^+ ions as projectiles, Cu and Zn can clearly be separated in the LEIS spectrum. Strong dose-dependence is apparent. Fig. 3.60 b shows the dose-dependent surface concentrations of Cu and Zn. At low doses ($<1.5 \times 10^{14}$ Ne cm^{-2}) the Zn concentration remains constant whereas the Cu concentration increases. At these low doses a hydroxyl layer on top of the catalyst is sputtered. The Zn signal stays constant despite removal of the adsorbate, indicating that at the virgin surface the Zn concentration was even higher.

As a final application, the analysis of a surface structure by TOF-SARS is discussed (a general discussion is given elsewhere [3.148]). Fig. 3.61 shows that the scattered intensity is increased when projectiles hit a surface at the critical angle of incidence, α_c, because of focusing collisions with one of the neighboring atoms in the surface (a) or below the surface (b). Note also that recoiling atoms can be emitted in the same direction. Fig. 3.62 a shows a TOF spectrum measured for bombardment of CdS(0001) by 4 keV Kr^+ ions [3.149]. A spectrum of Ne projectiles scattered by Cd in CdS is also shown as a function of the angle of incidence, α. The peak at approximately $\alpha \approx 20^\circ$ is because of focusing of the projectiles by first-layer Cd atoms on to their first-layer Cd neighbors, by which they are scattered into the detector. The peak at $\alpha \approx 65^\circ$ is because of focusing of the projectiles by first-layer Cd atoms on to their second-layer Cd neighbors, by which they are scattered into the detector. Analysis of these critical angles enables determination of the lateral first-layer interatomic spacings and the first–second interlayer spacing (see Fig. 3.61).

Finally, Fig. 3.61 c shows an azimuthal scan, again for CdS(0001) and Kr^+ ions incident under a grazing angle. This enables determination of the surface periodicity, because the scattered intensity is minimal for incidence along the crystallographic axes, again because of shadowing. The intensity increases when the direction of incidence is tilted relative to the crystallographic axis by a critical angle. The widths of the minima are related to the interatomic spacings along the particular directions. Wide deep minima are expected for short interatomic spacings, because of the large azimuthal tilt needed to move the neighboring atom out of the shadow cone.

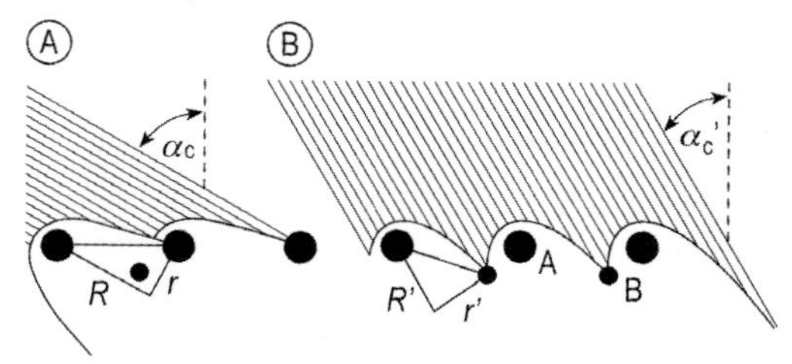

Fig. 3.61. Schematic illustration of projectile trajectories, showing focusing collisions when the projectiles impinge under a critical angle α_c [3.150].

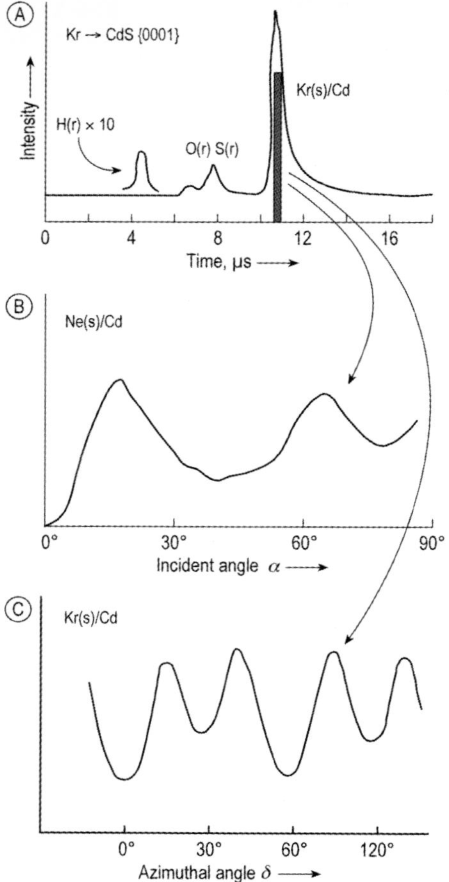

Fig. 3.62. Examples of TOF-SARS spectra for Kr scattering from CdS – time-of-flight spectrum, scans for the angle of incidence α, and the azimuthal angle δ [3.149].

3.7 Elastic Recoil Detection Analysis (ERDA)

Oswald Benka

3.7.1 Introduction

From methods based on elastic scattering of high-energy ions, one expects quantitative information about the depth distribution of elements in the surface region of solids. In Rutherford back-scattering spectrometry (RBS) projectiles scattered by angles larger than 90° are analyzed. Because projectiles with mass M_1 can only be back-scattered from a target atom with mass M_2 if $M_1 < M_2$, light projectiles such as protons and He ions are usually used in RBS. High back-scattered energies and large back-scattering cross-sections are found for heavy target atoms. RBS is, therefore, well suited to the analysis of heavy target elements, but its sensitivity for light elements is poor.

In elastic recoil detection analysis (ERDA) target atoms, which are recoiled in the forward direction by projectiles with energies in the MeV range, are analyzed, but not the scattered projectiles. Ecuyer et al. [3.151] described this method for the first time in the mid seventies. Because the sensitivity of ERDA is approximately the same for all target atoms, the technique is mainly used for light element profiling, which is hardly possible by RBS. An important application is hydrogen profiling, information that cannot be obtained from RBS and other standard techniques such as Auger and photoelectron spectroscopy.

Because of momentum conservation in forward scattering, both scattered projectiles and recoiled target atoms emerge from the sample in the forward direction and will be detected simultaneously in an ERDA experiment. The energy spectra of recoiled atoms and of scattered projectiles overlap. The measured spectrum is, therefore, very complex and a useful evaluation is usually not possible unless additional information is simultaneously obtained for particle identification so that the ERDA spectra can be split up into contributions of individual recoiled elements and of scattered projectiles. For particle identification, two methods are usually applied – the $\Delta E - E$ method and the TOF method. The $\Delta E - E$ method uses the specific energy loss ΔE of the detected particles in a thin solid or gas layer to distinguish between different ion species with the same energy E; this yields the atomic number ("effective charge") of the particles. The TOF method uses the velocity to distinguish between particles of the same energy; this yields the particle masses (see below).

ERDA, like RBS, is based on the following physical concepts:

(1) the *kinematic factors* describe the energy transfer from the projectile to target atoms in an elastic two body collision;
(2) the *differential scattering cross-section* gives the probability of the scattering event occurring; and
(3) the *stopping power* gives the average energy loss of projectiles and recoiled atoms as they traverse the sample, and define the depth scale.

By applying these concepts, individual recoil energy spectra can be transformed quantitatively into the corresponding concentration–depth profiles. The *depth* from which recoils originate correlates with the energy of these recoils. The *concentration* of analyzed atoms at this depth is obtained from the measured intensity of recoils at this energy. Analytical expressions which give the depth profile as a function of the measured energy spectrum are less straightforward than for RBS, because the energy loss of both, projectiles and recoil atoms, must be taken into account. Thus, measured energy spectra are most conveniently evaluated by use of computer codes. Several programs are available which calculate the energy spectra of projectiles and recoils for a given sample and experimental set-up (geometry).

In ERDA, different regimes have been developed with a broad range of projectiles and energies, which can roughly be separated into three groups:

(1) light projectiles (He, C ions) with energies between 2 and 10 MeV using the $\Delta E - E$ method, for the analysis of very light elements, mainly hydrogen;

(2) medium heavy projectiles (Cl) with energies of approximately 30 MeV using TOF, for the analysis of light elements; and
(3) heavy projectiles (Au) with energies larger than 100 MeV using mainly the $\Delta E - E$ method, to analyze a broad range of light and intermediate elements.

The sensitivity and depth resolution of ERDA depend on the type of projectile, on the type of particle, and on energy measurement. Because of the broad range of particles and methods used, general statements about sensitivity and depth resolution are hardly possible. Recent reviews of ERDA techniques are available [3.152–3.154].

3.7.2 Fundamentals

In ERDA the particle yield is measured in forward scattering geometry, i.e. at angles of detection $<90^\circ$ relative to the beam. Typical scattering geometry is shown in Fig. 3.63. Projectiles impinge at an angle of incidence α between the ion beam and the sample surface on a target. Recoils and scattered projectiles, which leave the sample at an exit angle β relative to the sample surface, are observed at a recoil and scattering angle θ.

When a projectile of mass M_1, energy E_1, and atomic number Z_1 collides with a target atom of mass M_2 and atomic number Z_2, it will transfer energy E_2 to the target atom at a recoil angle θ, which is given by:

$$E_2 = K_R E_1 \tag{3.45}$$

where K_R is the kinematic factor for elastic recoil. It can be derived from laws of conservation of energy and momentum to be:

$$K_R = \frac{4 M_1 M_2}{(M_1 + M_2)^2} \cos^2 \theta \tag{3.46}$$

The projectiles which are also scattered with scattering angle θ will have energy:

$$E_S = K_S E_1 \tag{3.47}$$

where K_S is the kinematic factor for elastic scattering:

$$K_S = \left[\frac{(M_2^2 - M_1^2 \sin^2 \theta)^{1/2} + M_1 \cos \theta}{M_1 + M_2} \right]^2 \tag{3.48}$$

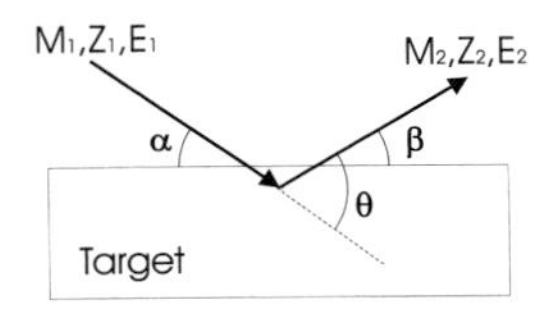

Fig. 3.63. Schematic diagram of the ERDA geometry.

Both kinematic factors K_R and K_S are functions of the mass ratio M_2/M_1 and the recoil and scattering angle θ. Because the recoil kinematic factor is symmetric in the masses M_1 and M_2, target atoms with masses $M_{2,1}$ and $M_{2,2}$ with $\frac{M_{2,1}}{M_1} = \frac{M_1}{M_{2,2}}$ will have the same recoil energy. The largest recoil energy is found for target atoms with $M_2 = M_1$ and is given by $E_2^{max} = E_1 \cos^2\theta$.

The scattered projectiles will have a higher energy than recoils for $M_2 > M_1$. If, therefore, in the surface region of a sample elements are present with atomic masses larger than that of the projectile, scattered projectiles will have the highest detected energy. When $M_2 < M_1$ for all masses M_2, the recoil atoms with highest masses have the highest energy. It is also important to remember that there is a critical angle for scattering. No scattered projectiles are found for scattering angles, θ, larger than arcsin (M_2/M_1). This has stimulated the use of heavy projectiles in ERDA.

A depth scale can be obtained from the energy of recoiled ions. If ions recoiled from a depth x are lower in energy by ΔE compared with ions recoiled from the surface, a simple relationship between ΔE and x can be found for thin layers, when constant stopping power is assumed:

$$\Delta E = x N \varepsilon_R \tag{3.49}$$

where N is the atomic density and ε_R is the recoil stopping cross-section factor of the target:

$$\varepsilon_R = K_R \frac{\varepsilon_{in}}{\sin\alpha} + \frac{\varepsilon_{out}}{\sin\beta} \tag{3.50}$$

ε_{in} and ε_{out} are the stopping cross-sections of the incident projectiles and of the recoiled atoms.

The recoil cross-section (in cm^2) for Rutherford scattering is:

$$\sigma_R = 5.18 \times 10^{-27} \left[\frac{Z_1 Z_2 (M_1 + M_2)}{M_2 E_1}\right]^2 \frac{1}{\cos^3\theta} \tag{3.51}$$

for projectiles of energy E_1, given in MeV.

For atomic masses $M_2 \ll M_1$ the recoil cross-section is almost independent of the atomic number, because the cross-section becomes proportional to $(Z_2/M_2)^2$ and the ratio Z_2/M_2 is close to 0.5 for all elements. ERDA with heavy projectiles thus has the advantage of almost constant sensitivity for all elements. Only for hydrogen the ratio Z_2/M_2 is equal to 1, hence the intensity of hydrogen recoils is enhanced by roughly a factor of four.

For quantitative evaluation of ERDA energy spectra considerable deviations of recoil cross-sections from the Rutherford cross-section (Eq. 3.51) must be taken into account. Light projectiles with high energy can penetrate the Coulomb barrier of the recoil atom; the nuclear interaction generally leads to a cross-section that is larger than σ_R, see Eq. (3.51). For example, the H recoil cross-section for MeV ^{4}He projectiles is increased by approximately a factor of three. Bozoian et al. [3.155] published a means of estimating, for any ion-target system, the critical projectile energy above which deviations from the Rutherford cross-sections because of nuclear interaction are expected. In this non-Rutherford regime, measured cross-sections should be used in the spectrum evaluation (for example, H recoil cross-sections for ^{4}He impact have been published by several authors [3.156–3.159]).

Deviations from Rutherford cross-sections are also found for heavy projectiles at lower impact energies, when the projectile can bind inner shell electrons which screen the nuclear charge. These deviations are usually small and can easily be taken into account by use of a theoretical correction [3.160].

3.7.3
Particle Identification Methods

As already mentioned, particle identification is achieved by energy-loss measurement (the $\Delta E - E$ method) or by velocity measurement (TOF method).

In the $\Delta E - E$ method, particles with the same energy, E, are identified by their energy loss, ΔE, in a thin solid or gas layer in front of the energy detector. This method can be used for particle energies $E > E_{Br}$, where E_{Br} is the Bragg energy, i. e. the energy of maximum electronic stopping power. For particles with a fixed energy $E < E_{Br}$, the stopping power depends only weakly on the atomic number of the ion, because in this regime stopping is roughly proportional to the velocity of the ion and at a given energy a heavier ion has a lower velocity. Hence, particle identification by ΔE measurement is hardly possible at energies below E_{Br}, which is approximately 100 keV for protons, approximately 0.5 MeV for He, 2.7 MeV for C, 7 MeV for O, and approximately 25 MeV for Si ions. For small accelerators (1–10 MeV ion energy), therefore, the $\Delta E - E$ method is applicable for identification of very light elements only (mainly H), while for large accelerators and ion energies >100 MeV it enables the analysis of elements up to Cu.

The simplest arrangement for $\Delta E - E$ measurement uses a stopper foil in front of the E-detector, which is usually a surface barrier detector (SBD) [3.161–3.163]. This system is mainly used for H profiling with low-energy light ions. The thickness of the foil must be such that H ions are transmitted through the foil (with a certain energy loss) whereas scattered projectiles and heavier recoiled atoms are stopped in the foil. A considerable improvement over the stopper foil method is the use of electron emission for particle identification [3.164] – when ions pass the foil, electrons are emitted from both surfaces, the number of emitted electrons, N_e, being roughly proportional to the ΔE deposited in the foil by the ions. Thus particles can be identified by measurement of the number of electrons emitted from a set of thin foils in front of the SBD. Because the emitted electrons originate from a layer close to the surface, foils as thin as possible can be used, so that energy loss and energy loss straggling are almost negligible. This results in a better depth resolution compared with the conventional stopper-foil technique.

For ERDA arrangements using high-energy heavy projectiles, mainly gas telescope detectors, are used for ΔE and E measurements [3.153, 3.165, 3.166]. In an ionization chamber individual particles ionize gas atoms. The emitted electrons are accelerated by an electric drift field (which is usually aligned perpendicular to the particle path) and detected by means of a split anode. The shorter section of the anode next to the entrance window yields the energy ΔE deposited in the front region next to the window. The remaining energy $E - \Delta E$ is then measured in the second part of the anode. The total energy, E, is obtained from the total charge accumulated in both sections of the anode. The second part of the ionization chamber, which measures the energy $E - \Delta E$, can be replaced by an SBD [3.167], and the first part, which measures the energy loss ΔE, by a transmission SBD [3.168, 3.169]. When SBDs are used to measure heavy ions, radiation damage of the detector by the ions must be taken into account.

In TOF systems, particle energies are usually determined by SBDs in addition to particle velocities being obtained with a TOF set-up which primarily measures the time needed by a particle to pass the distance between two thin foils 0.5–1 m apart [3.170, 3.171]. The first foil delivers a start signal, the second a stop signal. The stop signal can also be obtained from the SBD, but usually foils provide better timing signals. The timing signal from the foils is obtained from electrons emitted when the particle pass the foil. Usually carbon foils are used, because these are the most stable ultra-thin foils. The emitted electrons are focused to a micro-channel plate, which gives the time signal. The efficiency for detection of light particles (hydrogen) is low, because of the low probability of electron emission from the foil [3.172]. Particle identification by the TOF method can also be achieved at particle energies below the Bragg energy E_{Br}, in contrast with $\Delta E - E$ systems. For low-energy particles the particle energy, which is a measure for the emission depth, can sometimes be calculated more precisely from the measured time-of-flight than directly from the SBD spectrum [3.173]. In these circumstances the SBD is used to identify particles with same time-of-flight but different energy.

3.7.4 Equipment

In ERDA, the projectiles are high-energy ions produced by an ion accelerator. The lowest ion energies used for ERDA are several MeV for detection of very light elements, mainly hydrogen. For analysis of a wider mass range of elements, higher projectile energies are favorable. Higher projectile energies also increase the information depth. The largest projectile energies used in ERDA are hundreds of MeV for measuring the light-element composition in surface layers. Most of the ion accelerators used are of the tandem-Van de Graaff type.

The pressure in the scattering chamber should be below approximately 10^{-4} Pa (high vacuum). Lower pressures in the UHV range are usually not necessary, because ERDA is not surface-sensitive.

The energies of recoil atoms and scattered projectiles are usually measured by solid state SBDs. For identification of particles with same energy but different atomic number an additional quantity (TOF, ΔE, or N_e) must be measured in coincidence with the energy. Usually, both quantities (energy and the identification quantity) are then stored in a two-dimensional multichannel analyzer [3.164]. Only for the simplest case (ERDA with a stopper foil) can a one-dimensional multichannel analyzer be used. High efficiency can be obtained by use of detectors with a large solid angle. Here recoils are measured with recoil angles in an interval, $\Delta\theta$, around a mean recoil angle θ. The different recoil angles will lead to different recoil energies. This effect is called kinematic line broadening and can be minimized by use of a position-sensitive energy detector, here the energies for particles with recoil angles $\theta + \Delta\theta$ are corrected to the nominal energy at the recoil angle θ [3.153, 3.174].

The exit angle, β, and the angle of incidence, α, of the beam (Fig. 3.63) determine the depth resolution and information depth. Small angles increase the depth resolution but reduce the depth probed. To optimize both quantities recoil angles θ between 30° and 45° are often used, with $\alpha = \beta = \theta/2$.

3.7.5
Data Analysis

With particle identification, individual energy spectra of the different recoil elements and of scattered projectiles are obtained from the stored two-dimensional multichannel analyzer data. For a given recoil element of a given energy, the corresponding emission depth is deduced from the energy, and the concentration of the element is obtained from the number of particles detected. A simple evaluation of the energy spectra can be made by dividing the target into thin slabs and assuming a constant energy of the projectiles within one slab to calculate the recoil cross-sections (including deviations from the Rutherford values for light elements) and the energy losses [3.152].

More precise and rapid evaluation of the individual energy spectra can be performed by use of available computer programs, which simulate energy spectra of recoils and scattered projectiles for a given target composition and experimental arrangement. In these programs, SIMNRA [3.175] or DEPTH [3.176], non-Rutherford cross-sections can also be used. Multiple scattering of incident projectiles and recoils in the target can also be taken into account in the DEPTH program. Multiple scattering becomes important for analysis of deeper layers in ERDA geometries with small recoil angles and low energies. To make the evaluation procedure simple and efficient, energy spectra for a variety of target compositions may be simulated and compared to measured spectra, to determine which target composition gives best agreement between simulated and measured spectra.

3.7.6
Sensitivity and Depth Resolution

Reasonable estimates of ultimate sensitivity and depth resolution in ERDA can hardly be given because of the large range of projectiles and energies (from He ions of several MeV up to 200-MeV Au ions), and the use of different detection systems. In addition, stability of the sample under irradiation (which, of course, depends on the target material) is also important in the discussion of sensitivity and detection limits. The sensitivity is mainly determined by the recoil cross-section, the solid angle of the detector system, and the total number of incident projectiles which can be used for analysis.

For H profiling with light MeV projectiles, a mean detection limit of approximately 10^{14} hydrogen atoms cm^{-2} in a near surface region can be deduced from measurements of implanted H in Si [3.177]. Similar detection limits are obtained for analysis of light impurities (C, N, and O) by 200-MeV Au projectiles [3.153]. By use of position-sensitive detectors one can obtain large solid angles and, consequently, high sensitivity with minor deterioration of the depth resolution because of kinematic broadening [3.153, 3.174].

TOF detector systems usually have smaller solid angles and sensitivity than $\Delta E - E$ systems, because of the long TOF system in front of the energy detector and the limited size of the stop detector. They also have worse detection limits for very light elements (hydrogen), because of the low probability of obtaining start and stop signals for particles of very low atomic number [3.172].

The depth resolution of ERDA is mainly determined by the energy resolution of the detector system, the scattering geometry, and the type of projectiles and recoils. The depth resolution also depends on the depth analyzed, because of energy straggling and multiple scattering. The relative importance of different contributions to the depth resolution were studied for some specific ERDA arrangements [3.161, 3.163]. It was found that at the surface the depth resolution is mainly determined by the energy resolution of the detector, the recoil geometry, and the type of projectile, whereas in deeper layers the depth resolution is mainly defined by energy loss straggling and multiple scattering of projectiles and recoils. The optimum resolution is obtained at the surface. For H profiling with MeV He ions using a standard SBD approximately 40-nm depth resolution is obtained [3.177]. For 5-MeV Ne projectiles impinging on a polymer sample a depth resolution down to 8 nm was obtained for surface hydrogen in an arrangement optimized for depth resolution [3.163]. For heavy ion projectiles and $\Delta E - E$ telescope detectors depth resolutions of the order of several nanometers can be obtained in surface layers [3.153]. For layers as deep as several 100 nm the resolution increases to approximately 50 nm [3.178]. TOF systems with good timing signals enable good depth resolution of few nanometers, when the energy spectra are obtained from the timing signals [3.179]. TOF systems have the advantage of larger information depth for most elements compared with $\Delta E - E$ systems [3.180]. Extremely high depth resolution corresponding to atomic layer resolution can be obtained if the ion beam and detector system are optimized. Monolayer resolution has, for example, been realized for 60-MeV iodine projectiles impinging on HPOG graphite with a special magnetic spectrograph for energy determination [3.181].

3.7.7 Applications

An example of depth profiling of hydrogen implanted into Si is shown in Fig. 3.64 [3.177]. Measured energy spectra of H recoils are given for impact of 6-MeV C ions. H identification was achieved by the $\Delta E - E$ technique and use of ion-in-

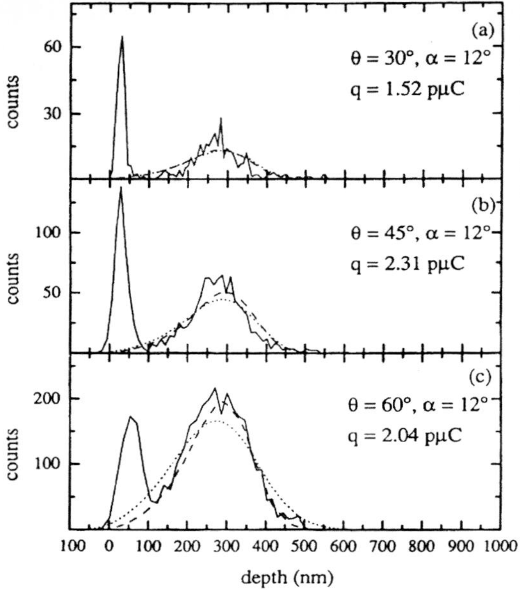

Fig. 3.64. H depth profile of an H-implanted Si sample obtained with 6-MeV C projectile ions for different recoil angles θ. q gives the charge of the incident ions. The experimental depth profiles (full line) are compared with simulated spectra (dashed line – SIMNRA, dotted line – DEPTH) [3.177].

duced electron emission [3.164]; the energy scale is converted to a depth scale for comparison with simulated spectra. To analyze the data, ERDA spectra were simulated by the code SIMNRA, for which an H depth profile calculated by the TRIM program was used as input. In Fig. 3.64, the surface contamination peak can clearly be separated from the implanted H peak. Spectra are given for different recoil angles. The depth resolution can be estimated from the width of the surface peak. At low recoil angles the depth resolution is better whereas at large recoil angles the sensitivity is higher. Good agreement was obtained between measured and calculated depth profiles. The number of implanted H atoms was found by this measurement to be 1.35×10^{16} atoms cm^{-2}.

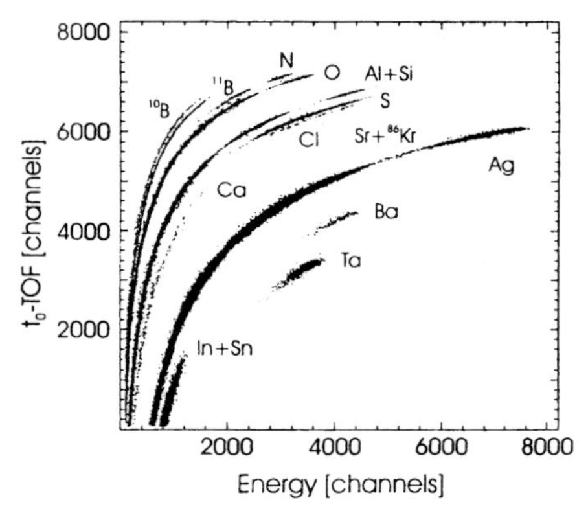

Fig. 3.65. Two-dimensional spectrum showing dependence of TOF on energy for a multilayer sample and impact of 120 MeV Kr ions [3.171].

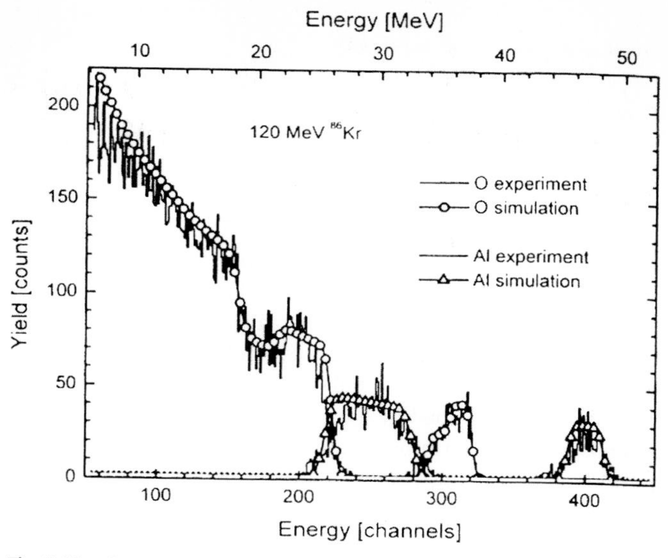

Fig. 3.66. Energy spectra of O and Al, calculated from the measured TOF of Fig. 3.65 Experimental results are compared with a simulation (SIMNRA) [3.171].

As a second example, results from a TOF ERDA measurement for a multi-element sample are shown in Fig. 3.65 [3.171]. The sample consists of different metal–metal oxide layers on a boron silicate glass. The projectiles are 120-MeV Kr ions. It can be seen that many different recoil ions can be separated from the most intense line, produced by the scattered projectiles. Figure 3.66 shows the energy spectra for O and Al recoils calculated from the measured TOF spectra, together with simulated spectra using the SIMNRA code. The concentration and thickness of the O and Al layers are obtained from the simulations.

3.8 Nuclear Reaction Analysis (NRA)*

Oswald Benka

3.8.1 Introduction

Nuclear reaction analysis (NRA) is a technique for the analysis of surface layers using light ion beams with energies in the range 1–10 MeV. When an ion 'a' hits an atom 'A' it might cause a nuclear reaction. An intermediate excited nucleus I* is then formed, which decays to the ground state 'B' by emission of a particle 'b'. The emitted particle 'b' can be a γ photon, or an ion (H^+, He^{2+}), and will have a characteristic energy determined by the Q value of the reaction. Such a reaction is described in an abbreviated form by A(a,b)B. For a constant number of incident projectiles the number of emitted reaction products 'b' depends on the number of target atoms 'A'. Therefore one can derive the number of atoms 'A' present in an analyzed sample from the measured number of reaction products 'b'. If the mean life-time of the excited state I* is reasonably long, reaction products can be measured after irradiation, and this technique is called *charged-particle activation analysis* (by analogy with *neutron activation analysis*). If the mean lifetime is small, or emission is prompt, the reaction products are measured during ion bombardment. This technique is then called *nuclear reaction analysis.*

Because a projectile which hits a target nucleus must overcome the Coulomb barrier to excite a nuclear reaction, only light projectiles with energies in the MeV range are used in NRA for analysis of light elements, taking advantage of the low Coulomb barrier. The probability of a nuclear reaction, which is given by the corresponding cross-section, depends on the energy and type of projectile and on the type of target atom, and there is no simple means of predicting its magnitude. For proton and He projectiles with energies in the MeV range, few elements have nuclear reactions with reasonable probability for analysis. Therefore, NRA is not a technique to probe the composition of an unknown sample by use of high-energy ions, like particle-induced X-ray emission (PIXE) or Rutherford back-scattering spectroscopy (RBS), but NRA is a powerful technique for quantification of the concentration or the depth profile of a certain element in a sample for which the qualitative chemical composition is known. In addition, NRA is a sensitive means of measuring depth profiles of individual isotopes of an element; these cannot be measured by techniques such as RBS, AES, or XPS.

Because the cross-sections for nuclear reaction are usually lower than the cross-sections for elastic scattering of projectiles used in RBS or in elastic recoil detection analysis (ERDA), higher currents must be used to obtain comparably high intensity in

* In most nuclear reactions A (a,b) B the emitted radiation 'b' used for analysis consists of charged particles, e. g. α-particles and protons. This is why NRA is covered in this section on ion detection. The emitted radiation can, however, also consist of γ-rays or neutrons. In accordance with the chosen structure of this book we ask for the reader's understanding of our omission of an additional NRA chapter dealing with these γ-ray- and neutron-producing reactions.

NRA as in RBS or ERDA, and possible modification of the target composition as a result of irradiation must be considered. Nuclear reaction cross-sections are also usually not available in analytical form for direct evaluation of measured data. Concentrations are, therefore, often obtained by comparison of the measured data with results from standard samples of known concentration.

The experimental requirements for NRA can be roughly divided into two groups depending on the emitted reaction products – gamma photons or particles. If the emitted γ photons are used for analysis, this method is also called PIGE, *particle-induced gamma emission*. In this method the gamma detector that measures the intensity of the photons is usually outside the scattering chamber. Only the concentration of a specific isotope in a thin layer can be deduced from the number of photons. To obtain information about the depth distribution also, the number of gamma photons must be measured as a function of the beam energy, E_0, for an equal number of impinging projectiles, which is called *excitation curve* $N(E_0)$.

For analysis of emitted particles, solid state surface barrier detectors (SBD) are used inside the scattering chamber to measure the number and energy of the reaction products. Stopper foils are used to prevent scattered projectiles from reaching the detector. Depth profiles can be obtained from the energy spectra, because reaction products emitted in deeper layers have less energy than reaction products emitted from the surface. The concentration in the corresponding layer can be determined from the intensity of reaction products with a certain energy.

The cross-section curve $\sigma(E)$ gives the dependence of the nuclear cross-section on the projectile energy, E. The measured energy spectra of emitted particles or the excitation curve $N(E_0)$ will depend on the depth profile $N(x)$ of the analyzed isotope and on the cross-section curve $\sigma(E(x))$, where $E(x)$ gives the energy of the projectiles at a depth x. Evaluation of the depth profile $N(x)$ from measured energy spectra or excitation curves often requires a tedious evaluation procedure if the cross-section curve has a complex structure. It is simplified for two special types of behavior of the cross-section curve:

(1) if $\sigma(E)$ is *constant*, i.e. independent of the projectile energy in an energy interval ΔE, the measured *energy spectrum* of the reaction products directly reflects the concentration profile in a depth interval corresponding to ΔE; and
(2) if the cross-section curve has a *strong resonance* at an energy E_R, i.e. it is large in a small energy interval around E_R and negligibly small for all other energies, the measured *excitation curve* will directly reflect the concentration profile.

An excellent review has recently been published on the various techniques of NRA [3.182]. Reviews of depth profiling with narrow resonances are also available [3.183, 3.184].

3.8.2
Principles

For a non-resonant nuclear reaction with emission of an ion, a depth scale can be obtained from the measured energy of the emitted ions. If ions emitted from a depth x are lower in energy by ΔE than ions emitted from the surface, a relationship between ΔE and x can be found, similarly to RBS and ERDA analysis:

$$\Delta E = x N \varepsilon_{nr} \tag{3.52}$$

where N is the atomic density of the sample and ε_{nr} is the nuclear reaction stopping cross-section factor, given by:

$$\varepsilon_{nr} = \alpha\, \varepsilon_{in} + \frac{\varepsilon_{out}}{\cos\varphi} \tag{3.53}$$

when the projectiles impinge perpendicularly on the surface and the reaction products are emitted at an angle φ relative to the surface normal. ε_{in} and ε_{out} are the stopping cross-sections for the incoming projectiles and the emitted ions. The reaction factor α weights the energy loss of the incident projectiles, in the same way as the kinematic factor in RBS, with:

$$\alpha = \frac{\mathrm{d}E_r}{\mathrm{d}E_p} \tag{3.54}$$

where E_p is the projectile energy when it produces a nuclear reaction and E_r is the energy of the reaction product. The reaction factor α can be calculated from the reaction kinematics [3.182] and depends on the masses of projectile, the reaction product, and the reacting atom, the Q value of the reaction, and the angle of emission. For light nuclei and backward emission α can become negative. In these circumstances case contributions of deeper layers extend the spectrum towards higher energies and the depth resolution can become worse.

For slowly changing reaction cross-sections the depth profile of the analyzed isotope can be calculated using the reaction cross-sections, by analogy with RBS, for which the scattering cross-section is the corresponding quantity. To obtain a simple estimate of the depth profile, the target can be divided into thin slabs, with constant cross-section and stopping power within each slab. The standard procedure is to simulate energy spectra for different target compositions, by use of a computer program [3.185, 3.186], and to deduce the depth profile by comparison of simulated and measured spectra.

When, in NRA, energy spectra of emitted particles are analyzed, a sufficiently thick foil in front of the detector is usually used to absorb the scattered projectiles. This reduces the depth resolution of NRA, because of energy loss straggling of the reaction products in the foil.

When, in NRA, resonances are used, and the depth profile of an isotope A is obtained from the excitation curve $N(E_0)$, the reaction depth x is given by the requirement that projectiles incident with energy E_0 are slowed down to the resonance energy E_R at x, which leads to:

$$x = \int_{E_R}^{E_0} \frac{1}{N\,(m\,\varepsilon_A\,(E) + (1-m)\,\varepsilon_B\,(E))}\,\mathrm{d}E \tag{3.55}$$

for a sample with atomic density N and composition $A_m B_{(1-m)}$, where m is the atomic fraction of the isotope A and ε_A and ε_B are the stopping cross-sections of the constituents A and B. For $m \ll 1$ and for thin layers, the stopping cross-sections can be assumed to be constant and Eq. (3.55) reduces to:

$$x = (E_0 - E_R)\frac{1}{N\varepsilon} \tag{3.56}$$

where ε is the stopping cross-section of the sample at the mean projectile energy, $(E_0 + E_R)/2$.

The easiest way to obtain the concentration m of the isotope A is to use a reference sample ('standard'), that contains the isotope A with a known atomic fraction f, for comparison. Knowing the NRA yield of the standard Y_{St} and its stopping cross-section ε_{St}, the atomic fraction m in the sample can easily be evaluated from the yield Y_A of the sample for the same projectile energy, taking the different stopping cross-sections into account:

$$m = \frac{f\,Y_A\,\varepsilon_B}{Y_{St}\,\varepsilon_{St} + f\,Y_A\,(\varepsilon_B - \varepsilon_A)} \tag{3.57}$$

For $m \ll 1$ Eq. (3.57) can again be approximated by:

$$m = f\frac{Y_A}{Y_{St}}\frac{\varepsilon}{\varepsilon_{St}} \tag{3.58}$$

where ε is the stopping cross-section of the sample.

3.8.3
Equipment and Depth Resolution

Nuclear reactions are excited when projectile energies are typically in the MeV range. Medium size ion-accelerators are, therefore, necessary to obtain these projectile energies. Protons and α projectiles, typical projectiles in other ion-beam analysis techniques as RBS or PIXE, have few useful nuclear reactions. Deuteron beams excite many more nuclear reactions, but the use of deuteron beams instead of standard beams is more hazardous, because of efficient neutron production. Strict safety rules are necessary when high-energy deuteron beams are used.

The main requirement for accelerators used in resonance NRA are reasonably good energy resolution and the possibility of changing the beam energy easily. The beam energy is usually increased stepwise by adjusting the magnetic field used to select the energy and to stabilize the terminal voltage. More sophisticated energy scanning systems [3.187, 3.188] have, however, been developed to change the beam energy while keeping the analyzing magnet constant.

The nuclear reaction products are usually measured in a high-vacuum scattering chamber. At resonance NRA, where the beam energy is varied, at each energy the number of incident projectiles must be known. A good beam-current measurement and integration system is necessary to determine the total incident beam charge.

In NRA depth profiling using the energy spectra of ion reaction products, an absorber foil is usually used to filter out the scattered projectiles [3.182]. The main disadvantage of this method is that the reaction products will suffer energy straggling when they pass the foil, resulting in degradation of depth resolution. The main limiting factors of depth resolution are:

(1) the energy resolution of the detector;
(2) kinematic broadening, because of the solid angle of the detector; and
(3) energy straggling of the incident beam and of the reaction products in the sample and in the absorber foil [3.189].

Which of these contributions dominates depends on the nuclear reaction, the energy of the projectiles, the analyzed depth, and the geometry of the equipment used.

The depth resolution in resonance NRA close to the surface is mainly determined by the stopping power of the projectiles in the target. For deeper layers, there are different contributions:

(1) energy loss straggling;
(2) energy resolution of the beam;
(3) resonance width; and
(4) Doppler broadening because of vibrations of the target atoms [3.182–3.184].

Doppler broadening is usually a small contribution, which becomes important only for nuclear reactions of heavy projectiles and light target atoms [3.184, 3.192]. Close to the surface of the sample, energy loss straggling can be neglected and for narrow resonances high depth resolution can be obtained for beams with good energy resolution (typically 100 eV). The corresponding depth resolutions for proton and He projectiles are in the range of several nanometers [3.182]. H depth profiling is best achieved with an ^{15}N beam which has a resonance at 6.5 MeV [3.182, 3.184]; for H depth profiles close to the surface the depth resolution is limited by Doppler broadening. The corresponding depth resolution in Si is approximately 25 nm [3.182, 3.192]. The depth resolution in deeper layers is mainly determined by energy-loss straggling. Depth resolution in NRA is, therefore, similar to that for the other high-energy ion-beam profiling techniques, e. g. RBS.

In resonance NRA with gamma emission, the emitted gamma rays are usually measured outside the scattering chamber by means of an NaI or a Ge(Li) detector. NaI scintillation counters are used when high efficiency is needed and energy resolution is not critical. This is often so when few light elements are present in the sample and only one well isolated resonance is measured. BGO (bismuth–germanium–oxide) detectors recently became available; these are similar to NaI detectors, but are more efficient [3.190]. Solid-state Ge(Li) detectors are used when high gamma-energy resolution is necessary to distinguish between adjacent gamma lines of nuclear reactions from different target elements. If an element present at a low concentration must be analyzed, high sensitivity is required. To achieve this, the background radiation must be minimized [3.191], by careful shielding of the gamma detector and by reducing the gamma rays produced by the beam in the beam line, e. g. on beam collimators in front of the target.

3.8.4
Applications

The major application of NRA is determination of the concentration of selected light elements as a function of depth in thin surface layers. The accessible depth is typically several microns, with a depth resolution of approximately 10–100 nm. A review of useful nuclear reactions, with applications, is available [3.182].

For protons with energies in the MeV range (p,γ) reactions for the isotopes ^{7}Li, ^{15}N, ^{18}O, and ^{19}F can be used for depth profiling, where only ^{7}Li, and ^{19}F are the most abundant isotopes of the natural elements. With MeV deuteron projectiles many nuclear reactions can be used for NRA; this subject has been reviewed [3.193]. Deuterons, for example, have useful nuclear reactions with the isotopes ^{12}C, ^{14}N, and ^{16}O, which are abundant in natural elements and have no useful reactions with other projectiles. An example of a recent application [3.194] is shown in Fig. 3.67. The constituent light elements in a Cu patina of an historical object were determined with an external 2 MeV deuteron beam. The proton energy spectrum from different (d,p) reactions was measured; the different reactions are denoted by the reacting target atom and by the number of proton reactions, according to different final states of the target atom; p_0 corresponds to the transition with highest proton energy. The scattered deuterons were stopped in an absorber foil. In addition, an RBS spectrum was measured to analyze Cu using a 3 MeV proton beam. Figure 3.67 also shows simulated spectra fitted to the measured spectra together with the concentration profiles, when the patina is separated into four layers of different thickness. Within each

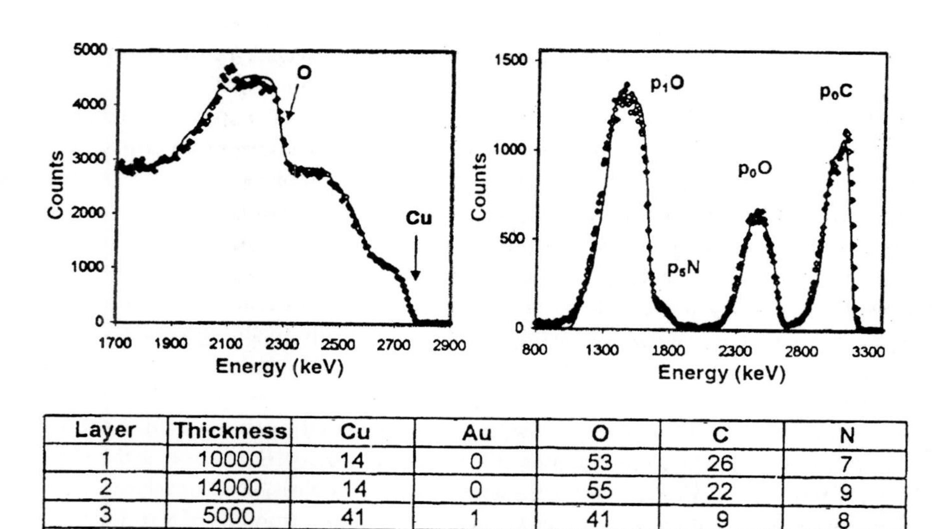

Layer	Thickness	Cu	Au	O	C	N
1	10000	14	0	53	26	7
2	14000	14	0	55	22	9
3	5000	41	1	41	9	8
4	5000	68	2	25	5	0

Fig. 3.67. RBS spectrum for impact of 3 MeV protons and NRA proton spectrum for impact of 2 MeV deuterons. The full line gives the fit result (SIMNRA) of a four-layer simulation. The thickness is expressed in 10^{15} atoms cm^{-2} and the element content in at% [3.194].

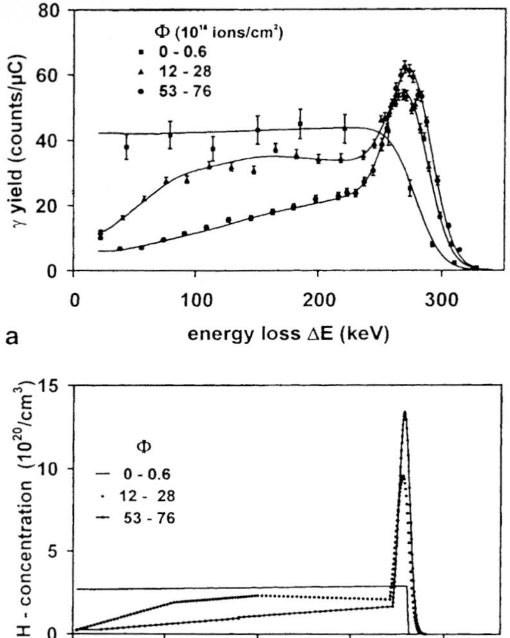

Fig. 3.68. (a) Dependence of measured γ yield on ΔE for impact of $(6.385 + \Delta E)$ MeV N^{15} ions on an SiO_2–Si sample and for different fluence intervals. (b) Deconvoluted H concentration profiles corresponding to (a) [3.195].

layer constant concentrations of elements were assumed. The layer thickness is expressed in 10^{15} atoms cm^{-2}.

NRA with gamma emission is mainly used for depth profiling of medium light isotopes, i. e. N, F, Na, Mg, Al, and Si. Some useful reactions, with detection limits, were recently given by Elkes et al. [3.191]. An important application of NRA with gamma emission is the determination of H depth profiles, because hydrogen is probably the most common elemental contamination in thin-film materials and it is invisible to many analytical methods. In principle all inverse p-induced nuclear reactions can be used, but the ^{15}N reaction is most often used; it has a strong narrow resonance at 6.385 MeV (laboratory energy). The main problem is usually obtaining a good ^{15}N ion beam, because the ^{15}N abundance in natural N is below 1% and N does not form a stable negative ion as necessary in tandem accelerators. A recent application [3.195] of H depth profiling is shown in Fig. 3.68. Figure 3.68 a gives the gamma yield as a function of projectile energy; here the projectile energy $E_p = E_r + \Delta E$, where E_r is the resonance energy (6.385 MeV) and ΔE is the abscissa. The sample is an SiO_2/Si layer structure and gamma yield measurements are given for different amounts of N irradiation. Figure 3.68 b shows the deconvoluted hydrogen concentration profiles corresponding to measurements of Fig. 3.68 a. It can be seen that an increasing ^{15}N fluence, Φ, degrades the original shape of the H profile: H is enriched to a depth of 173 nm which corresponds to the depth of the SiO_2/Si transition region.

3.9
Other Ion-detecting Techniques

John C. Rivière

3.9.1
Desorption Methods

3.9.1.1 Electron Stimulated Desorption (ESD) and ESD Ion Angular Distribution (ESDIAD)

Electron irradiation of a surface, particularly one covered with one or more adsorbed species, can give rise to many types of secondary particle, including positive and negative ions. In ESD and ESDIAD, the surface is irradiated with electrons of energies in the range of 100–1000 eV, and the ejected positive-ion currents of selected species are measured in a mass spectrometer. If the angular distribution of the secondary ions is also measured, either by display on a screen or by using position-sensitive detection, then electron-stimulated desorption (ESD) becomes electron-stimulated desorption ion angular distribution (ESDIAD).

The electron desorption techniques are not used, and probably cannot be used, for compositional analysis, but they provide valuable information about the nature of electronic interactions leading to the breakage of bonds and, in the angle-resolved form, about the geometry of surface molecules and the orientation of broken bonds. The primary electrons do not, at the energies employed, succeed in breaking molecular or surface-to-molecule bonds or in knocking ions out of the surface directly, but the process is one of initial electronic excitation. An electron is absorbed by a surface–adsorbate complex or an adsorbed molecule itself, leading to excitation to an excited state by a Franck–Condon process. If the excited state is antibonding and the molecule or radical is already far enough from the surface, desorption can occur. Because return to the ground state after excitation is a much more probable process, the cross-sections for ion desorption are low, 10^{-20}–10^{-23} cm^2. If core-level ionization is involved in the initial interaction with the incident electron, rather than valence levels, as in Franck–Condon-type excitation, a desorption mechanism based on Auger decay has been proposed. The core level left behind after interatomic Auger decay creates a positive ion which is then expelled by the repulsive Madelung potential.

The most common ions observed as a result of electron-stimulated desorption are atomic (e.g., H^+,O^+,F^+), but molecular ions such as OH^+, CO^+, H_2O^+, and $CO_2{}^+$ can also be found in significant quantities after adsorption of H_2O, CO, CO_2, etc. Substrate metallic ions have never been observed, which means that ESD is not applicable to surface compositional analysis of solid materials. The most important application of ESD in the angularly resolved form ESDIAD is in determining the *structure and mode of adsorption* of adsorbed species. This is because the ejection of positive ions in ESD is not isotropic. Instead the ions are desorbed along specific directions only, characterized by the orientation of the molecular bonds that are broken by electron excitation.

3.9.1.2 Thermal Desorption Spectroscopy (TDS)

TDS, sometimes called temperature-programmed desorption (TPD), is simple in principle. A gas or mixture of gases is adsorbed on a clean metal foil for a chosen time; then, after the gas is pumped away, the foil is heated, at a strictly linear rate, to a high temperature, during which the current of a particular ion or group of ions is monitored as a function of temperature. The ion masses are selected in a quadrupole mass spectrometer. As the binding energy thresholds of the adsorbed species on the surface are crossed, peaks in the desorbed ion current appear at characteristic temperatures. From the characteristic temperatures and the shape of the desorption peak above the threshold, the activation energies for desorption can be obtained, along with information about the nature of the desorption process. The mass spectrum from the mass spectrometer, of course, provides information about the species that actually occur on the surface after adsorption.

Although simple in principle, experimental artifacts that are possible in TDS must be avoided. Thus, ions accepted by the mass spectrometer must originate from the surface of the foil only, and the temperature distribution across the foil should be uniform to avoid the overlapping of desorption processes occurring at different temperatures. To ensure these experimental requirements are met, the angle of acceptance into the spectrometer is restricted by placing a drift tube with an aperture between the foil and the spectrometer, and the foil itself is usually in the form of a long thin ribbon, and only its center section contributes to detected ions. In addition, the heating rate must be sufficiently fast so that the desorbed species accepted by the mass spectrometer is characteristic of the desorption process, but not so fast that a sudden pressure increase occurs around the foil.

With correct experimental procedure TDS is straightforward to use and has been applied extensively in basic experiments concerned with the nature of reactions between pure gases and clean solid surfaces. Most of these applications have been catalysis-related (i. e. performed on surfaces acting as models for catalysts) and TDS has always been used with other techniques, e. g. UPS, ELS, AES, and LEED. To a certain extent it is quantifiable, in that the area under a desorption peak is proportional to the number of ions of that species desorbed in that temperature range, but measurement of the area is not always easy if several processes overlap.

3.9.2 Glow-discharge Mass Spectroscopy (GD-MS)

Glow discharge mass spectrometry (GD-MS) is a very promising technique for direct ultra-trace analysis of conducting and semi-conducting solids [3.196, 3.197]. The physical principles for bulk and surface analysis are identical to those of GD-OES except that the operating conditions of the GD source are optimized for maximum ionization efficiency. The advantages in comparison with GD-OES are convincing. In principle, all elements of the periodic table can be detected with improved detection limits below 1 ng g^{-1}. Mass spectra are easier to interpret than line-rich emission spectra and, therefore, concepts for quantification are much more straightforward. The only available commercial instrument, the VG 9000, a double focusing sector field device, was launched in 1985, and has been used in many different applications for ultra-trace analysis of pure metals, semiconductor materials and even for non-conducting powders. It has been discussed in more detail in a recent review article [3.198].

In principle GD-MS is very well suited for analysis of layers, also, and all concepts developed for SNMS (Sect. 3.3) can be used to calculate the concentration–depth profile from the measured intensity–time profile by use of relative or absolute sensitivity factors [3.199]. So far, however, acceptance of this technique is hesitant compared with GD-OES. The main factors limiting wider acceptance are the greater cost of the instrument and the fact that no commercial ion source has yet been optimized for this purpose. The literature therefore contains only preliminary results from analysis of layers obtained with either modified sources of the commercial instrument [3.200, 3.201] or with homebuilt sources coupled to quadrupole [3.199], sector field [3.202], or time-of-flight instruments [3.203]. To summarize, the future success of GD-MS in this field of application strongly depends on the availability of commercial sources with adequate depth resolution comparable with that of GD-OES.

3.9.3
Fast-atom Bombardment Mass Spectroscopy (FABMS)

Fast atom bombardment mass spectrometry (FABMS) is very similar to SSIMS in practice, the only difference being that instead of using positively charged ions as the primary probe, a beam of energetic neutral atoms is used. Secondary ions are emitted as in SSIMS and analyzed in a mass spectrometer, usually of the quadrupole type. The beam of fast atoms is produced by passing a beam of ions through a charge-transfer cell, which consists of a small volume filled with argon to a pressure of approximately 100 Pa. Charge transfer occurs by resonance between fast argon ions and argon atoms with thermal energy, with 15–20% efficiency if geometric and pressure conditions are optimized. Residual ions are removed by electrostatic deflection.

FABMS has two advantages over SSIMS, both arising from the use of neutral rather than charged particles. Firstly, little or no surface charging of insulating materials occurs, so organic materials such as polymers can be analyzed without the need to employ auxiliary electron irradiation to neutralize surface charge. Secondly, the extent of beam damage to a surface, for the same particle flux, is much lower using FABMS than SSIMS, thus enabling materials such as inorganic compounds, glasses, and polymers to be analyzed with less worry about damage introducing ambiguity into the analysis.

3.9.4
Atom Probe Microscopy (APM)

The *atom probe field-ion microscope* (*APFIM*) and its subsequent developments, the *position-sensitive atom probe* (*POSAP*) and the *pulsed laser atom probe* (*PLAP*), have the ultimate sensitivity in compositional analysis (i.e. single atoms). FIM is purely an imaging technique in which the specimen in the form of a needle with a very fine point (radius 10–100 nm) is at low temperature (liquid nitrogen or helium) and surrounded by a noble gas (He, Ne, or Ar) at 10^{-2}–10^{-3} Pa. A fluorescent screen or a microchannel plate is situated a few centimeters from the needle. A high positive voltage is applied to the needle, which causes noble-gas atoms approaching the needle to be ionized over points of local field enhancement (i. e. over prominent atoms at the surface); the ions are then repelled from these points and travel in straight lines to the screen or plate, where an atomic image of the needle tip is formed. This is called a field-ion image, and the voltage at which the image is optimized is called the *best imaging voltage*. The latter is in the range 5–20 kV.

3.9.4.1 Atom Probe Field Ion Microscopy (APFIM)

If the voltage is increased further above the imaging voltage, the cohesive energy that binds atoms to the surface can be exceeded, and atoms can be removed by what is termed *field evaporation*. The evaporation field required is a function of the sample material and the crystallographic orientation of the needle. The removal of atoms is the basis of APFIM and its daughter techniques. If the evaporation field is pulsed with very short ($\leq$ 10 ns) voltage pulses of extremely sharp rise times (~1 ns), and the time taken for a field-evaporated atom to travel from the tip to a detector is measured, the mass-to-charge ratio of the ionized atom can be established – i. e. elemental identification of the individual atom is possible. The region on the tip from which atoms are removed is selected by rotating or tilting the tip until the desired region, as viewed in the FIM image, falls over an aperture of approximately 2-mm diameter in the fluorescent screen. Most field-evaporated atoms will strike the screen, but those from the selected area will pass through the aperture into the time-of-flight mass spectrometer. For an aperture of 2 mm, the area analyzed on the tip is ca. 2 nm in diameter.

3.9.4.2 Position-sensitive Atom Probe (POSAP)

Pulsed operation of the APFIM leads to analysis of all atoms within a volume consisting of a cylinder of ca. 2 nm diameter along the axis of the tip and aperture, with single atomic layer depth resolution but no indication of just where within that volume any particular atom originated. The advent of position-sensitive detectors has enabled APFIM to be extended so that three-dimensional compositional variations within the analyzed volume can be determined. The development is called *position-sensitive atom probe* (*POSAP*). The aperture and single-ion detector of APFIM are replaced by a wide-angle double channel plate, with a position-sensitive anode just behind the plate. Field- or laser-evaporated ions strike the channel plate, releasing an electron cascade which is accelerated toward the anode. The impact position is located by the division of electric charge between three wedge-and-strip electrodes. From this position the point of origin of the ion on the tip surface can be determined, because the ion trajectories are radial. Thus after many evaporation pulses, leading to removal of a volume of material from the tip, both the identities and the positions of all atoms within that volume can be mapped in three dimensions. Because an evaporated volume can contain many thousands of atoms, the data collection and handling capabilities must be particularly sophisticated.

POSAP is the only technique available for identifying and locating precipitates, second phases, particles, and interfaces on an atomic scale, and has therefore found considerable application in metallurgical and semiconductor problems.

4 Photon Detection

4.1 Total Reflection X-ray Fluorescence Analysis (TXRF)

Laszlo Fabry and Siegfried Pahlke

4.1.1 Principles

Less than two decades after the first practical application of X-ray fluorescence [4.1], Moseley established the relationship between the reciprocal wavelength of emitted characteristic *X-ray fluorescence* (XRF) and the atomic number Z of the elements [4.2]. The XRF arrangement was not, however, sufficiently sensitive for trace or ultratrace analyses. Compton reported in 1923 that the reflectivity of a flat target increases below a critical angle of 0.1° under conditions of total X-ray reflection. The high reflectivity of the sample support reduced the spectral background of the support and improved the detection limit (DL) down to picogram levels in the early 1970s when, in 1971, Yoneda and Horiuchi applied the principle of TXRF to, mainly, ultratrace elemental microanalysis of biological samples [4.3]. More details of the history of TXRF are given elsewhere [4.4–4.6].

Since then, TXRF has become the standard tool for surface and subsurface microanalysis [4.7–4.11]. In 1983 Becker reported the angular dependence of X-ray fluorescence intensities in the range of total reflection [4.12]. Recent demands have set the pace of further development in the field of TXRF – improved detection limits [4.13] in combination with subtle surface preparation techniques [4.14, 4.15], analyte concentrations extended even to ultratraces (pg) of light elements, e. g. Al [4.16], speciation of different chemical states [4.17], and novel optical arrangements [4.18] and X-ray sources [4.19, 4.20].

For information on both the history and principles of, and future trends in, XRF, please refer to the article by Jenkins in this book. For TXRF see the outstanding handbook by Klockenkämper [4.21] and current reviews [4.22–4.24]. This contribution relies extensively on these referenced works.

The basis of XRF analysis is the photoelectric absorption and the subsequent emission of X-ray photons characteristic of the fingerprints of analyte atoms in the sample. Element composition can be quantified by the relative intensities of the indivi-

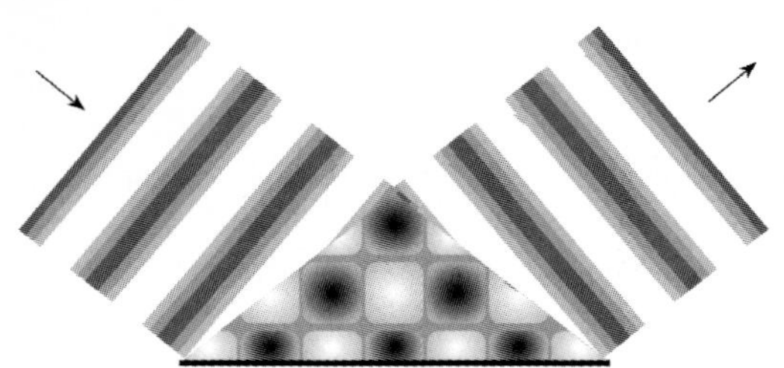

Fig. 4.1. Interference of incoming and the reflected X-ray waves in the triangular region above a flat and thick reflecting substrate. The strength of the electromagnetic field is represented on the gray scale by instantaneous crests (white) and troughs (black). In the course of time, the pattern moves from the left to the right [4.21].

dual element fingerprints. In a TXRF arrangement, however, the primary beam interacts with the surface analyte atoms as a standing wave field (Fig. 4.1) and excites surface atoms rather than the bulk, in contrast with XRF. The emitted intensity is directly proportional to the standing wave field intensity. In the subsurface the primary beam generates an evanescent wave field. Thus the characteristic fluorescence of analyte atoms is a response to these varying excitations by the standing and evanescent wave fields, and its dependence on the glancing angle is an opportunity to be used for stratigraphy of parallel-layered structures and to distinguish particulates from film-type surface contamination.

In X-ray spectrometry, attenuation, deflection and interference must be considered. Attenuation is described by the well-known Lambert–Beer law and the mass attenuation coefficient as given for conventional XRF.

Practical X-ray energies do not exceed 100 keV. The primary beam is mainly attenuated by the photoelectric effect. Scattering, both elastic (Rayleigh) and inelastic (Compton), represents a minor contribution to attenuation at energies below 100 keV.

Reflection and refraction of X-rays follow the laws of optics (Fig. 4.2): The glancing angles of incidence (ϕ_1) and reflection (ϕ_1^*) are equal. The respective refraction angles follow Snell's law for different phase velocities v in mediums 1 and 2:

$$v_2 \cos\phi_1 = v_1 \cos\phi_2 \tag{4.1}$$

When the glancing angle ϕ_2 of the refractive beam becomes zero, the refractive beam runs tangential to the boundary surface. Below that "critical" glancing angle of incidence reflectivity increases to 100% and the penetration – i.e. information – depth is limited to a few nanometers (nm). Reflection below the critical angle is defined as total reflection. The critical angle ϕ_c depends on the photon energy E and

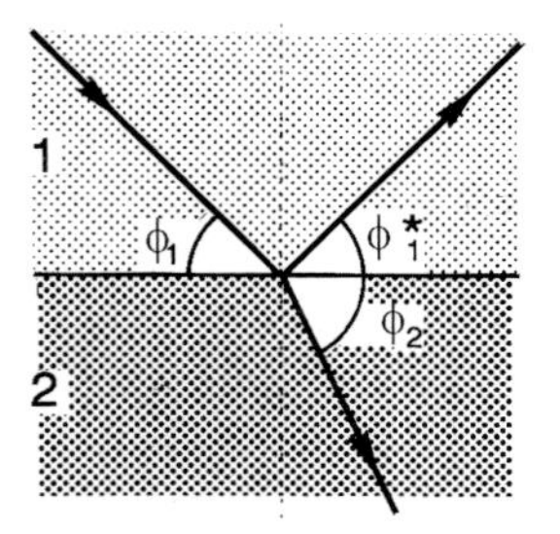

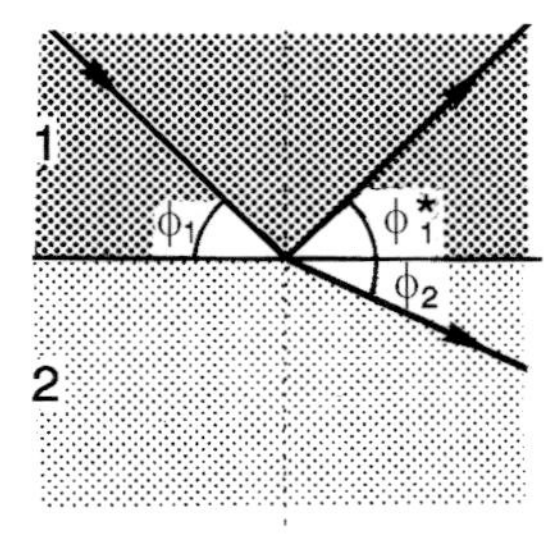

Fig. 4.2. Incident, reflected, and refracted beams at the interface between two media 1 and 2. On the left, medium 2 is optically denser than medium 1 ($n_2 > n_1$); on the right ($n_1 > n_2$) [4.21].

the atomic mass M of the medium, its atomic number Z and density ρ. The signal intensity is also a function of surface roughness.

$$\phi_c \approx \frac{1.65}{E}\sqrt{\frac{Z}{M}\rho} \tag{4.2}$$

At ϕ_c the penetration depth approaches a minimum, particularly for reflective surfaces such as chemi-mechanically polished Si. Total reflection disappears on rough surfaces (Fig. 4.3). Below ϕ_c the penetration depth is in the range of a few nanometers (Fig. 4.4).

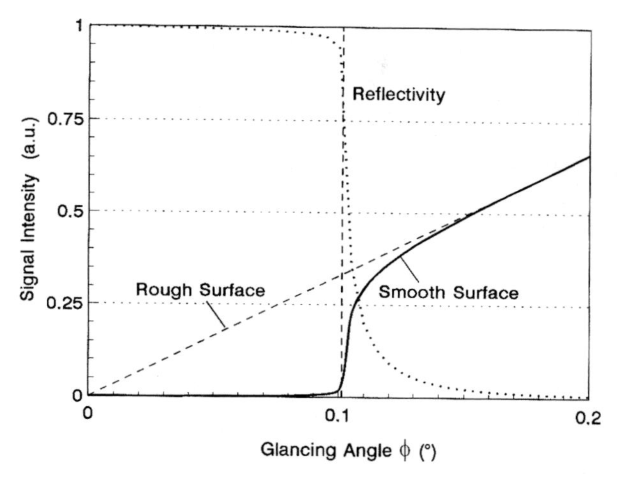

Fig. 4.3. Signal intensity of a thick, flat, and smooth Si-substrate (–), calculated for an impinging Mo-Kα beam. The reflectivity R (···) is shown; it depends on the glancing angle ϕ. Below $\phi_c = 0.102°$, total reflection occurs with a stepwise increase in reflectivity and a stepwise decrease in signal intensity. The oblique dashed line represents the intensity from a rough Si substrate [4.21].

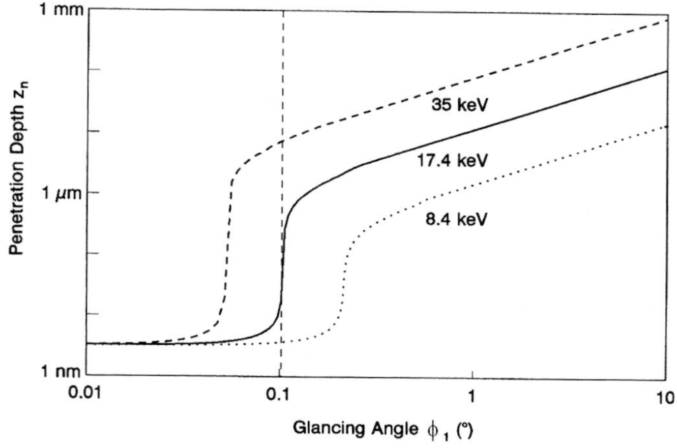

Fig. 4.4. Penetration depth z_n of X-rays striking silicon at a variable glancing angle ϕ_1. The curves were calculated for three different photon energies. The dashed vertical line signifies the respective critical angle [4.21].

4.1.2
Instrumentation

The construction of a TXRF system, including X-ray source, energy-dispersive detector and pulse-processing electronics, is similar to that of conventional XRF. The geometrical arrangement must also enable total reflection of a monochromatic primary beam. The totally reflected beam interferes with the incident primary beam. This interference causes the formation of standing waves above the surface of a homogeneous sample, as depicted in Fig. 4.1, or within a multiple-layered sample. Part of the primary beam fades away in an evanescent wave field in the bulk or substrate [4.28].

A TXRF instrument (Fig. 4.5) consists of an X-ray source, low-pass filter, collimator, monochromator, sample holder, detector and an electronic registration unit. A variety of target materials e.g. Mo, W, or Mo/W alloys are used in fine-focus, sealed-anode X-ray tubes powered by a water-cooled 3.5 kW generator. Maximum permissible power for high-Z anodes is approximately 2–3 kW. During their lifetime of 2000–4000 h with high uptime beyond 90% they gradually lose intensity. Rotating anodes are warranted for 2000 h operation at higher operating power up to 30 kW and more intense water cooling (15 L min^{-1} instead of 5 L min^{-1}).

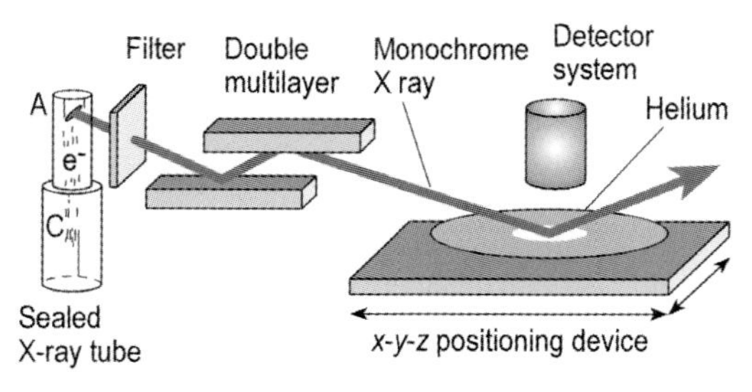

Fig. 4.5. Schematic diagram of a TXRF system equipped with a fine-focused anode for wafer detector system.

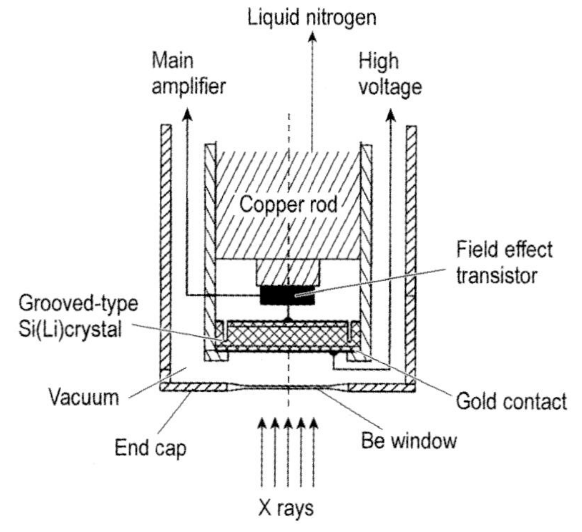

Fig. 4.6. Cross section of the front end of an SSD (solid-state detector), here with a grooved Si(Li) crystal. Crystal and preamplifier are connected with a cooled copper rod and shielded by a case with an end cap and Be window [4.21, 4.29].

Currently, a double reflector is used as a low-pass filter [4.9]. Tunable monochromators in combination with an alloy anode are also commercially available. For sensitive surface analysis monochromators are necessary. They are made of LiF, highly oriented pyrolytic graphite (HOPG), or W/Si, W/C, or Mo/B_4C multilayers, which provide intense and monochromatic radiation of, preferentially, WLβ or MoKα. The sample-positioning device must operate at high geometric reproducibility in all three dimensions with automated sample loading and unloading stations. The sample must be in an evacuated or He-flushed chamber to avoid disturbing absorption by air.

The energy-dispersive (EDX) solid state detector (SSD, Figs 4.6, 4.7) is made of lithium-drifted Si crystal (Si(Li)). Between a thin p-type and an n-type layer lies a high-resistivity Si crystal of centimeter dimensions. The front and end planes of the crystal are coated with Au and serve as electrodes. The crystal, cooled to 77 K by liquid nitrogen, represents a p–i–n diode (Fig. 4.7). An incident X-ray photon with

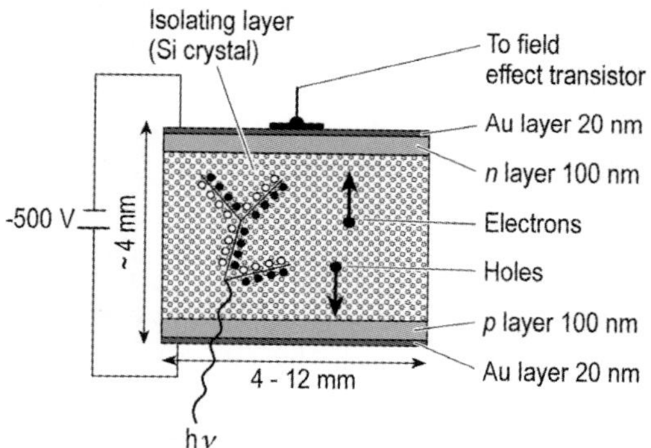

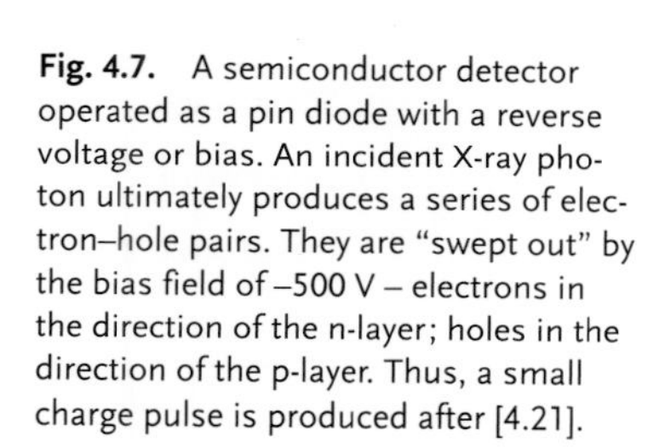
Fig. 4.7. A semiconductor detector operated as a pin diode with a reverse voltage or bias. An incident X-ray photon ultimately produces a series of electron–hole pairs. They are "swept out" by the bias field of –500 V – electrons in the direction of the n-layer; holes in the direction of the p-layer. Thus, a small charge pulse is produced after [4.21].

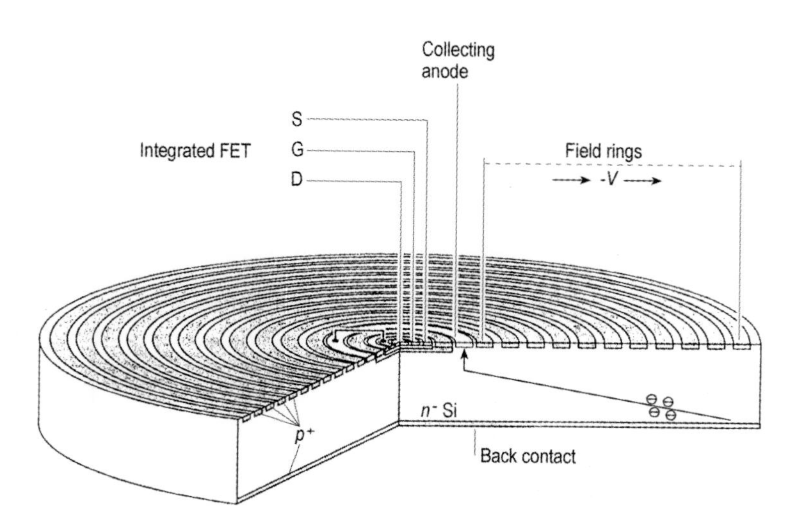

Fig. 4.8. Schematic view of a cylindrical silicon drift detector (SDD) [4.30].

an energy above 2 keV causes a cascade of photoelectron–hole pairs into which the initial photon energy is completely converted. Because of the reverse bias, electrons rapidly drift to the positively polarized n-type electrode, and holes to the negative p-type electrode. The number of electron–hole pairs is proportional to the energy of the impinging photon; thus, the charge pulse is a measure of the characteristic

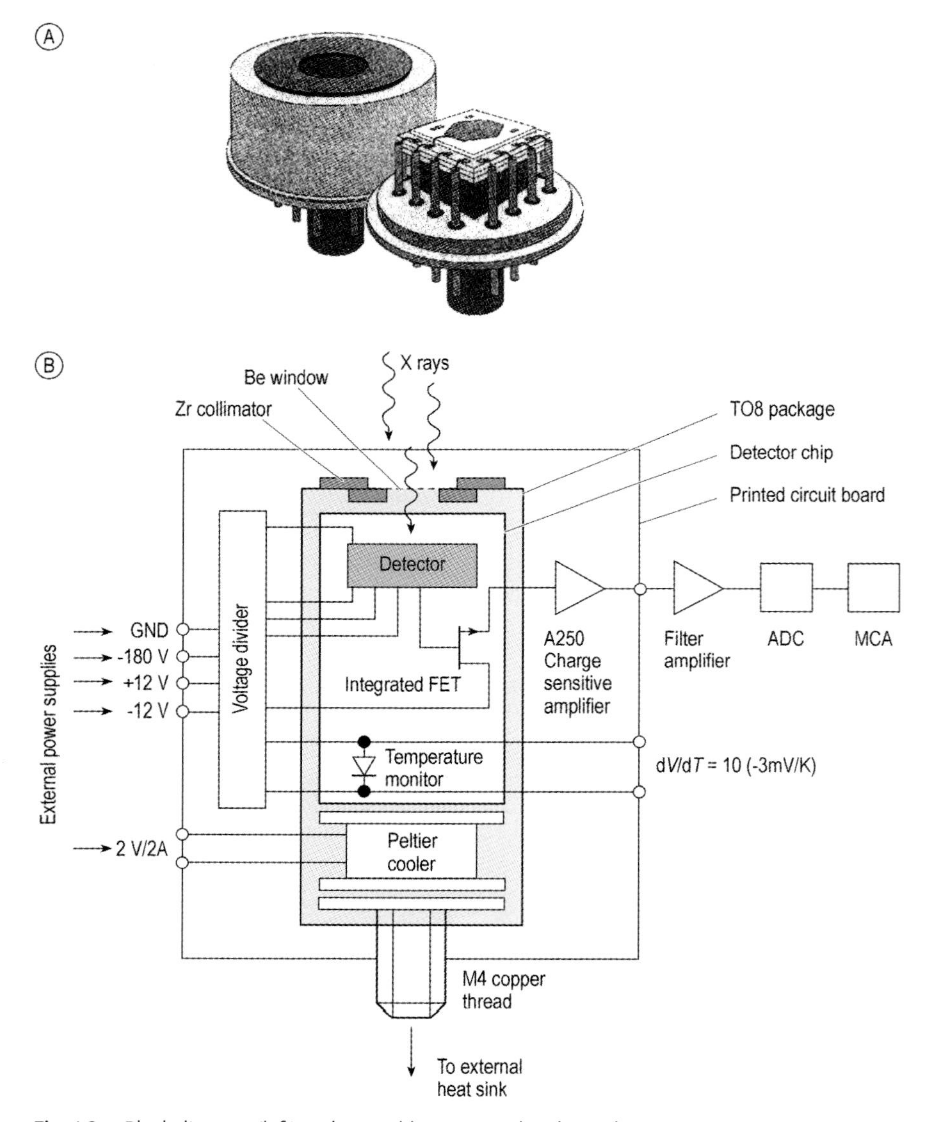

Fig. 4.9. Block diagram (left) and assembly image (right, also with cap removed) of a silicon drift detector (SDD) module, including electronics [4.30].

photon energy. Finally, the X-ray spectrum is established by a multichannel analyzer which distributes the pulses according to their height into a series of channels, each being reserved for a small range of pulse heights (= energy), and sums them to give the measured intensity.

Optimum resolution, i. e. low full width at half-maximum (*FWHM*), is a trade-off between high count rate, i. e. low dead-time, and good spectral resolution.

Silicon drift detectors (SDD, Figs 4.8 and 4.9) now also provide sufficient resolution ($FWHM = 0.175$ keV) above a sample spot sized 2×2 to 100×100 mm^2, and enable high-speed operation ($> 10^5$ counts s^{-1}). SDD can be combined with microelectronics and applied in portable TXRF models for microanalytical applications [4.30]. They must be cooled by a Peltier element.

4.1.3 Spectral Information

Spectral information from TXRF spectra is similar to that from conventional XRF-EDX spectra. The analyst must be always aware of potential spectral interferences that cannot be resolved with standard EDX detectors [4.16]. Spurious peaks can be assigned to escape peaks in energy-dispersive spectra. An escape peak emerges when intense fluorescence of sufficiently high energy (giving rise to the mother peak) impinges on the SSD expelling photoelectrons from Si core shells. Excited Si atoms emit X-ray photons that are usually reabsorbed by the crystal. Photoelectrons generated by these photons initiate additional electron–hole cascades, i. e., charge pulses in the same way as the sample fluorescence. Photons near the SSD crystal surface, however, also escape without producing a pulse, thus carrying off the difference between the energy of the mother peak and the generated core photoelectron. This "escaped" energy gives rise to a separate, low-energy, daughter peak (Table 4.1).

Spurious peaks can also appear as a result of energy-doubling, when charge collection induced by the first fluorescence photon has not been completed when a second photon from the same fluorescence source impinges. The detector then registers twice the number of charges giving rise to a sum-up peak. Sum-up peaks can be reduced by use of a pulse pile-up-rejector (PUR). Contamination along the beam path,

Tab. 4.1. Certain elements whose Kα peaks interfere with escape peaks of other elements using solid state detector SSD or silicon drift detector SDD [4.16].

Elements	*Escape peaks*
V	Co-Kα, Mn-Kβ, Dy-Lα, Ho-Lα, Er-Lα, Gd-Lβ
Cr	Co-Kα, Fe-Kβ, Er-Lα, Tm-Lα, Tb-Lβ, Dy-Lβ
Mn	Ni-Kα, Co-Kβ, Yb-Lα, Lu-Lα, Hf-Lα, Ho-Lβ
Fe	Cu-Kα, W-Lα, Ta-Lα, Hf-Lα, Ni-Kβ, Tm-Lβ, Er-Lβ
Co	Zn-Kα, W-Lα, Cu-Kβ, Yb-Lβ, Lu-Lβ, Re-Lα, Os-Lα
Ni	Ga-Kα, Pt-Lα, Ir-Lα, Ta-Lβ, Hf-Lβ
Cu	Zn-Kβ, Au-Lα, W-Lβ, Hg-Lα, Ge-Kα

e.g. Fe in the detector window (Be), can also result in spurious peaks and limit the detection capability of certain elements involved [4.31].

4.1.4 Quantification

Reliable quantification is based on peak-search software that combines peak location, peak identification, and element deduction. Element deduction means that, for unambiguous detection, at least two of the principal peaks must be detected for each analyte of interest. In trace analysis, only the strongest _ peaks can be detected and special attention must be paid to interfering satellites and spurious peaks.

By using an optically flat plate of a pure element such as Ni, TXRF systems can be calibrated in the sense of absolute calibration [4.32]:

$$I_X = K_X c_X \tag{4.3}$$

I_X is the background-corrected net intensity of the principal peak of analyte X, K_X a proportionality factor for the absolute sensitivity of the standard reference, e.g. an Ni plate, and c_X the concentration of X. Multielement analyses are based on known relative sensitivities S:

$$S_j = \frac{I_j / c_j}{I_{ref} / c_{ref}} S_{ref} \tag{4.4}$$

where the subscript "ref" denotes reference values. Values of relative sensitivities can be determined by theoretical calculations including, e.g., mass absorption [4.33]. The fluorescence yield is angle-dependent. Because matrix effects are absent for minute amounts of sample ($<\mu g\ cm^{-2}$), quantification by internal standardization is possible. Equation (4.4) is valid correspondingly.

The bulk type response curve depends also on surface roughness [4.34]. Reference materials must, therefore, be carefully investigated by angle-scan before use. Angle-scan characteristics of the sample, i.e. the fluorescence intensity recorded at more than one glancing angle near ϕ_c, should not deviate from those of the reference. The measurement must be performed under similar optical conditions.

With monochromatic Au-L_β (11.44 keV) excitation, the variation of Cu depth distribution in monocrystalline Si with time was studied by means of TXRF [4.35]. Such, and similar, effects must be carefully investigated when using reference samples of unknown behavior. Complementary analytical techniques might be helpful. The propagation of errors must, however, also be taken into consideration, and, moreover, the calibration must cover the concentration range of interest [4.36] because the accuracy of reference concentrations strongly affects the accuracy of the results [4.37, 4.38]. External calibration with reference materials can yield deviations of 3–20% from the true value [4.39]. Reliable preparation of controlled spiking has, however, recently been reported [4.40]. For general issues of external calibration of TXRF systems, please refer to Diebold [4.41] and Hockett [4.42].

4.1.5 Applications

TXRF is an ideal tool for microanalysis [4.21]. The analytical merits are that TXRF has a broad range of linearity (10^{13}–10^{8} atoms cm^{-2}) and it is extremely surface-sensitive and matrix-independent. TXRF can be applied to a great variety of different organic and inorganic samples such as water, pure chemicals, oils, body fluids and tissues, suspended matters, etc., down to the picogram range.

4.1.5.1 Particulate and Film-type Surface Contamination

The penetration depth for impinging X-rays is very limited under the condition of total reflection. For light substrates such as Si, quartz, or poly(methyl methacrylate) the spectral background of TXRF is six orders of magnitude smaller than that of XRF. The fluorescence signal arises from the uppermost layer; therefore the scattered intensity, i.e. background, is lower than in a conventional XRF arrangement. If metallic smear, organic tissues, or thin (< 100 nm), dried residues of a solution, suspension, or dispersion are present on a substrate within the standing wave front, strong oscillation of signal intensity results (Fig. 4.10). Such oscillations vanish when grain size or film thickness increase to approximately 1000 nm [4.43].

One can clearly distinguish thick films from particulate analytes by angle-characteristics (compare curves b and c in Fig. 4.11). Although the particulate and thin-layer types of angle-characteristics (Fig. 4.11) do not differ at large angles [4.44, 4.45], surface roughness crucially modifies peak height [4.34] and can simulate a stratified structure [4.46].

4.1.5.2 Semiconductors

Metrology and contamination analysis in particular have been decisive factors for profitable semiconductor production [4.47]. Semiconductor applications of TXRF go back to the late nineteen-eighties and were introduced by Eichinger et al. [4.48, 4.49]. Because of its high sensitivity, wide linear range, facile spectrum deconvolution, and

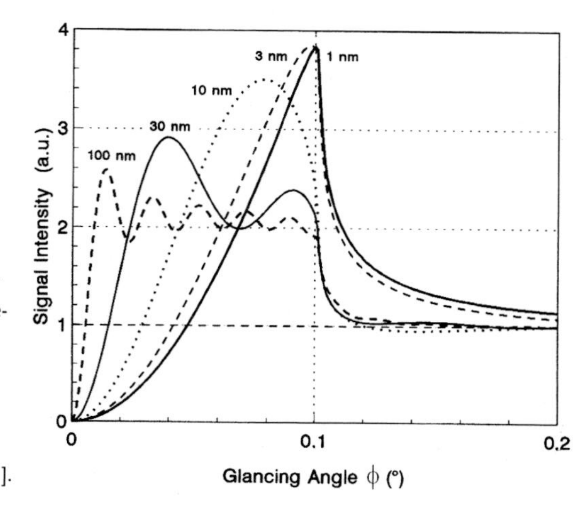

Fig. 4.10. Fluorescence signal from small particles or thin films deposited on a silicon substrate used as sample carrier. The intensity was calculated for particles, thin films, or sections of different thickness but equal mass of analyte, and plotted against the glancing angle ϕ. A Mo-Kα beam was assumed for excitation. Particles or films more than 100 nm thick show double intensity below the critical angle of 0.1° [4.21].

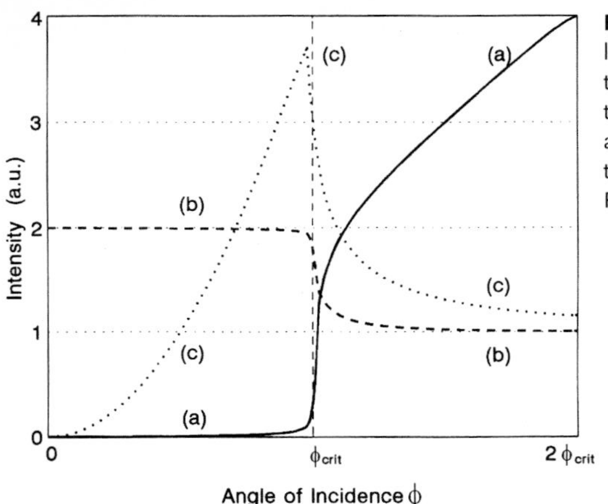

Fig. 4.11. Characteristic intensity profiles for three different kinds of concentration: (a) bulk type; (b) particulate type; and (c) thin-layer type. The critical angle ϕ_c is determined by total reflection at the substrate ([4.21], after Ref. [4.44]).

ease of calibration, TXRF has rapidly become the economic workhorse for monitoring cleaning efficiency and front-end wafer cleanliness [4.50, 4.51]. TXRF provides the mandatory quality control with reliable data which are easily integrated in the fab data management [4.52, 4.53]. Automatic operation with repeated one-point calibration every 8 h and automatic data management and transfer has led to abundant applications of TXRF (Figs. 4.12, 4.13).

Detection limits for various elements by TXRF on Si wafers are shown in Fig. 4.13. Synchrotron radiation (SR) enables bright and horizontally polarized X-ray excitation of narrow collimation that reduces the Compton scatter of silicon. Recent developments in the field of SR-TXRF and extreme ultra violet (EUV) lithography nurture our hope for improved sensitivity down to the range of less than 10^7 atoms cm^{-2}

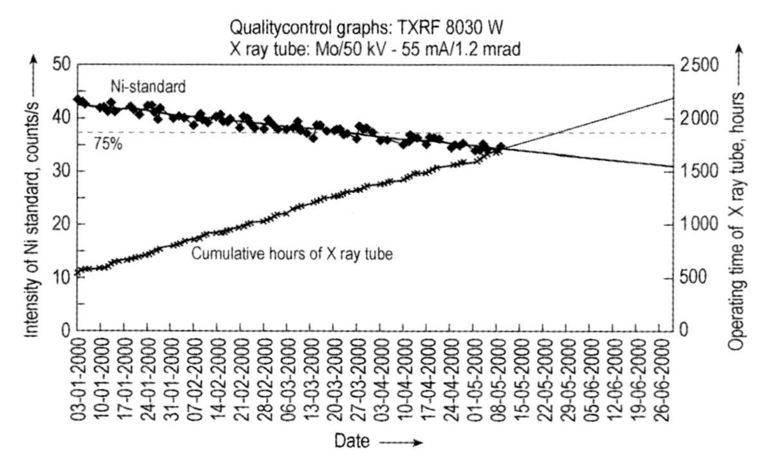

Fig. 4.12. Statistical process control chart for TXRF measurement systems. The sensitivity of the system can be controlled by daily calibration with an Ni standard reference sample. When the X-ray intensity sinks below 75% of the original intensity, the tube must be replaced with a new one.

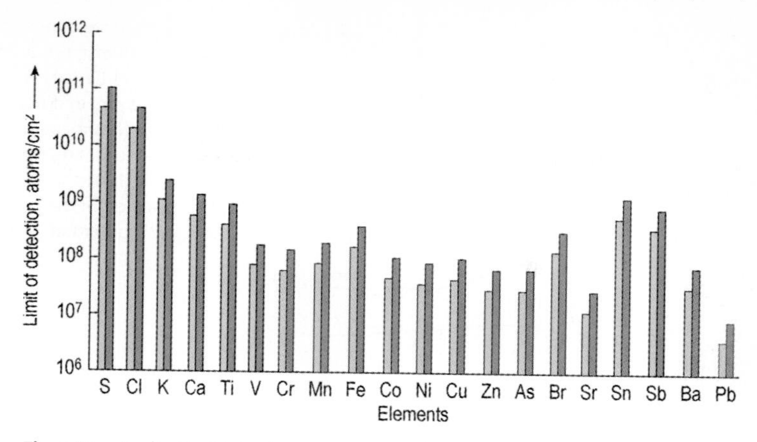

Fig. 4.13. 3σ limit of detection [atoms cm^{-2}] of VPD-TXRF measurements for 200 and 300-mm wafers using the wafer surface preparation system (WSPS) for sample preparation.

[4.13, 4.54–4.57]. Practical applications of SR-TXRF currently focus on fast mapping of wafers up to a diameter of 300 mm and reference measurement capabilities in the range of 10^5–10^7 atoms cm^{-2} [4.58]

4.1.5.2.1 **Depth Profiling by TXRF, and Multilayer Structures**

Buried layers are important parts of microcircuits. TXRF is a sensitive microanalytical tool for inspection of buried layers (Fig. 4.14). On a thick substrate, the angle characteristics are a function of layer thickness. [4.59, 4.60].

Recently, first-principle calculations have been reported on model-free TXRF of stratified structures [4.61].

Shallow doping profiles, particularly those of As, require nanoscale information on dopant distribution. Although SIMS can be reliably applied for layers below 5 nm

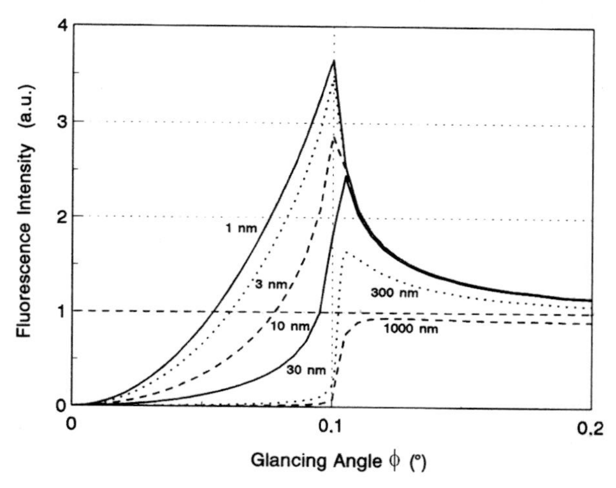

Fig. 4.14. Fluorescence intensity from layers buried in a thick substrate. The dependence of intensity on the glancing angle was calculated for layers of different thickness but with a constant analyte area density. Silicon was assumed as substrate and Mo-Kα X-rays as primary beam. Total reflection occurs in the region below 0.1°. Without total reflection, the dashed horizontal line would be valid throughout [4.21].

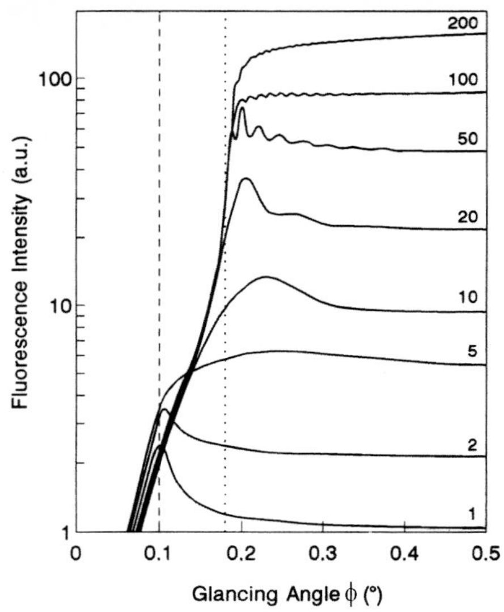

Fig. 4.15. Angular dependence of the fluorescence radiation emitted from a Co-layered Si substrate. The Co-Kα intensity is plotted semi-logarithmically for layers of different thickness (mm). The maxima for the ultra-thin Co-layers are located at the critical angle of Si (dashed vertical line). They are shifted to the critical angle of Co (dotted vertical line) if the layer is more than 10 mm thick ([4.21], after Ref. [4.41]).

depth, the technique of choice for the upper subsurface is angle-dependent TXRF [4.62]. Using angle-scan TXRF, stratified microstructures can be accurately analyzed for element composition, layer thickness, and density (Fig. 4.15) [4.41]. A combination of TXRF with layer-by-layer chemical etching provides reproducible results on the stratigraphy of metallic implants [4.15, 4.63–4.65].

4.1.5.2.2 **Vapor Phase Decomposition (VPD) and Droplet Collection**

Vapor-phase decomposition and collection (Figs 4.16 to 4.18) is a standardized method of silicon wafer surface analysis [4.11]. The native oxide on wafer surfaces readily reacts with isothermally distilled HF vapor and forms small droplets on the hydrophobic wafer surface at room temperature [4.66]. These small droplets can be collected with a scanning droplet. The scanned, accumulated droplets finally contain all dissolved contamination in the scanning droplet. It must be dried on a concentrated spot (diameter approximately 150 μm) and measured against the blank droplet residue of the scanning solution [4.67–4.69]. VPD-TXRF has been carefully evaluated against standardized surface analytical methods. The user is advised to use reliable reference materials [4.70–4.72].

The accuracy of VPD-TXRF depends largely on the reliability of the reference material (cf. Sect. 4.1.1.3) and on the reproducibility of the droplet drying procedure (Fig. 4.17) [4.75, 4.76]. Mass absorption sets the upper limit of reliable analyses to 10^{14} atoms cm^{-2} [4.77].

VPD-TXRF is also a facile technique for interface analysis [4.78, 4.79]. Automated VPD equipment (Fig. 4.16) improves both the detection limit (upper range 10^7 atoms cm^{-2}) and the reliability (by > 50%) of the VPD-TXRF measurement [4.14]. Current research focuses on sample holders [4.80, 4.81] and light-element detection capability [4.82–4.84].

Fig. 4.16. Wafer surface preparation system WSPS: Automated VPD system for wafers of diameter from 100 to 300 mm.

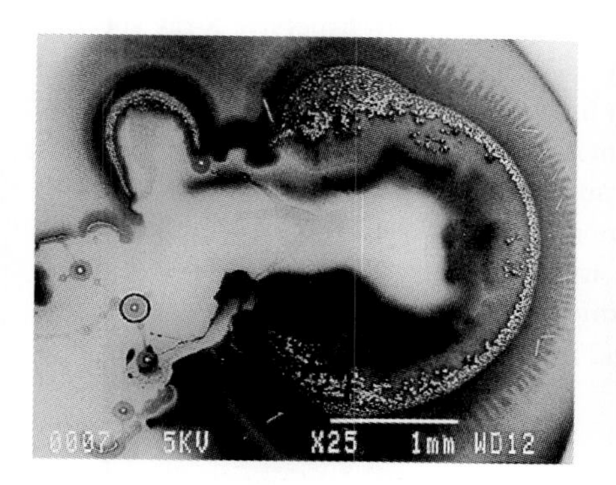

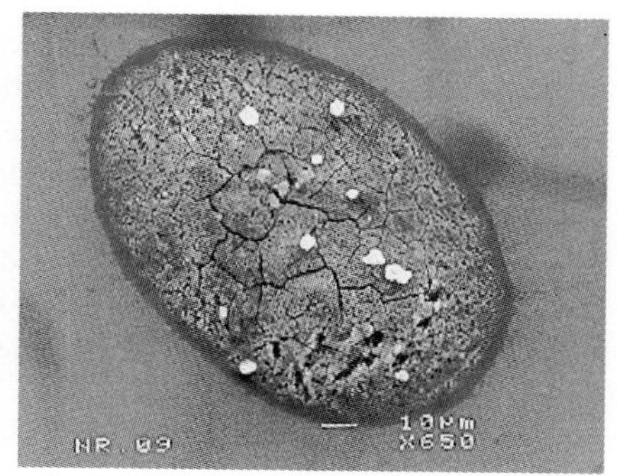

Fig. 4.17. Dried VPD-droplet (left) worst case, the VPD solution exploded under fast drying using an infrared lamp, droplet size of a few mm; (right) best case (WSPS),VPD solution dried under controlled conditions using vacuum and carrier gas (L. Fabry, S. Pahlke, L. Kotz, *Fresenius' J. Anal. Chem.* **354** (1996) 267, Figures 1 and 2 [4.36]).

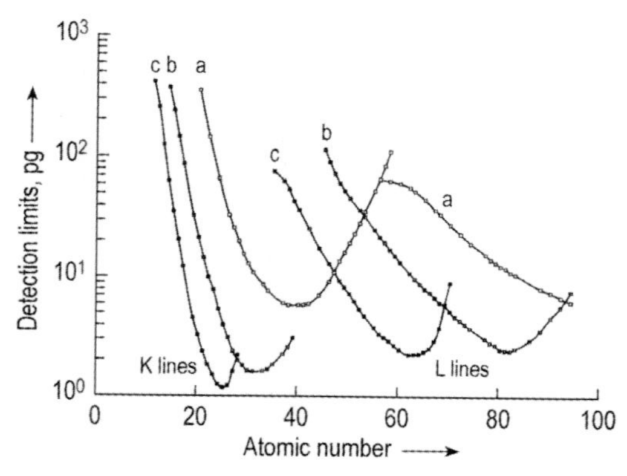

Fig. 4.18. Detection limit of TXRF for the residues of aqueous solutions, depending on the atomic number of the analyte element. Three excitations modes were used: (a) W-tube, 50 kV; Ni-filter, cutoff 35 keV; (b) Mo-tube; 50 kV; Mo filter, cutoff 20 keV; (c) W-tube, 25 kV, Cu-filter. The three curves to the left were determined by the detection of K-peaks; the three curves to the right, by that of L-peaks [4.21], after Refs [4.73] and [4.74].

4.2 Energy-dispersive X-ray Spectroscopy (EDXS)

Reinhard Schneider

Physically the generation of X-rays is often a secondary process preceded by the ionization of an atom. There are, therefore, several possibilities of X-ray generation depending on the type of the exciting medium – neutrals or charged particles such as electrons and ions and high-energy photons, i. e. X-rays themselves.

The occurrence of X-rays can, therefore, also be observed in all electron–optical instruments where samples are bombarded by electrons with energies high enough to ionize matter (cf. Sect. 2.3). The following discussion is mainly restricted to the use of X-rays for microanalysis in transmission electron microscopy (TEM). Since X-ray spectroscopy is widely used in scanning electron microscopy (SEM) some fundamental considerations must be made if the techniques are applied in combination. In general, two main X-ray spectroscopic techniques can be distinguished depending on the physical property measured – the wavelength or the energy of the emitted X-rays. The corresponding methods are wavelength-dispersive X-ray spectroscopy (WDXS) and energy-dispersive X-ray spectroscopy (EDXS), where WDXS is almost exclusively used in conjunction with SEM. In the following discussion merely an introduction is given to the fundamentals, instrumentation, and application of energy-dispersive X-ray spectroscopy. Because of its importance, peculiarities, and application possibilities, a wealth of monographs [4.85–4.91] and original papers have been devoted to EDXS/WDXS and many concern TEM/SEM. Likewise, the fundamentals and principles of electron microscopy itself including such details as electron–target interaction, electron optics, image and contrast generation are presented elsewhere.

4.2.1 Principles

When an atom has been ionized by electron bombardment, for instance by transition of a K-shell electron into the vacuum, after a finite dwell time the generated hole can be filled by an electron of the L shell (Sect. 2.3, Fig. 2.31). There is a certain probability that the energy set free by this process leads to the excitation of characteristic X-rays, viz. Kα radiation for the transition described. If an M-shell electron falls into the hole state, emission of so-called Kβ X-rays occurs. A process competing with the emission of X-rays is the emission of Auger electrons, where the sum of both the fluorescence yield of X-rays and probability of Auger electron emission is always equal to 100%. The fluorescence yield of X-rays increases with atomic number Z, while the portion of emitted Auger electrons behaves in the opposite manner. For ionization of the L-shell and filling of the remaining hole by electrons from the M or N shells, the emission of Lα or Lβ radiation, respectively, is observed. For heavier elements electrons from more outer shells can also occupy the holes, although not all conceivable transitions between the existing electronic energy levels are allowed. The

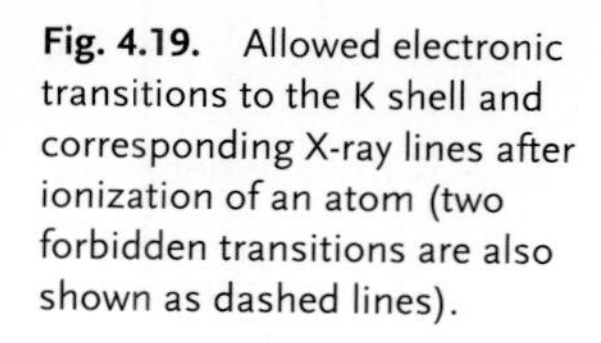

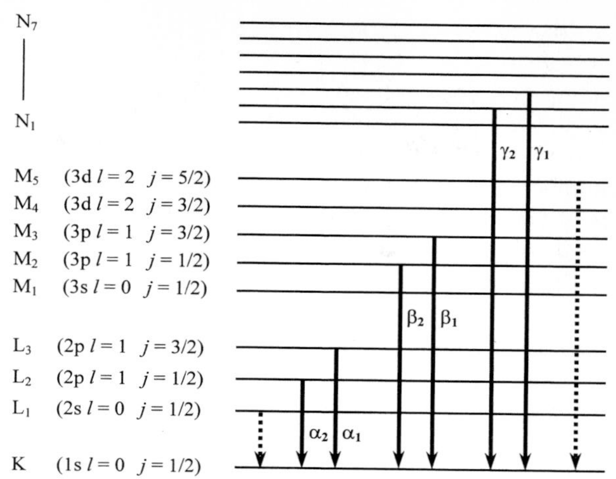

Fig. 4.19. Allowed electronic transitions to the K shell and corresponding X-ray lines after ionization of an atom (two forbidden transitions are also shown as dashed lines).

physically allowed transitions are determined by the following quantum-mechanical selection rules [4.92]:

(1) $\Delta n > 0$, i. e. L $\rightarrow$ K, M $\rightarrow$ K transitions but not, e. g., $L_3 \rightarrow L_2$
(2) $\Delta l = \pm 1$ and $\Delta j = -1, 0, +1$,

where n is the principal quantum number, l the angular momentum quantum number, and $j = l \pm s$ is that of the total angular momentum, i. e. the combination of orbital momentum and spin. These so-called dipole selection rules enable prediction of all possible transitions between electronic levels and thus the corresponding X-ray lines, also. This is illustrated in Fig. 4.19 in which some allowed transitions contributing to the K series of X-rays are shown as full lines, whereas forbidden transitions are drawn dashed.

In accordance with spin-orbit coupling [4.93], higher energy levels are split, enabling additional transition processes. Thus, for the Kα radiation two individual Kα_1 and Kα_2 lines can be differentiated which are attributed to transitions from $2p_{3/2}$ and $2p_{1/2}$ levels to the $1s_{1/2}$ state, because $\Delta l = \pm 1$ and $\Delta j = 0, \pm 1$. In practice, because of an average energy resolution of approximately 130 eV of EDXS detectors this splitting of energy levels is only measurable for heavier elements, whereas WDXS can usually be used to resolve the individual sub-lines (Fig. 4.22). The Kα_1/Kα_2 line splitting amounts to approximately 2 eV for potassium ($Z = 19$), 75 eV for yttrium ($Z = 39$), and 1.815 keV for gold ($Z = 79$).

The energies of the X-ray quanta of the K and L series can be calculated by use of the equations [4.94]:

$$E(\mathrm{K}\xi(n)) = E_{\mathrm{A(n+1)}} - E_{\mathrm{K}} = hcR(Z-1)^2(1/n^2 - 1) \tag{4.5 a}$$

and

$$E(L\xi(n)) = E_{A(n+2)} - E_L = hcR(Z-7.4)^2(1/n^2 - 1/2^2) \quad (4.5\,b)$$

where $\xi(n) = \alpha, \beta, \gamma, \ldots$, $A(n) = K, L, M, \ldots$, and $n = 1, 2, 3, \ldots$, R is the Rydberg constant, h is Planck's constant, and c is the velocity of light.

In these approximations for the K series the value 1 is subtracted from the atomic number Z to correct for the screening of the nuclei by the remaining K-shell electron. For the L series the screening effect of the two K-shell electrons and the seven remaining L-shell electrons must be taken into consideration by subtracting 7.4.

The energy E of the characteristic X-rays within a given series of lines, i. e. Kα, Kβ, etc., increases regularly with the atomic number Z. This dependence is called Moseley's law of X-ray emission:

$$\sqrt{E} \propto (Z-1) \quad (4.6)$$

In an electron-excited X-ray spectrum the discrete X-ray lines are superimposed on a continuous background; this is the well-known bremsstrahlung continuum ranging from 0 to the primary energy E_0 of the electrons. The reason for this continuum is that because of the fundamental laws of electrodynamics, electrons emit X-rays when they are decelerated in the Coulomb field of an atom. As a result the upper energy limit of X-ray quanta is identical with the primary electron energy.

4.2.2
Practical Aspects of X-ray Microanalysis and Instrumentation

For electron-beam illumination of an object the spatial extension of the volume from which a specific interaction signal is gained depends strongly on the material itself, the diameter of the electron beam, the primary electron energy, and specimen thickness. In bulk materials many more interaction processes occur between incoming electrons and target atoms in comparison with a thin film. Each individual interaction process can change the direction of the electron path within the specimen, leading to broadening of the average diameter of the electron beam. In general, electron propagation directly determines the distribution of the energy within the target and, therefore, also the size of the interaction or excitation volume, because the absorbed energy promotes the generation of secondary information carriers like secondary electrons, electron–hole pairs, and X-rays. The thicker the specimen the more extended is the interaction volume essentially limiting the achievable lateral resolution. Consequently, for thin films ca. 10 nm thick the resolution can be much better than for massive samples.

The phenomena of beam broadening as a function of specimen thickness are illustrated in Fig. 4.20; each figure represents 200 electron trajectories in silicon calculated by Monte Carlo simulations [4.91, 4.95–4.97] for 100-keV primary energy, where an infinitesimally small electron probe is assumed to enter the surface. In massive Si the electrons suffer a large number of elastic and inelastic interactions during their paths through the material, until they are finally completely stopped. The resulting penetration depth of the electrons is approximately 50 µm and in the

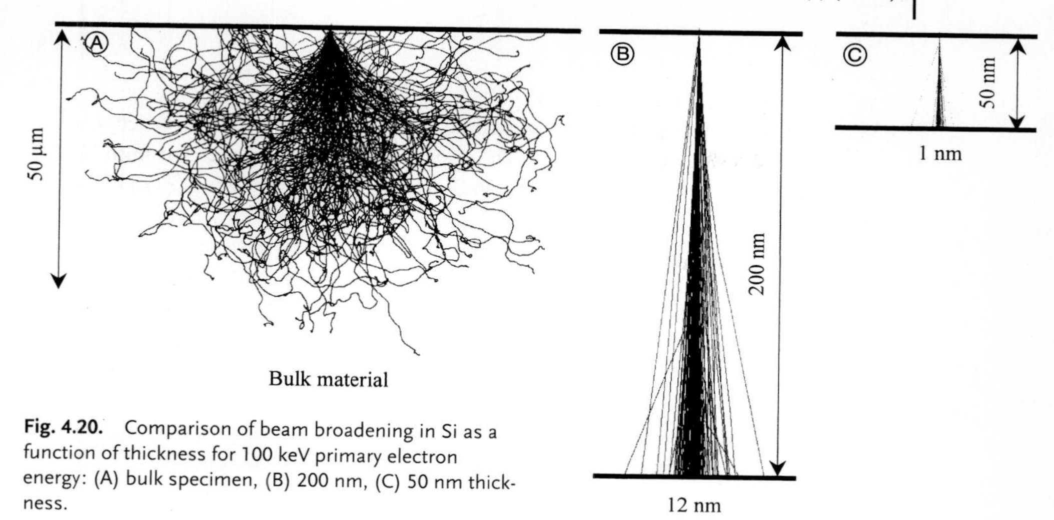

Fig. 4.20. Comparison of beam broadening in Si as a function of thickness for 100 keV primary electron energy: (A) bulk specimen, (B) 200 nm, (C) 50 nm thickness.

perpendicular direction their ranges are nearly twice that, owing to back-scattering. This situation meets the conditions present in scanning electron microscopes (SEM) and electron probe microanalyzers (EPMA). For the foil specimens applicable in transmission and scanning transmission electron microscopy (TEM/STEM) the beam-broadening effect is usually defined by the diameter of the disc at the exit surface containing 90% of the emerging electrons [4.96, 4.97]. In Fig. 4.20 this diameter is approximately 1 nm for the 50-nm thick foil and reaches as much as 12 nm for 200 nm thickness. From these considerations it is obvious that analytical information of extremely high lateral resolution can only be attained for very thin specimens in TEM/STEM.

Before the development of semiconductor detectors opened the field of energy-dispersive X-ray spectroscopy in the late nineteen-sixties crystal-spectrometer arrangements were widely used to measure the intensity of emitted X-rays as a function of their wavelength. Such wavelength-dispersive X-ray spectrometers (WDXS) use the reflections of X-rays from a known crystal, which can be described by Bragg's law (see also Sect. 4.3.1.3)

$$n\lambda = 2d\sin\theta \tag{4.7}$$

where λ is the wavelength of the X-rays, θ the scattering angle, n the order of reflection, and d the interplanar spacing of the single crystal chosen. The point of the electron probe on the specimen, the surface of the analyzing crystal, and the detector slit are placed on the circumference of a focusing circle, also called the Rowland circle (see also Sect. 2.1, and 2.2). The movement of the crystal and the detector is mechanically coupled such that the detector forms an angle θ with the crystal surface while it moves an angular amount 2θ. The wavelength, λ, measured is given by the relationship:

$$\lambda = \frac{d}{R} L \qquad (4.8)$$

where R is the radius of the Rowland circle and L the variable distance between the specimen and the analyzing crystal. The experimental arrangement for both WDXS and EDXS attachable to SEM/EPMA and STEM/TEM is shown schematically in Fig. 4.21.

A major disadvantage of the WDXS arrangement is the precise mechanical adjustment needed to meet the focusing conditions. Because of this WDX spectroscopy can be complicated or even fail for samples of high surface roughness. To cover the range of wavelengths of interest a set of three or four crystals differing in their lattice spacing are usually employed. For example, for angular scanning, $\theta = 15$–$65°$ LiF crystals cover a wavelength range from 0.35 to 0.1 nm, which is equal to X-ray energies of 3.5–12.5 keV. X-ray detection is usually achieved by use of a sealed proportional counter for shorter wavelengths and a gas-flow counter for longer wavelengths (> 0.7 nm).

Owing to the particular kind of X-ray detection, different experimental requirements have to be met for interfacing an EDXS system to an electron microscope, compared with those of a WDXS arrangement (cf. Fig. 4.21). Because of the specific geometry and dimensions of specimen chambers different experimental arrangements can be found for SEMs and TEM/STEMs – the free space for inserting a semiconductor detector enabling EDXS is extremely limited for the latter. For combined EDXS/(S)TEM the detector should be attached to the microscope in an orthogonal position relative to the tilt axis of the specimen; this enables alignment of interface structures of cross-section specimens parallel to the detector axis and hence enables X-ray analysis at high lateral resolution. It is important to know the exact position and angle of the front side of the detector relative to the region of the specimen contributing to the X-ray signal measured. The angle between the plane of the specimen and the detector axis is called the X-ray take-off angle α. For a detector arrangement in the plane of the specimen and perpendicular to it the take-off angle is simply the goniometer tilt. The ideal configuration is, however, a positive take-off angle (Fig. 4.21) providing X-ray detection without tilting of the specimen. In transmission electron microscopy

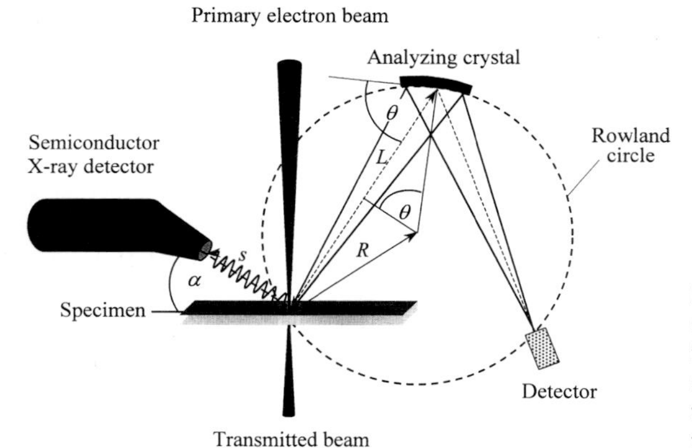

Fig. 4.21. Schematic diagram of spectrometer arrangements for wavelength-dispersive and energy-dispersive X-ray spectroscopy (WDXS/EDXS) in electron microscopy.

EDXS detectors are differentiated into two groups according to the take-off angle – low take-off systems with approximately 10–20° and high take-off systems the angle of which can reach up to 70°. A second important setting determining detector efficiency is the collection angle, Ω, which is the solid angle subtended by the front face of the detector. In practice, it is given by the approximation [4.91, 4.101]:

$$\Omega \approx \frac{A}{s^2} \tag{4.9}$$

where A is the active area of the detector and s the distance from the specimen to the detector. For a detector with 30 mm^2 area in 15 mm distance from the specimen (typical values for EDXS/(S)TEM) the collection angle is approximately 0.13 sr. The collection angles in EDXS range from approximately 0.01 to 0.2 sr, and are thus more than one order in magnitude wider than for WDXS.

As already mentioned above, energy-dispersive X-ray spectroscopy is based on the use of a semiconductor detector, which can be a lithium-drifted Si crystal (a so-called Si(Li) detector) or a Ge crystal of high purity, respectively [4.89–4.91, 4.100–4.102]. For Si (Li) the lithium is introduced to compensate for impurity effects in Si thus producing an intrinsic zone inside the crystal that is effective for X-ray detection. In general, inside the detector material electron–hole pairs are generated by X-ray excitation (Fig. 4.22). The energy necessary to produce one electron–hole pair is approximately 3.8 eV for Si (Li) and only 2.9 eV for Ge. Because the energy of X-rays can be several keV, depending on the primary electron energy and on the bombarded material, a single photon can generate thousands of electron–hole pairs. Thus, the number of electrons and holes is a direct measure of the energy of the incoming X-rays. The electron–hole pairs are separated by the effect of an electric field caused by a negative bias of approximately 0.5–1 keV applied across the Si crystal. Gold layers deposited on the front and back of the crystal serve as electrical contacts. The resulting electrical pulses are amplified, analyzed for pulse height and finally stored in a multichannel analyzer (MCA) or computer. The detector crystals are cooled with liquid nitrogen to minimize the portion of thermally activated electron–hole pairs, to reduce the noise level of the associated field-effect transistor acting as pre-amplifier,

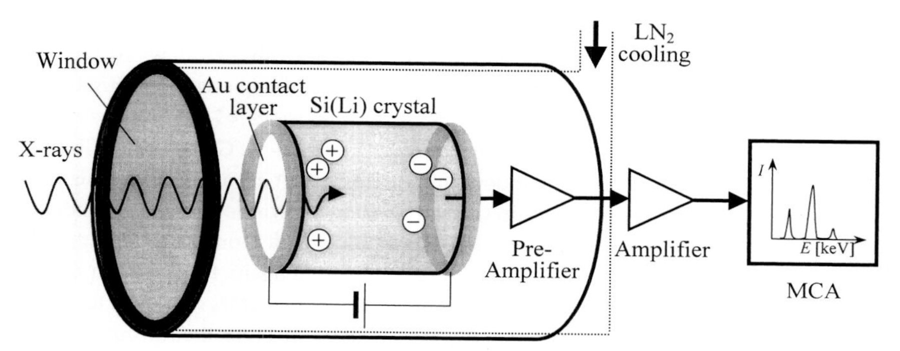

Fig. 4.22. Principal set-up of an X-ray detector with Li-drifted Si crystal.

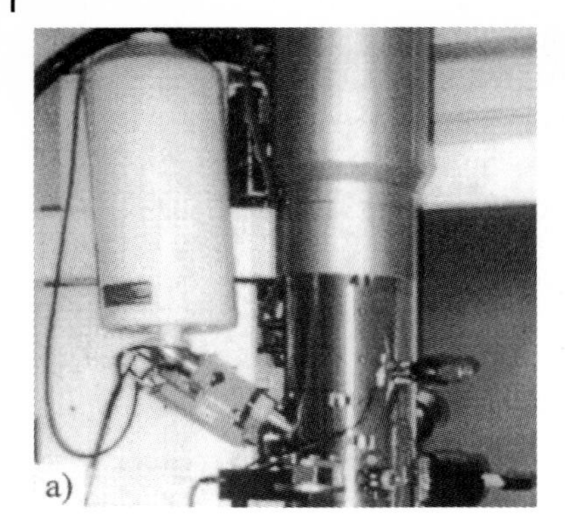

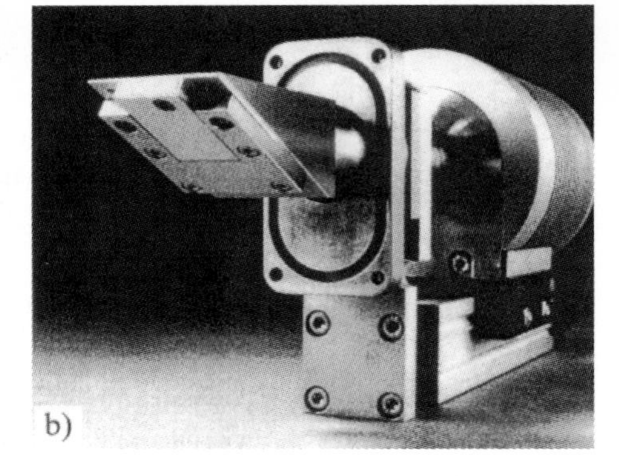

Fig. 4.23. EDXS instrumentation for TEM and SEM: (a) conventional EDX detector attached to a TEM/STEM; (b) EDX system with Si drift-chamber detector (XFlash, Röntec) for SEM [4.105].

and, for Si(Li), to prevent diffusion of Li. Thus the Dewars needed for cooling are characteristic of such X-ray detector assemblies (cf. Fig. 4.23 a).

The photographs presented in Fig. 4.23 should give an impression of the set-up of EDXS systems in practice. Fig. 4.23 a shows the column of a 200 kV TEM/STEM from the section of the objective lens (below) to the segment containing the electron gun and accelerator (above) with a high take-off spectrometer attached to the microscope. In contrast, Fig. 4.23 b shows a special energy-dispersive X-ray spectrometer based on the use of a thermoelectrically cooled Si drift chamber detector [4.103, 4.104]. This system is particularly suitable for handling high count rates in SEM, therefore enabling the acquisition of element maps with up to the four times the speed of conventional EDXS detectors. Many companies offer EDXS and WDXS systems, attachable to all commercially available electron microscopes. A selection of these companies is cited in the list of suppliers of surface and thin-film analytical equipment (Chap. 7).

To prevent condensation of hydrocarbons and even ice from the microscope environment on the cold surface of Si(Li) or Ge detector crystals they are commonly sealed in a pre-pumped tube. A window in front of the crystal enables the transmission of X-rays. In electron microscopes with particularly good oil-free vacuum, however, X-ray detectors can also be used without windows. Historically, windows made from sheets of beryllium, typically about 12 µm thick or even thicker, were the first to be used. Because of this window thickness X-rays with energies <1 keV were strongly absorbed, preventing the analysis of light elements such as B, C, N, and O. To overcome these drawbacks ultrathin windows of less absorbent advanced materials have been developed, for example very thin C (diamond) or BN windows. The dependence of the absorption effect on the type of detector window is demonstrated in Fig. 4.24. In this figure the transmitted fraction of X-rays is plotted against the photon energy for both the Si(Li) and the Ge crystal, where windowless detectors and detectors with windows (Be and diamond coated with Al) are assumed. Besides the influence of the window used the individual curves also have the absorption edges of the detector

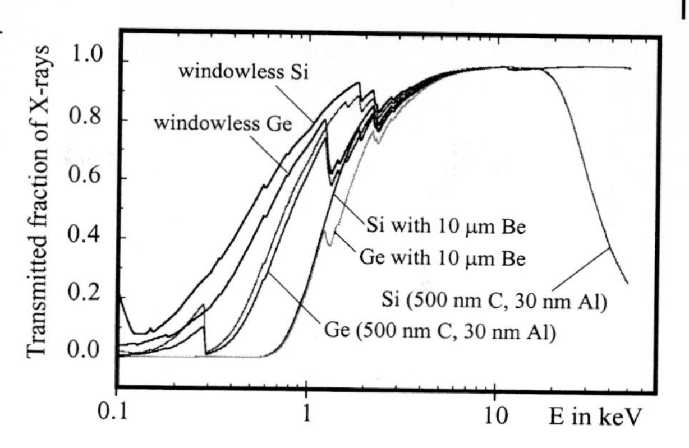

Fig. 4.24. Fraction of transmitted X-rays calculated for EDXS detectors with windowless Si(Li) and Ge crystals, and for different windows.

material itself, i.e. Si and Ge, respectively. The transmission of low-energy X-rays is insignificantly higher for Si(Li) compared with Ge, whereas the efficiency of Si(Li) detectors drops remarkably above 20 keV. For the high-energy range the application of Ge crystals is more useful; this enables detection of X-rays up to approximately 50 keV and higher with 100% efficiency. The dependence of the portion of transmitted X-rays on the type of window is clearly visible – the windowless configurations and, partly, those with Al-coated diamond window are feasible for detection of light-element X-rays.

The energy resolution of an X-ray detector is experimentally defined by the full width at half maximum (*FWHM*) of the Mn-Kα line. The *FWHM*, in eV, can also be calculated by use of the relationship:

$$FWHM = (\Delta n^2 + 2.35^2 F \varepsilon E + \Delta c^2)^{1/2} \tag{4.10}$$

where Δn is the electronic noise contribution, Δc is that of incomplete charge collection, F is the Fano factor, ε the energy to create an electron–hole pair, and E the initial X-ray energy [4.106]. Owing to the lower ionization energy for Ge the number of charge carriers is approximately 25% higher than that in Si. Because, in addition, the Fano factor for Ge is approximately 0.06 compared with 0.087 for Si, Ge detectors intrinsically afford better energy resolution. There resolution also depends on the area of the detector surface – broadening is observed with increasing area. It should, moreover, be noted that it is also a function of output count rates and the setting of the pulse processor; the best resolution values are obtained at rates less than 3000 counts per second and at medium pulse processor times [4.107]. Typically, Si(Li) detectors have an average energy resolution of approximately 140 eV measured at the Mn-Kα line, whereas that of Ge detectors can be better than 120 eV.

Because of the limited energy resolution in EDX spectra an overlap of peaks can often occur, depending on the composition of the material to be analyzed. The situation is much improved in WDXS, for which the energy resolution is approximately 10 eV and better. This is demonstrated in Fig. 4.25, in which the WDX and EDX spectra recorded from $BaTiO_3$ are compared. Here, WDXS enables easy resolution of the Ba-Lα and Ti-Kα lines; this is impossible by EDXS. In addition, for WDXS the

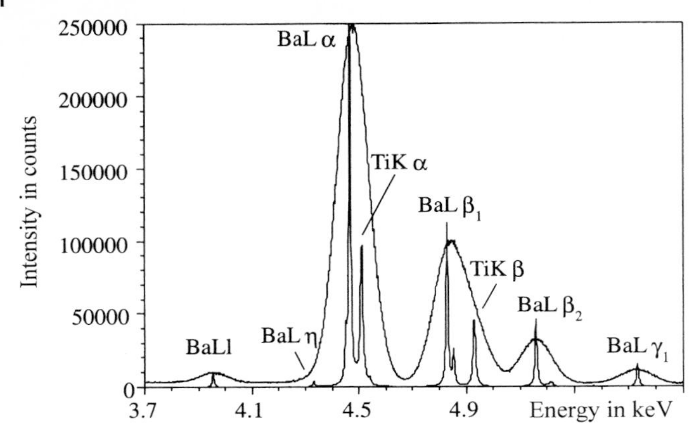

Fig. 4.25. A WDX spectrum of $BaTiO_3$ plotted against energy and compared with the corresponding EDX spectrum [4.91].

peak-to-background ratio is also better. But, owing to the low X-ray collection efficiency compared with EDXS, the wavelength-dispersive technique is commonly not combined with TEM/STEM. Some more advantages of the EDXS technique are the compact and stable instrumentation, its ease of use, and the short acquisition times as a result of parallel detection.

4.2.3
Qualitative Spectral Information

In EDX spectra the X-ray intensity is usually plotted against energy. They consist of several approximately Gaussian-shaped peaks being characteristic of the elements present in the transmitted volume. The characteristic peaks are superimposed on a background (bremsstrahlung continuum) of distinctly low intensity if the specimen is very thin. Most of the chemical elements can be identified by EDXS. The only limit is whether the particular type of detector window registers soft X-rays of the light elements (Sect. 4.2.2). Thus detectors equipped with Be windows only enable detection of elements of atomic number $Z \geq 11$ (Na), whereas detectors with ultrathin light-element windows can even be used for analysis of boron ($Z = 5$). Problems in element detection can also arise as a result of peak overlap caused by the restricted energy resolution discussed above.

An EDX spectrum typical of thin-film analysis in TEM/(S)TEM is shown in Fig. 4.26. It was obtained from a polycrystalline TiC/ZrO_2 ceramic by use of an Si(Li) detector at 100 keV primary electron energy. For spectrum recording the electron probe of approximately 1 nm in diameter was focused on the triple junction between the grains in the STEM mode (Fig. 4.26 a). Besides the elements expected for the material under investigation, viz. Ti and Zr, Si, Fe, and Co were also detected, hinting at the presence of a (Fe, Co) silicide as an impurity. For ceramic materials it is known that

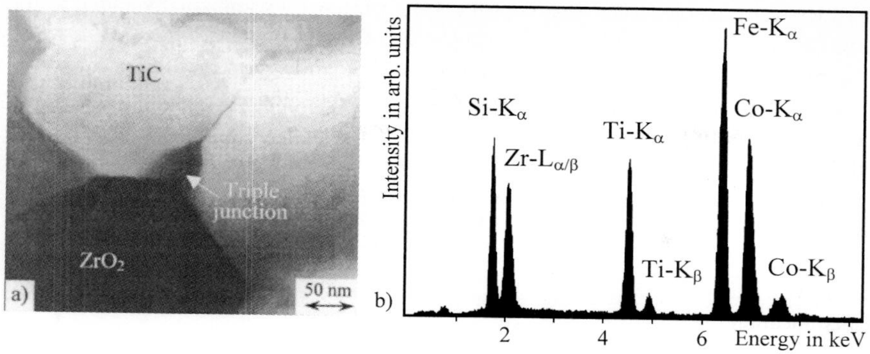

Fig. 4.26. Typical X-ray spectra: (a) STEM bright-field image of a polycrystalline ZrO_2/TiC ceramic with a triple junction; (b) corresponding EDX spectrum.

such foreign phases are predominantly present in triple regions and in interlayers between the individual grains.

Depending on the measurement conditions chosen there is some probability that artifacts will contribute to an EDX spectrum. They comprise effects intrinsically produced by signal detection, viz. in the form of escape peaks and internal fluorescence peaks, and, in addition, those of signal processing, for example pile-up peaks. Escape peaks appear because an X-ray photon entering the Si(Li) detector can lose part of its energy by exciting the Si to fluoresce Si-Kα radiation (1.74 keV). Thus, an extra peak is produced at an energy of the characteristic peak reduced by 1.74 keV. In Fig. 4.27 a this phenomenon is demonstrated for the Cu-Kα line, where an escape peak of low intensity is also observed for the Cu-Kβ line. Of course, high-purity Ge crystals do also generate escape lines, but several peaks (Ge-Kα and Ge-Lα) can occur because of the electronic structure of Ge. The software used to process EDX spectra commonly

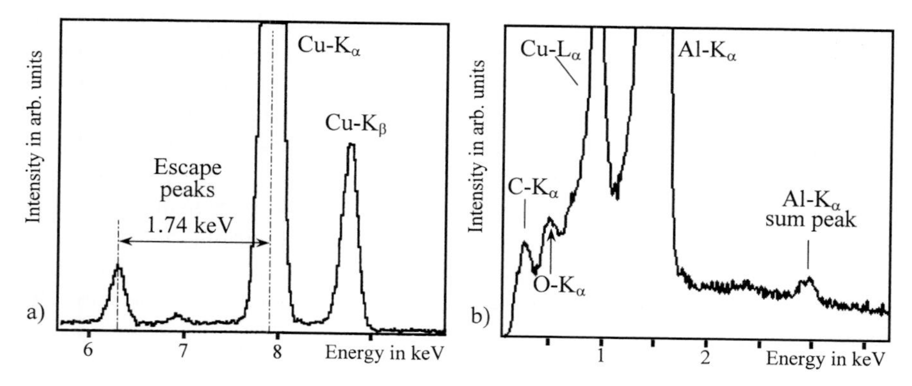

Fig. 4.27. Artifacts in energy-dispersive X-ray spectra. Occurrence of (a) escape and (b) sum peaks.

uses routines to correct for escape peaks – their intensities are usually added to the original characteristic peak. As a consequence of the internal fluorescence a corresponding peak can be seen in the spectrum at high count rates and long acquisition times. When the number of incoming X-ray photons is too high (more than 10 000 counts per second) the signal electronics can no longer differentiate between two single photons of a certain energy. This leads to the occurrence of so-called pile-up or sum peaks, respectively, with exactly twice of the energy of a characteristic X-ray line as shown in Fig. 4.27 b for the Al-Kα peak.

4.2.4 Quantification

As it is usual in all spectroscopic methods for quantification, the original signal characteristic of an individual element inside the excited volume that can be correlated with the content of this element must be extracted from the raw-data spectrum. Hence, quantitative EDXS analysis involves the determination of the background contribution, background subtraction, and counting the net intensities of the characteristic X-ray peaks. There are different possible ways of fitting the background curve underlying the whole EDX spectrum. For thin-foil analysis it is nearly a horizontal line of rather low intensity. Thus, one procedure used to determine the background signal below an X-ray peak is to assume a straight-line contribution that is subtracted. Another method of background extrapolation comprises averaging of the background intensities in two windows of identical width just below and above the characteristic peak. It is also possible to model the background by Kramers' law [4.108] defining the number N_{Back} of bremsstrahlung photons as a function of the photon energy E:

$$N_{\mathrm{Back}} = CZ\,\frac{E_0 - E}{E} \tag{4.11}$$

where Z is the mean atomic number of the material and E_0 is the primary energy of the electrons. The factor C must be found by modeling and includes the original Kramers constant and corrections for the efficiency of collection and processing by the detector and the effects of absorption of the excited X-rays within the specimen. In Fig. 4.28 an EDX spectrum of a layered Al/Cu sample is shown with the modeled background marked by a full line. After background subtraction the net X-ray peaks are usually fit by Gaussian peaks and their intensities are determined by integration.

The procedure commonly used to quantify EDX spectra was originally outlined by Castaing [4.109], although for the general situation of investigating bulk materials. To a good approximation it can be assumed that the concentration C_{Sp} of an element present in an unknown sample is related to the concentration C_{St} of the same element in a standard specimen by

$$\frac{C_{\mathrm{Sp}}}{C_{\mathrm{St}}} = K\,\frac{I_{\mathrm{Sp}}}{I_{\mathrm{St}}} \tag{4.12}$$

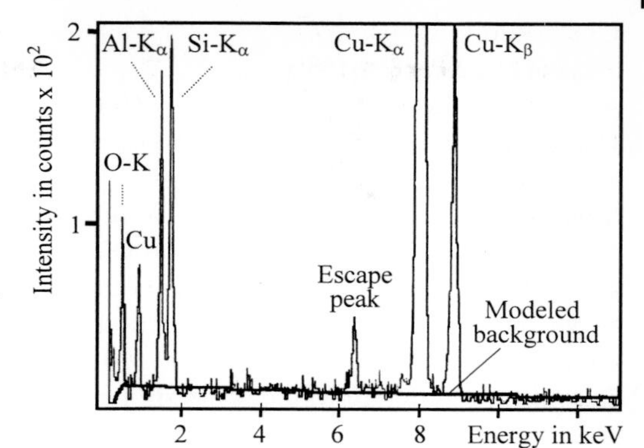

Fig. 4.28. Quantification of EDS spectra (background fit and subtraction).

where the corresponding X-ray intensities I_{Sp} and I_{St} emerging from the sample and standard, respectively, must be measured under identical experimental conditions. The factor K accounts for the different X-ray intensities generated inside a material and measured by the detector for both the standard and the specimen under investigation. Three contributions to K which must be considered are usually termed the *ZAF* correction (Z – atomic number, A – absorption, F – fluorescence) [4.89–4.91, 4.95, 4.110–4.117]. In detail, this procedure involves:

Z – correction for the different inelastic scattering properties introduced by differences between the mean atomic numbers of the specimen of interest and the standard;

A – correction for differences between X-ray absorption; and

F – correction for corresponding X-ray fluorescence differences.

For an electron-transparent specimen the absorption and fluorescence correction parts can often be neglected, this is the so-called thin-film criterion introduced by Cliff and Lorimer [4.118]. Thus, for a thin specimen containing two elements A and B yielding the net X-ray intensities I_A and I_B, the concentration ratio reduces to:

$$\frac{C_{\mathrm{A}}}{C_{\mathrm{B}}} = k_{\mathrm{AB}} \frac{I_{\mathrm{A}}}{I_{\mathrm{B}}} \tag{4.13}$$

where the atomic-number correction factor, k_{AB}, is also called Cliff–Lorimer factor. This factor varies with the accelerating voltage and depends on the particular microscope and EDXS detector system used. The k_{AB} factors needed for quantification are either calculated from first principles or determined experimentally by use of reference materials; the latter procedure is more accurate. To obtain quantitative results of high accuracy and reliability the influence of absorption and fluorescence effects should, however, also be taken into account for thin films; this procedure is described in detail elsewhere [4.91, 4.119–4.121]. It should be noted that the minimum element concentration detectable by EDXS is approximately 0.01–1 atom% – the value depends on the element to be analyzed in a matrix and is a function of instrumental settings determining the peak-to-background ratio.

4.2.5 Imaging of Element Distribution

The combination of EDXS and scanning or scanning transmission electron microscopy (SEM/STEM) enables imaging of element distribution either along a line or two-dimensionally in a rectangular field of view. Thus, facilities for generating a fine electron probe and scanning it are essential for obtaining information on element distribution. Usually, before mapping of the element distribution, a spectrum of the whole area of interest must be acquired; the peaks of the elements sought are then selected by defining energy windows. While the electron probe is scanned the signals from the different windows are simultaneously stored on a computer, creating a line profile or even building up element-specific images. Careful interpretation of these element-selective profiles and images is sometimes advisable, because artifacts can arise from thickness variations, peak overlap, and the effects of absorption and fluorescence of X-rays. The lateral resolution achievable in EDXS line-profile analysis and maps is strongly determined by such factors as the size of the S(T)EM probe, and the thickness and composition (mean atomic number) of the specimen. For very thin specimens investigated in TEM/STEM microscopes resolution values of several nanometers are attainable when electron probes 1 nm in diameter or even smaller are used.

In EDXS the so-called spectrum-image method [4.122] can also be employed. A series of spectra is taken from a scanned rectangular field resulting in a data cube with its upper plane as the scanned x–y area and the third axis as the X-ray spectrum. Comprehensive information about the chemical composition and element distribution is extractable from this data set by subsequent processing.

An example of interface analysis by EDXS line profiling at high lateral resolution is given in Fig. 4.29. It is of particular importance, because the distribution of light elements like carbon, nitrogen, and oxygen is also revealed. This was possible by means of an Si(Li) EDXS detector (Kevex) with an ultrathin window attached to a dedicated STEM HB 501, from Vacuum Generators, with a cold field-emission cathode.

The TEM bright-field image (Fig. 4.29, upper) shows an interfacial region of a fiber-reinforced SiC composite including the carbon fiber with a pyrolytically deposited graphitic layer, approximately 30 nm thick, and the SiC matrix [4.123]. EDXS line profiles were taken across this layer system in the STEM mode (1 nm probe diameter, 100 kV accelerating voltage). The individual line profiles (Fig. 4.29, lower) furnished detailed insight into the particular processes occurring during processing of the composite. Nitrogen was detected inside the graphitic fiber coating; this was introduced from the environment during chemical vapor deposition. The profile of the Si signal from the matrix in the direction of the fiber indicates, moreover, that silicon diffused towards the fiber. The local maximum in the Si content in the region of the graphitic fiber coating can be correlated to the formation of a compound, probably of the type Si_xN_y or $Si_xC_yN_z$.

Figure 4.30 is an example of X-ray mapping of an (In,Ga)As quantum wire structure using a TEM/STEM Philips CM20 equipped with a thermally-assisted field-emitter and a Ge EDXS detector (Tracor Northern) [4.124]. The cross-section STEM bright-

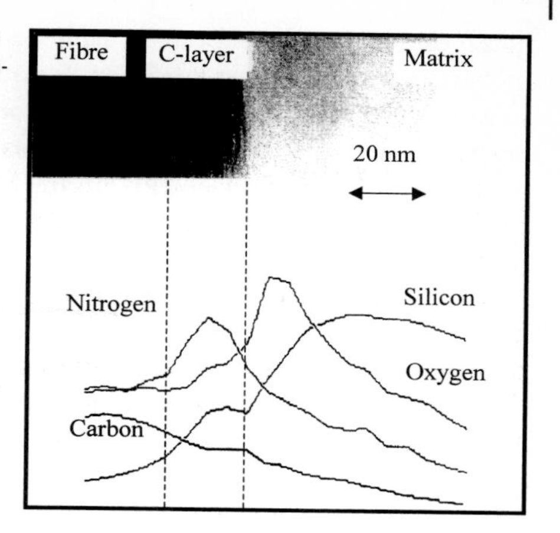

Fig. 4.29. EDXS line-profile analysis across the interfacial region of a C-fiber reinforced SiC composite and corresponding TEM bright-field image.

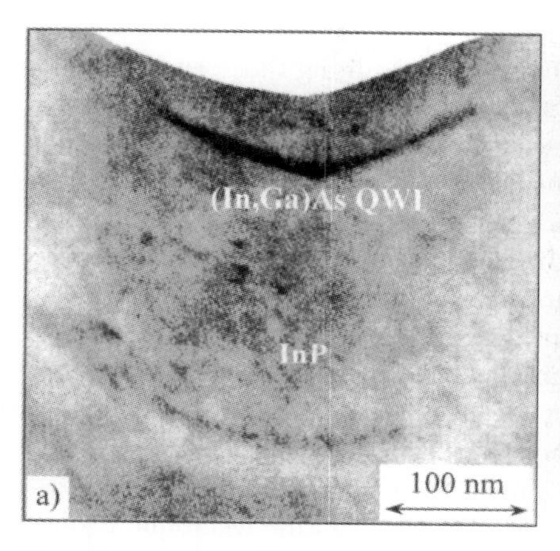

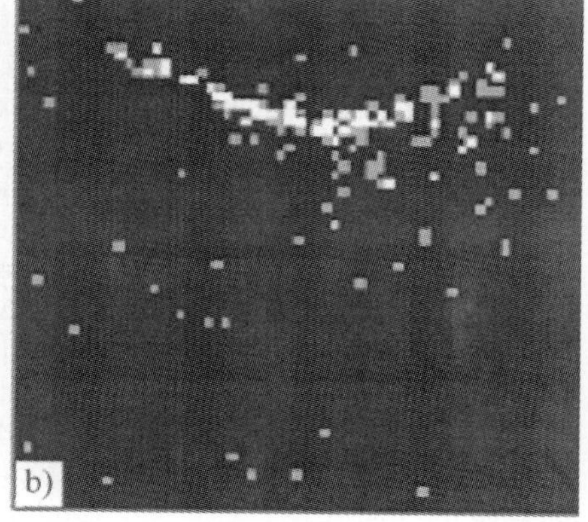

Fig. 4.30. Imaging of element distribution by X-ray mapping: (a) cross-section STEM bright-field image of an (In,Ga)As quantum wire (QWI) in InP; (b) corresponding As-Kα map.

field image (Fig. 4.30 a) exhibits the crescent-like shape of this wire, which was grown on (001) InP by metal–organic chemical vapor deposition (MOCVD). Its width is approximately 240 nm, whereas the extension in the growth direction varies from approximately 10 to 20 nm. Despite of the low signal-to-noise ratio of the corresponding As-Kα map the position of the quantum wire is clearly revealed.

4.2.6
Summary

The combined use of energy-dispersive X-ray spectroscopy and TEM/STEM is a routine method of analytical electron microscopy enabling both qualitative and quantitative chemical analysis of interfaces and interlayers with high lateral resolution. Resolution of few nanometers can be achieved, depending, in particular, on the size of the electron probe and on specimen thickness. Because EDXS is a particular highly sensitive means of detecting elements of medium and high atomic number, it is a useful complement to electron energy loss spectroscopy (EELS), which is predominantly sensitive to light elements. In the STEM mode the element distribution can be imaged as line profile or element map, by use of characteristic X-ray lines.

4.3
Grazing Incidence X-ray Methods for Near-surface Structural Studies

P. Neil Gibson

4.3.1
Principles

X-ray diffraction (XRD) is the classical method for determining the crystalline structure of solid materials. Although the basic theory of X-ray diffraction was developed several decades ago, instrumentation has evolved continuously, with new diffraction geometries opening up possibilities for structural determination in different applications. In particular, as research into surface modification has led to more applications in industry, "glancing angle" or "grazing incidence" geometries have been developed to render X-ray diffraction more surface-sensitive. All major XRD equipment manufacturers now offer systems, or system attachments for *glancing angle X-ray diffraction* (GAXRD). The main purpose of this chapter is to provide a broad outline of the two major glancing angle X-ray diffraction geometries and the differences between these and the standard θ–2θ diffraction geometry (where θ is the incident angle). The technique of *grazing incidence X-ray reflectivity* (GXRR) is also described, because the theory of GXRR is of central relevance here and the technique is often applied on the same systems as used for GAXRD. To complete the picture of glancing angle X-ray structure analysis, *X-ray absorption spectroscopy* is briefly mentioned in this section, even though this technique can be effectively applied only at synchrotron radiation sources, and a full description of it, and its enormous range of applications, would require a lengthy article in itself. GXRR and one of the glancing angle diffraction techniques are described in a little more detail, because these can be applied on normal laboratory X-ray systems. The chapter will not deal in detail with theory, and the reader is referred to the appropriate literature for details about the underlying theory of the techniques described.

4.3.1.1 Glancing Angle X-ray Geometry (GAXRD)

The basic idea of the glancing angle X-ray geometry is rather straightforward. To reduce the penetration of X-rays into a surface and thus limit the depth from which information will be gathered one simply reduces the angle of incidence ϕ of the beam on the sample surface. At angles of incidence higher than a few degrees the intensity I of X-rays within a material as a function of the distance z from the surface is given by $I = I_0 \exp(-\mu z/\sin\phi)$ where I_0 is the intensity at the surface and the absorption coefficient, μ, depends on the composition and density of the material. The depth at which the intensity has fallen to $1/e$ of its value at the surface is often defined as the penetration depth $z_{1/e}$ and is given by:

$$z_{1/e} = \sin(\phi)/\mu. \tag{4.14}$$

At angles of incidence of less than a few degrees the above relationships are not valid. The refractive index of solid and liquid materials at typical X-ray wavelengths is slightly less than unity. Thus at low angles of incidence refraction becomes significant, and below a critical angle, ϕ_c, total external reflection occurs [4.125] and the effective penetration of the beam into the material decreases to a few nanometers. This only occurs for extremely flat and smooth surfaces; the situation for rough surfaces is discussed below. The complex refractive index, n, of a material is usually written $n = 1 - \delta - i\beta$, where δ and β are given by:

$$\delta = (\lambda^2 e^2/2\pi mc^2)\sum_i N_i (Z_i + f_i') \tag{4.15}$$

and

$$\beta = (\lambda^2 e^2/2\pi mc^2)\sum_i N_i (f_i'') \tag{4.16}$$

where e and m_e are the charge and mass of the electron, λ is the wavelength of the X-rays, c is the velocity of light, N_i is the atomic number density of atoms of type i, Z_i is the corresponding atomic number, and f' and f'' are the real and imaginary parts of the anomalous dispersion correction to the atomic scattering factor. The absorption coefficient is related to β by $\mu = \beta\lambda/4\pi$ and ϕ_c is related to δ by $\phi_c \approx (2\delta)^{1/2}$.

The equation for the coefficient of reflection of a flat surface in air at an incident angle ϕ is given by:

$$R(\phi) = \frac{I_r}{I_0}(\phi) = \frac{(\phi - A)^2 + B^2}{(\phi + A)^2 + B^2} \tag{4.17}$$

where

$$2A^2 = \left[(\phi^2 - 2\delta)^2 + 4\beta^2\right]^{1/2} + (\phi^2 - 2\delta) \tag{4.18}$$

and

$$2B^2 = \left[(\phi^2 - 2\delta)^2 + 4\beta^2\right]^{1/2} - (\phi^2 - 2\delta) \tag{4.19}$$

For low angles of incidence, i. e. angles of incidence of a few degrees the $1/e$ penetration depth is given by:

$$z_{1/e} \approx \lambda/4\pi B \tag{4.20}$$

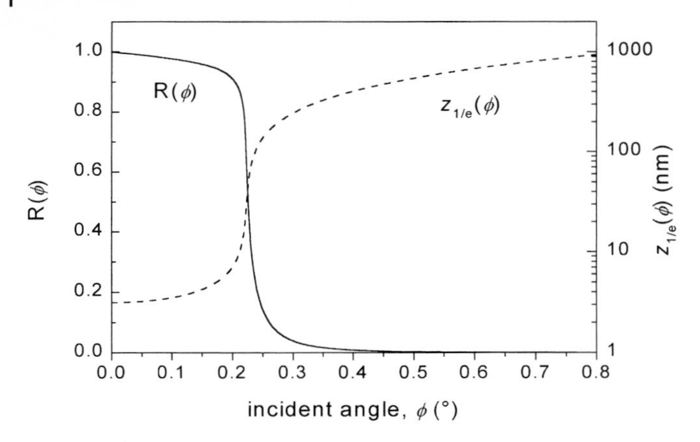

Fig. 4.31. Variation of the coefficient of reflection and penetration depth for X-rays of 1.5405 Å incident on a perfectly flat silicon surface.

Figure 4.31 shows how R and $z_{1/e}$ vary for a perfectly flat silicon surface. The rapid change in penetration depth near the critical angle shows how X-ray techniques become very surface-sensitive at very low incident angles, and can theoretically be used to study extremely thin films, including atomic monolayers. For diffraction studies, however, the intensity of the diffracted beam is drastically reduced at incident angles below ϕ_c, and the range of materials that can be studied in this way with conventional X-ray sources is limited. In addition, to study the structure of the first few atomic layers of a material, the sample must be exceptionally flat and smooth – single crystal semiconductor wafers are ideal. Some examples of such studies are given below.

For slightly rough surfaces (e.g. metals polished mechanically to a mirror finish, with an rms roughness of several tens to hundreds of nanometers) the intensity of the reflected beam below ϕ_c is lower, and as the roughness increases the reflectivity decreases. The non-reflected part of the beam penetrates the surface. It is difficult to derive a formula for calculating average penetration depth at this point, because it will be sensitive to details of the surface topography. It is, however, likely that for the purposes of diffraction analysis Eq. (4.14) rather than Eq. (4.20) can usually be taken as a more accurate estimate, even if a significant fraction of the beam is being reflected. If analysis is made of the reflected beam, as for ReflEXAFS (see below), the surface sensitivity will still be rather good even for slightly rough surfaces, although poorly reflecting surfaces introduce significant experimental difficulties and are best avoided.

4.3.1.2 **Grazing Incidence X-ray Reflectivity (GXRR)**

We have seen that partial reflection of an X-ray beam occurs from a surface at angles of incidence above ϕ_c with the intensity dropping off rapidly with ϕ. In fact partial or total reflection occurs at the interface between any two materials with different refractive indices. For a layered material these reflected beams can combine to form an interference pattern, and analysis of this can provide extremely accurate information about film thickness. In addition, the reflectivity coefficient at each interface depends on its roughness and on the difference between the densities on each side of the interface. Thus the technique of GXRR, i.e. the analysis of plots of R against ϕ, can provide information on all these properties. Kiessig [4.126, 4.127] was the first to note interference effects in X-ray reflectivity, and interference fringes seen in GXRR plots are often referred to as "Kiessig fringes". He derived a formula for the position of the interference maxima as a function of the film thickness, although nowadays this is not normally used in GXRR analysis, having been replaced by full mathematical models combined with least-squares fitting techniques that can be run on modern computers. The basic theory for modern GXRR analysis was developed by Parratt in 1954 [4.128].

4.3.1.3 Glancing Angle X-ray Diffraction (GAXRD)

For details about general X-ray diffraction theory the reader is referred to the mass of material on the subject produced over the last half century. Here we are concerned only with the application of the technique in the glancing angle geometry, and the appropriate considerations that must be made. The basic aim of a diffraction measurement is to detect diffraction peaks from the material under investigation. Diffraction only occurs when the incident angle of the X-ray beam on a set of planes is such that (for a particular wavelength) Bragg's law is satisfied and constructive interference occurs. The diffracted beam from that set of crystalline planes is at the same angle to those planes as the incident angle. Reversing this argument, it is immediately apparent that any diffraction peak will arise only from crystals or crystallites oriented in a particular manner. In the standard XRD geometry, also known as the θ–2θ or Bragg–Brentano, geometry, the incident angle on the sample surface is maintained at the same angle as the detector. Thus any crystalline planes detected will lie parallel to the surface of the sample. This is not true for the two main glancing angle X-ray geometries – the *grazing incidence angle asymmetric Bragg* (GIAB) geometry [4.129] and the *grazing incidence X-ray scattering* (GIXS) geometry [4.130]. These are shown schematically in Fig. 4.32, with the standard θ–2θ geometry. It is also worth mentioning, in passing, the asymmetric parafocusing Seeman–Bohlin geometry [4.131], developed as a surface-sensitive diffraction technique, but now not generally employed. This technique maintains a parafocusing geometry at low angles of incidence by moving the detector around a focussing circle defined by the X-ray source and the sample surface, the latter being at a tangent to the focussing circle. The focussing circle increases in radius as the angle of incidence is lowered, so the technique cannot be used below approximately $\phi = 5°$.

We consider first the GIAB geometry, as this is the most often applied using standard laboratory sources, and the most widely available commercially. The discussion above makes it quite clear that the angle of the crystalline planes that contribute to any particular diffraction peak are inclined relative to the sample surface, and that the angle of inclination changes as the detector is scanned. This makes this geometry much more suitable for studying polycrystalline surfaces than for studying single

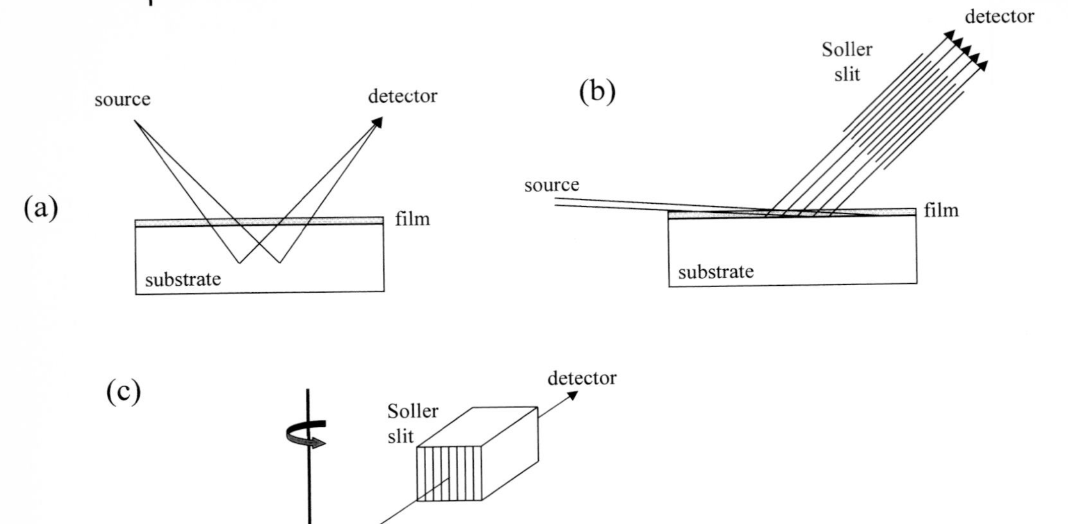

Fig. 4.32. Schematic diagrams of three diffraction geometries: (a) θ–2θ, (b) GIAB, (c) GIXS.

crystal surfaces (e.g. epitaxial films), because for single-crystal studies the orientation of the sample relative to the incident beam is an important variable. Where polycrystalline surfaces with no texture (or preferential orientation of the crystallites) are under investigation, GIAB will produce a diffraction pattern very similar to that of identical material in the bulk form examined by θ–2θ diffraction, apart from slight differences in peak intensities, because of absorption. Thin film materials are, however, usually textured to some extent, and often very strong preferential orientations are apparent. In these circumstances very different peak intensities will be recorded. Experimental techniques and procedures that can be employed for such work are described below. With regard to penetration depths, the GIAB geometry applied just above the critical angle for total external reflection is approximately two orders of magnitude more surface sensitive than the standard θ–2θ X-ray diffraction method, although it should be remembered that this occurs together with significant loss in intensity, because of the restricted dimensions of the incident beam, the lower penetration depth, and the non-focussing geometry. At angles of incidence below ϕ_c the penetration depth is only a few nanometers for very smooth surfaces and most materials can only be studied in this way with high intensity sources. As will be described below, however, slightly rough surfaces of some materials can be studied successfully below ϕ_c using laboratory sources.

In contrast to the classical θ–2θ and GIAB geometries, the GIXS geometry is ideal for studying sets of crystalline planes that are perpendicular to the sample surface.

Ideally the incident angle and the detector elevation angle should both be set at the critical angle to the surface material. In this way the refracted X-ray beam travels parallel to the surface and the surface evanescent wave intensity is maximized. The restrictions of GIXS geometry greatly reduce the beam intensity at the detector, and thus the GIXS is totally inappropriate for studying randomly oriented polycrystalline surfaces with a laboratory X-ray source. It can be used on strongly textured polycrystalline surfaces for determining texture or strain, but such measurements can be made in other ways. Where GIXS really comes into its own is in the study of single-crystal surfaces and epitaxial films. With typical sealed laboratory sources only strongly scattering materials can be studied in the GIXS geometry and it is desirable that several nanometers of material are present to achieve a good signal/noise ratio. With GIXS facilities at synchrotron radiation sources the structure of fractions of monolayers can be studied, even those of relatively weakly scattering materials.

4.3.1.4 **ReflEXAFS**

The technique of *extended X-ray absorption fine structure* (EXAFS) [4.132, 4.133] is based on analysis of variations in the absorption coefficient of a material observed in a range of several hundred eV above the absorption edge of one of the atomic components of the material. Because the incident energy must be scanned in an EXAFS experiment, high intensity over a wide energy range is required and EXAFS is normally applied only at synchrotron radiation sources, where it is now a commonplace technique. When an inner shell electron is ejected from an atom the outgoing photoelectron has a wavelength that depends on its energy. Scattering from surrounding atoms occurs and effectively leads to constructive or destructive interference effects that modulate the measured absorption coefficient. Analysis of these modulations involves background subtraction and then analysis with an appropriate computer program to extract information about the distance and type of the surrounding atoms. EXAFS is a very powerful technique for local structural analysis around particular atomic species of a sample. It is often combined with analysis of the near-edge region, this being called *X-ray absorption near-edge structure* (XANES) or *near-edge X-ray absorption fine structure* (NEXAFS), from which information is obtained on the oxidation state and the symmetry of the local environment of the atomic species under investigation.

Two main techniques are used to render EXAFS surface sensitive. Firstly, the current from surface photoelectron yield can be monitored [4.134]. Because the escape depth of the photoelectrons is only a few nanometers this is effectively the region from which information is obtained. This technique, usually called *surface* EXAFS, or SEXAFS, is normally performed with the sample in a vacuum chamber. Secondly, the technique can be employed in air or, indeed, in a special environmental chamber by using the grazing incidence geometry [4.135, 4.136] ; this technique is termed *reflection* EXAFS or ReflEXAFS. Typically an incident angle of $\phi_c/2$ is used and the intensity of the reflected beam as a function of incident beam energy is analyzed. The reflected beam contains information only from the surface of the sample, even for slightly rough surfaces. The technique can also be applied by monitoring X-ray fluorescence instead of the reflected intensity [4.137]. This has advantages in some situations, but care should be taken if the sample is slightly rough, because the fluorescence detection method will rapidly lose its surface sensitivity as the reflectivity coefficient decreases.

4.3.2 Experimental Techniques and Data Analysis

In this section we consider only GXRR and GIAB, these being techniques that can be applied using sealed laboratory X-ray sources, and that are available commercially from several suppliers of X-ray equipment. GIXS and ReflEXAFS are rather more specialized and need to be carefully studied before use. In addition they usually require high-intensity sources and are normally only employed at synchrotron radiation facilities, although rotating anode laboratory sources and even sealed tubes can be sufficiently intense for some GIXS applications.

4.3.2.1 Grazing Incidence X-ray Reflectivity (GXRR)

GXRR can be applied only to surfaces that are extremely smooth and flat. An rms roughness of greater than a few nanometers will render the technique unusable. It is thus usually applied only to films deposited on to single-crystal silicon substrates, or float glass, where the effective roughness can be from zero to a few Angstroms. It can also be applied to liquid surfaces. Because of absorption in the material under examination and/or the instrumental resolution of the measuring apparatus, for film thickness evaluation it can only be applied to thin films or multilayers of less than a few hundred nanometers thickness. Thus the range of applications is limited, and sample preparation is of critical importance. The information obtained from GXRR is, however, difficult or impossible to derive using other techniques, as the examples outlined below illustrate.

Several experimental methods can be employed for GXRR. Figure 4.33 shows a very generalized schematic set-up. For many applications it is of great importance that the intensity of the reflected beam can be measured over several orders of magnitude, so background radiation must be reduced to an absolute minimum. On dedicated GXRR systems a monochromator in the incident beam is normally used to remove unwanted X-ray wavelengths and, in combination with slits, to collimate the incident beam. Often the $K\alpha_1$ emission line of a Cu anode source (1.5405 Å) is employed. Two concentric goniometers are generally used with the sample mounted on

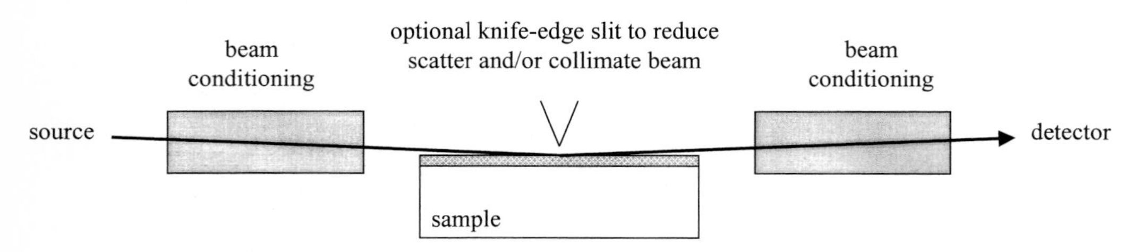

Fig. 4.33. Schematic diagram of the experimental arrangement for GXRR measurements. Beam-conditioning optics can include slits, monochromators, Goebel mirrors, and evacuated beam tubes, etc.

one and the detector on the other. A slit is placed in front of the detector to define the angle of the reflected beam. Alternatively a second monochromator–slit arrangement can be used for this purpose, with the advantage of further reducing unwanted background radiation. Special evacuated beam tubes can be inserted to reduce air scatter of the beam. Careful alignment of the whole system is required to ensure the results are reliable. Any distortion of the curve will make analysis almost impossible.

On several systems somewhat less rigorous beam "conditioning" is employed. These systems may be completely adequate for most studies but might not achieve good enough performance for particular applications. Monochromation might be limited to a monochromator crystal placed before the detector, and may enable both the $K\alpha_1$ and $K\alpha_2$ wavelengths to reach the detector. On at least one commercially available system an interesting solution is used to achieve both collimation and sample alignment – a simple "knife edge" is positioned near the sample center. No incident beam monochromator is used. This has the advantage of simplifying alignment, in particular for small samples. A relatively recent development is in the use of a Goebel mirror on some systems to achieve collimation and some degree of wavelength discrimination. As with any other expensive scientific instrument, before investing in a GXRR system it is important to consider its suitability for the types of sample to be investigated.

GXRR data analysis is usually based on theory developed by Parratt [4.128]. He derived a recursive formula, based on the Fresnel equation and Snell's law of refraction, for the coefficient of reflection of multilayered surfaces. This, however, could be applied only to perfectly flat surfaces or multilayers. Following the later work of Cowley and Ryan [4.138] and of Névot and Croce [4.139, 4.140] roughness can also be taken into account in the analysis. The approach of Cowley and Ryan was to apply a roughness-related Debye–Waller type factor to the reflectivity from each interface. Névot and Croce, on the other hand, developed a more rigorous method by generating sublayers at each interface to simulate the gradual change of optical constants. While the latter approach is occasionally slightly more accurate, and certainly easier to relate to the rms roughness, that of Cowley and Ryan is less intensive on computer processing time, and is more often used.

4.3.2.2 **Grazing Incidence Asymmetric Bragg (GIAB) Diffraction**

The discussion above on roughness and penetration depth makes it clear that ultrasmooth samples are not necessary for GIAB, and great improvements in surface sensitivity can be achieved even with rather rough surfaces. Very rough surfaces will cause some experimental problems and uncertainty, as will curved surfaces, because the incident angle will not be well defined and alignment might prove difficult. Samples should, therefore be, flat over an area of at least a few cm^2 and not too rough – preferably they should appear at least slightly shiny. As mentioned above the optimum incident angle and beam dimensions depend on the sample under investigation. A rough calculation should be made of the absorption coefficient of the material so the incident angle can be set to achieve the desired penetration depth. The beam dimensions can then be adjusted to maximize the X-ray "footprint" on the sample, bearing in mind that the beam should only hit the surface of the sample, and the footprint size should be less than the aperture of both the Soller slit and the detector-active area, to avoid distortion of the collected XRD pattern. Such distortion might occur because the Soller slit and detector "see" more of the sample at low 2θ angles than they do at high 2θ angles. It should be borne in mind that at low incident angles on very smooth surfaces one should make a correction to the measured 2θ value [4.129] to account for refraction.

Special experimental procedures are necessary for samples that have a strong texture, or preferred crystallite orientation. If the texture is very strong it might happen that no crystalline planes are aligned relative to the incident beam to satisfy the Bragg condition and no diffraction peaks from the surface layer are observed. The best procedure then is to try a θ–2θ scan, because a very strong texture will render the diffraction peak from the preferentially oriented planes intense enough to be detected, even from rather thin films. This might fail if the film is too thin or if the preferred orientation is associated with an unallowed diffraction peak. An example of the latter would be the preferred 111 orientation of a bcc metal such as iron or chromium, although the 222 peak should appear if the diffractometer can run to a high enough 2θ angle in the θ–2θ geometry. If the preferred orientation cannot be determined in the θ–2θ geometry the only procedure left is to use different incident angles to try to find an asymmetric geometry where the Bragg condition is satisfied for some set of planes. If the surface crystalline structure is not known, this must be found by trial and error. If the probable crystalline structure is known then it is not difficult to calculate the appropriate angle of incidence for different sets of planes under the assumption of a particular preferred orientation. Weak texture in a surface layer will manifest itself in different peak intensity ratios than those of a randomly oriented material. Such samples can prove to be the most difficult with regard to identification of the direction of preferred orientation, especially if the film is too thin for detection at high angles of incidence. It is possible [4.141], assuming a certain direction and spread of orientations, to calculate how the relative peak intensities will be modified at different glancing angles, although this is not always straightforward.

The GIAB geometry is a non-focussing geometry and requires a Soller slit between the sample and detector. This leads to instrumental resolution considerably poorer than that of modern θ–2θ diffraction systems. A typical GIAB system can have a resolution of 0.1–0.2°. For simple phase identification this causes no real difficulty. If, however, line profile analysis is used to determine crystallite size or average microstrain [4.142, 4.143] care should be taken in drawing conclusions, especially where the line broadening is of the same order of magnitude as instrumental broadening. The often noisy nature of GIAB patterns only serves to compound the problem, because separating crystallite size from microstrain broadening requires reliable determination of the peak shape and width.

Macrostrain is often observed in modified surfaces such as deposited thin films or corrosion layers. This results from compressive or tensile stress in the plane of the sample surface and causes shifts in diffraction peak positions. Such stresses can easily be analyzed by standard techniques if the surface layer is thick enough to detect a few diffraction peaks at high angles of incidence. If the film is too thin these techniques cannot be used and analysis can only be performed by assuming an unstrained set of lattice parameters and calculating the stress from the peak shifts, taking into account the angle of the detected sets of planes relative to the surface (see discussion above). If the assumed unstrained lattice parameters are incorrect not all peaks will give the same values. It should be borne in mind that, because of stoichiometry or impurity effects, modified surface films often have unstrained lattice parameters that are different from the same materials in the bulk form. In addition, thin film mechanical properties (Young's modulus and Poisson ratio) can differ from those of bulk materials. Where pronounced texture and stress are present simultaneously analysis can be particularly difficult.

4.3.3 Applications

In this section examples of the application of the various techniques described are presented. The works mentioned have been selected simply to illustrate the wide range of areas of application and in general early rather than more recent examples have been quoted. The reader will find a huge number of other examples in the literature.

4.3.3.1 Grazing Incidence X-ray Reflectivity (GXRR)

X-ray reflectivity is now routinely used for surface and thin film analysis. Applications include simple determination of film thickness, surface density determination, surface and buried interface roughness evaluation, structural studies of Langmuir–Blodgett films, and studies of ion-beam or thermal mixing, etc. The first GXRR studies of Kiessig [4.126, 4.127] showed how film thickness could be determined from X-ray reflectometry measurements. Parratt [4.128] developed the theory of how to apply the technique to multilayer samples to extract an electron density profile of the surface region. In 1959 X-ray reflectometry was used to determine the density of copper films [4.144], and in 1960 the annealing and oxidation of evaporated films of Cu, Ni, Ge, and Se was studied [4.145]. The technique remained relatively obscure, however, until studies in the nineteen-seventies [4.139, 4.140] enabled surface and interfacial roughness to be taken into account in the analytical procedures. It was at this point that the potential of GXRR became apparent and the technique started to attract more attention.

In 1979 diffusional alloying of Ag–Al thin film couples was studied [4.146], and the results supported a particular model for the alloying process and enabled the calculation of the atomic diffusion coefficients. In 1987 Cowley and Ryan studied the growth of thermal oxides on silicon wafers and presented their model for incorporating interfacial roughness effects into GXRR theory [4.138]. This model is now widely used in GXRR analysis, although it is not the only model available. In 1988 Le Boité et al. studied the ion beam mixing of Ni/Au and Ni/Pt multilayers [4.147], and in a separate paper compared GXRR with Rutherford back-scattering as a method for studying ion-beam mixing of thin films [4.148]. An example of GXRR patterns of Ti/BN multilayers before and after ion-beam mixing [4.149] is shown in Fig. 4.34. These curves can be analyzed by using an appropriate computer program to monitor how

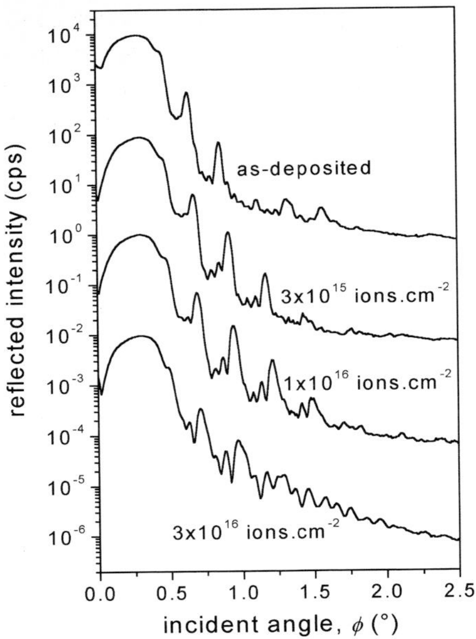

Fig. 4.34. GXRR patterns of Ti/BN multilayers before and after ion-beam mixing with various ion fluences.

the ion beam causes interfacial mixing and surface sputtering. GXRR has often been used to assess the quality or performance of multilayer optics for X-ray or X-UV systems. In 1988 Névot at al. [4.150] performed such a study on W/C multilayer stacks of up to 40 periods of between 3 nm and 6 nm.

Perhaps the most useful and potentially important application of GXRR is in the study of surface modifications that are either difficult or impossible to assess structurally by the use of other methods. Such systems include Langmuir–Blodgett films and grafting of molecules on to polymer surfaces for biocompatibility improvement. Examples of the former are studies of manganese stearate deposited on silicon wafers [4.151], of lead stearate films [4.152], of arachidic acid monolayers [4.153] directly on water, and of amphiphilic cyclodextrins on silicon [4.154]. Surface grafting of bioactive molecules, e.g. heparin, on to the surface of bio-implants, to improve biocompatibility, or perhaps to form specialized biosensors, has great potential for future health-care technologies. GXRR is one of the few methods by which the structure of such surfaces can be studied. Application of the method in this field is still rather new and it will be interesting to follow developments in the coming years.

4.3.3.2 Grazing Incidence Asymmetric Bragg (GIAB) Diffraction

Perhaps the most widespread use of GIAB glancing angle XRD systems is in the study of the structure of deposited solid films. It can also be employed in a variety of other studies – surface corrosion, ion-beam modification, etc. The technique was described by Lim et al. in 1987 [4.129]; they employed the high intensity of a synchrotron radiation source to apply it at incident angles near and below ϕ_c for studying thin iron oxide layers, and proposed a method of correcting the diffraction patterns for refractive index effects. GIAB studies of sputtered iron oxide layers had also been

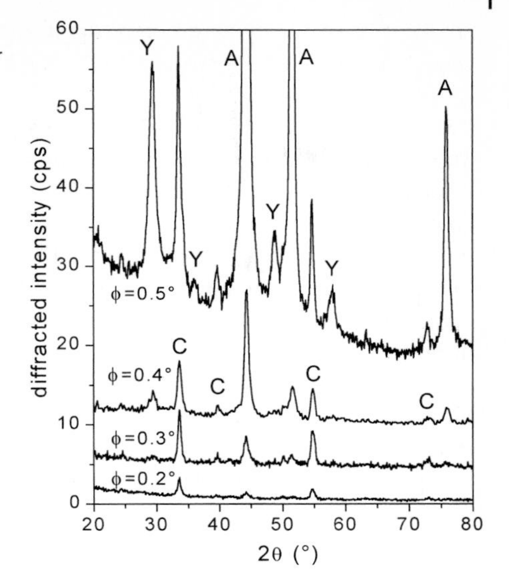

Fig. 4.35. GIAB depth profiling of yttrium ion-implanted NiCr that had been oxidized in air for 8 min at 700 °C. A = alloy substrate, C = Cr_2O_3, Y = Y_2O_3.

reported using a laboratory rotating anode system [4.155]. Subsequently the application of the technique at angles just above ϕ_c, to maximize the diffracted intensity while maintaining reasonable surface sensitivity with sealed laboratory sources became more widespread.

An example of the application of the technique for studying thin corrosion layers is that of yttrium ion implanted NiCr that had been subjected to oxidation in air for 8 min at 700 °C [4.141, 4.156]. Standard θ–2θ diffraction patterns on this sample revealed only diffraction peaks from the face-centered cubic alloy substrate. Figure 4.35 shows GIAB patterns obtained from the sample at four angles of incidence. At the higher angle of 0.5° oxide peaks can be observed; these indicate the presence of both Y_2O_3 and Cr_2O_3 near the surface. At 0.4° the relative intensities of the Y_2O_3 and Cr_2O_3 peaks change. Reducing the incident angle to 0.3° or even 0.2° shows that the hexagonal Cr_2O_3 is present as a surface layer whereas the Y_2O_3 is embedded in the alloy. Such low incident angles are actually below the critical angle for total external reflection of Cr_2O_3 and the mechanical polishing of the sample before oxidation would have enabled rather deeper penetration than for a perfectly smooth sample, thus increasing the diffracted intensity. Despite this, the reduction in diffracted intensity with decreasing incident angle is very apparent. Fifteen hours were needed to collect the spectra shown. Analysis of the relative peak intensities revealed that the Cr_2O_3 had a pronounced 001-preferred orientation [4.141]. A subsequent Auger depth profile of the sample revealed that the thickness of the Cr_2O_3 surface layer was only a few hundred Angstroms.

GIAB studies of sputtered thin films of different composition for tribological applications have been reported [4.157–4.159]. The technique has been used to study the structure of very thin CdS layers (deposited by chemical bath deposition) for photovoltaic applications; in combination with θ–2θ diffraction it enabled identification of their polytype structure [4.160]. Glancing angle diffraction in the GIAB geometry has been used for many different applications and the interested reader will find a large number of other examples in the literature.

4.3.3.3 **Grazing Incidence X-ray Scattering (GIXS)**

The first description of the GIXS technique was that of Marra et al. [4.130], who presented a structural study of the interface between a GaAs single crystal substrate and an epitaxial Al layer grown by molecular beam epitaxy (MBE). In this study a 60-kW rotating anode source was used and GIXS normally requires high-intensity sources. Most synchrotron radiation sources now have dedicated GIXS facilities, sometimes combined with specialized deposition chambers so that the growth of thin epitaxial films can be studied in-situ. A variety of such studies has been reported, with analysis even at the initial sub-monolayer growth stage. Many of these are studies of MBE film growth. An example of a GIXS study of monolayers formed by a different mechanism is the analysis by Samant et al. of Pb monolayers on Ag 111 and Au 111 electrode/electrolyte interfaces [4.161].

Another important application of GIXS is in the structural analysis of reconstructed surfaces; examples of such studies include analysis of Ge 001 and Au 110 surfaces [4.162, 4.163]. GIXS analysis of the melting of Pb monolayers on Cu 110 surfaces has been reported [4.164]. GIXS analysis of the Si–SiO_2 interface was reported by Fuoss et al. [4.165], and several other examples of GIXS applications have been described by Segmüller [4.166]. GIXS diffraction studies of Langmuir–Blodgett films [4.152] and also of oriented polycrystalline Cr_2O_3 layers [4.167] have also been reported (the latter study employed a sealed laboratory X-ray tube). Many other examples of GIXS surface diffraction analysis can be found in the literature.

4.3.3.4 **ReflEXAFS**

ReflEXAFS can be used for near-surface structural analysis of a wide variety of samples for which no other technique is appropriate. As with EXAFS, ReflEXAFS is particularly suited for studying the local atomic structure around particular atomic species in non-crystalline environments. It is, however, also widely used for the analysis of nanocrystalline materials and for studying the initial stages of crystallization at surfaces or interfaces. ReflEXAFS was first proposed by Barchewitz [4.135], and after several papers in the early nineteen-eighties [4.136, 4.168–4.170] it became an established (although rather exotic) characterization technique. Most synchrotron radiation sources now have beam-lines dedicated to ReflEXAFS experiments.

ReflEXAFS studies of the passivation of Ni electrodes were reported by Bosio et al. [4.171]. Studies of the effect of oxygen on interfacial reactions in Al–Ni bilayers were made by Chen and Heald in 1989 [4.172], and the technique has been applied in the field of long-term storage of nuclear waste by analysis of the local structure around uranium in leached waste-containing borosilicate glasses [4.173]. It has been used to study the oxidation of stainless steel [4.174] and to investigate the so-called "active element effect" in chromia-forming alloys by analysis of the local atomic structure around yttrium and chromium during the initial stages of oxidation of yttrium ion implanted NiCr and Cr [4.175, 4.176]. Fluorescence detection of surface EXAFS was described by Heald et al. in 1984 [4.137]; an example of the application of this technique is the study of the local atomic structure around iron dispersed in polymer coatings on steel [4.177]. As with the other techniques described in this article, few examples of the huge variety of applications of ReflEXAFS have been mentioned here; the reader will find many more in the scientific literature.

4.4 Glow Discharge Optical Emission Spectroscopy (GD–OES)

Alfred Quentmeier

GD–OES is becoming one of the most important techniques for the direct analysis of solids, at their surface, through their depth, and in their bulk. The ease and speed of operation, high sensitivity to all elements, accuracy of analysis and breadth of application make GD–OES a very versatile and powerful technique. GD–OES has proven its capability, in particular, for the rapid and reliable analysis of metallic coatings, oxide scales, diffusion and alloy layers, and can thus be used for the quality control of technical products during different stages of heat treatment, e. g. nitriding or nitrocarburizing steps, hard coating, or PVD processes.

4.4.1 Principles

Glow discharges as used in this context are low-pressure electrical discharges in a noble gas at pressures from 10^2–10^3 Pa [4.178]. It is well known that the glow discharge in the discharge chamber comprises several alternating dark and luminous zones. The complicated structure can be altered, depending on the geometry of the chamber, especially the distance between the two electrodes. In general, when the electrode distance is relatively long, the positive column (the bright zone near the anode) is the most prominent of the zones in the discharge. It is found that when the space between the electrodes is reduced, the positive column shrinks and finally disappears whereas the negative glow (the bright zone near the cathode) and the cathode dark space are hardly affected. Both negative glow and cathode dark space are essential to maintain the discharge and cannot be omitted. Under these conditions the discharge is called an "obstructed" glow and can be realized when the inter-electrode separation is just a few times the thickness of the cathode dark space.

Discharge tubes working under obstructed glow conditions are employed extensively as excitation sources for optical emission spectrometry. For this purpose the analytical sample of interest serves as the cathode. During the discharge, the cathode is bombarded by positive ions formed from the noble gas. This bombardment induces surface erosion called cathodic sputtering – atoms, electrons and ions are removed from the surface of the material constituting the cathode. This sputtered material then participates in different phenomena occurring in the discharge. The result is emission of photons corresponding to the characteristic spectral lines of the elements in the plasma. By this means the glow discharge furnishes information on the elemental composition of the analytical material being removed layer-by-layer from the cathode. The sputtering rates change from 0.1 nm s^{-1} up to more than 100 nm s^{-1}, depending on the cathode material and the excitation conditions used. The glow discharge thus enables rapid and detailed characterization of the near-surface layer, interfaces, and bulk composition, if the spectral line intensities of the elements of interest are recorded simultaneously.

4.4.2 Instrumentation

4.4.2.1 Glow Discharge Sources

If the potential applied across the cathode and the anode is constant (dc), conductive materials can be analyzed; if the potential is varying at radio frequency (rf), both conductive and non-conductive materials can be analyzed.

The most important dc-discharge source used to date in GD–OES is the Grimm-type source [4.179]. The principle of the Grimm source is illustrated in Fig. 4.36. In the Grimm source the anode is no longer a flat plate but a hollow tube, the annular face of the tube stopping only 0.1–0.2 mm from the surface of the sample. By this means the discharge is restricted to a sample area which is of equal size to the aperture of the anode tube (typically 4–8 mm). When the discharge current is sufficiently high, i.e. in the so called "abnormal" glow regime, the whole area is covered by the discharge, which results in very efficient cathodic sputtering. The sample (which acts as the cathode) is mounted against the discharge chamber and can thus be removed very easily. To maintain the low pressure conditions required, the sample must be vacuum tight and seal the discharge chamber by means of an O-ring. Normally flat samples with sufficiently low surface roughness will be used, although samples with other regular shapes (rods or wires) can also be used with special sample holders and a discharge chamber with a modified front plate. The chamber is evacuated, usually by means of a rotary pump, with application of a differential pumping scheme with a low pressure region in the narrow space between the anode tube and the front (cathode) plate. A noble gas (mostly argon) is flushed continuously through

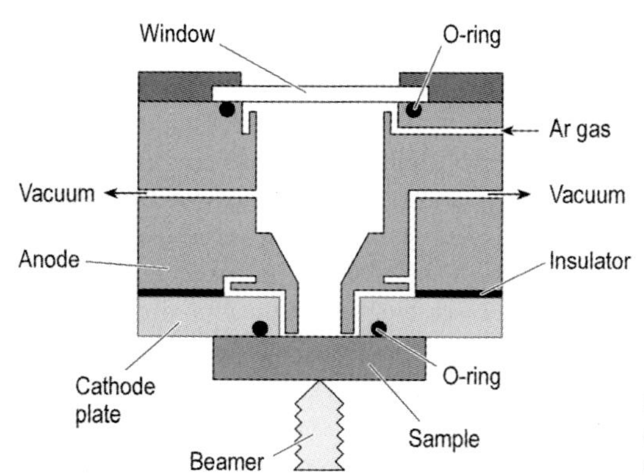

Fig. 4.36. Diagram of a typical glow discharge device used for GD–OES depth profiling [4.189].

the chamber. The pressure inside the source during sputtering is typically 0.3–1.3 × 10^3 Pa and is monitored carefully. The optical radiation of the excited sample atoms is observed end-on through a window with the required transmittance. MgF_2 is most commonly used as window material.

Besides the conventional Grimm-type dc source, which has dominated the GD–OES scene for approximately 30 years, other discharge sources are well known. Among those are various boosted sources which use either an additional electrode to achieve a secondary discharge, or a magnetic field or microwave power to enhance the efficiency of excitation, and thus analytical capability; none of these sources has, however, yet been applied to surface or depth-profile analysis.

The introduction of rf-powered sources has extended the capability of GD–OES to non-conductors, and several rf sources of different design have become commercially available. This is of the greatest importance for surface and depth-profile analysis, because there exists a multitude of technically and industrially important non-conductive coating materials (e. g. painted coatings and glasses) which are extremely difficult to analyze by any other technique.

The first commercially available rf GD sources were based on the Grimm design concept, and enabled both rf and dc operation if parts of the source were interchanged. A leading instrument manufacturer recently introduced a dedicated rf source designed by Marcus [4.180]. The Marcus-type source is operated at pressures similar to those of dc discharge sources but the essential characteristic is the conventional 13.56 MHz radio frequency power applied to the back of the sample by an impedance-matching device. For proper operation the sample must be electrically isolated from all metallic parts of the source and ancillary components to reduce rf radiation and conduction losses. In this way the sample no longer acts as the cathode, and the second electrode or counter-electrode (corresponding to the anode in the dc device) is no longer a tube but is represented by the metallic walls of the discharge chamber. The sample area exposed to the discharge is confined by an orifice disk and is much smaller than the counter-electrode, thereby ensuring most of the drop in rf potential occurs at the sample so that no significant sputtering of the counter-electrode occurs.

Continuous discharge and sputtering of the sample are accomplished by the negative dc self-bias voltage, or offset, which is acquired at the surface of the insulator without any external dc potential supplied. This phenomenon is explained by the different mobility of positive ions and electrons which reverse their direction of movement toward the sample with every half-cycle of the bipolar rf potential. The positive charge of the insulator surface during bombardment with ions is easily neutralized and shifted to negative charge by electrons in the next half-cycle, resulting in the observed steady-state negative offset potential and thus the effective sputtering of the sample by positive ions.

A practical drawback of the rf sources is the influence of sample thickness. Basically, the thickness of a dielectric sample directly affects the propagation of rf energy through the sample and thus thicker samples, run at a fixed rf power level, are characterized by reduced sputtering rates and thus reduced analytical signals. This difficulty is overcome by a new rf source design based on the Grimm device and which enables active control of rf power by means of current and voltage probes [4.181].

4.4.2.2 Spectrometer

Apart from the glow discharge source with its associated gas and power supplies, vacuum pumps, and controls, the optical emission spectrometer is the most important part of a GD–OES instrument. To make use of the analytical capability of the discharge source, the spectrometer must have a sufficient spectral resolving power and an adequate spectral range. Depending on the application, a commercial spectrometer can cover the wavelength range from 110–800 nm, which contains the most sensitive lines of all elements including the light elements (Li, Be, B, C, etc.) and the gases (H, O, N, Cl). The high spectral resolution is required to avoid spectral interference with lines of other elements, in particular of the discharge gas argon.

The most widely used spectrometer designs are the Paschen–Runge polychromator configuration for simultaneous spectrometers and the Czerny–Turner monochromator for sequential analyzers. In the Paschen–Runge mount the concave grating focuses the spectral line intensities of preselected elements on to fixed exit slits, which are positioned on the Rowland circle. Up to 64 elements are detected simultaneously in this way. In the Czerny–Turner mount quasi-monochromatic radiation is observed at the exit slit. By turning the plane grating around its axis it is possible to scan a spectral range and to profile an emission line including background radiation and neighboring spectral lines.

4.4.2.3 Signal Acquisition

The optical radiation passing the exit slit(s) of the spectrometer is recorded by a detector system. The best known detector is still the photomultiplier tube (PMT) which is implemented in most commercial spectrometers. In the common polychromator system each detection channel is equipped with a PMT and the corresponding high-voltage supply. Each supply can be adjusted individually to afford appropriate sensitivity for each element required to match the analytical conditions. Because of the linear characteristic of the PMT the output signal is proportional to the radiation power of the spectral line recorded with each detection channel. The analog output signals are amplified, converted into digital information, and transferred to a PC for further data processing. For depth profiling and continuous sample erosion it is necessary to use a fast A/D converter, because the acquisition rate must be high (~1 kHz) to ensure sufficient depth resolution.

The rapid development in the performance of the CCD (charge coupled device) and related detection techniques, e. g. the CID (charge-injection device) makes these detector systems very attractive, especially when combined with the well known compact échelle spectrometer. This configuration results in a small multi-channel spectrometer with high spectral resolving power [4.182]. When CID are coupled with échelle spectrometers these systems have very desirable properties – high sensitivity and a wide dynamic range with very low read noise and almost non-existent dark current. The random access of individual detector sites enables flexible selection of spectral lines [4.183], whereas the new collective readout mode will promise faster readout and improved signal-to-noise ratios as required for depth-profile analysis.

4.4.3 Spectral Information

The emission process in a glow discharge source depends on several obvious factors – the number of atoms in the plasma, the availability of energetic particles such as high-energy electrons and metastable argon atoms which provide the excitation energy by numerous collision processes, and the excitation cross-sections of the competing energy levels of the elements. It can be estimated that the number of argon atoms in the discharge exceeds the number of sputtered sample atoms by a factor of approximately 10^3. Because only a small fraction of sputtered atoms is charged and energetically excited, the discharge plasma is (for a given sample material) dominated mainly by argon atoms. This explains why calibration curves obtained with GD–OES are usually linear over a wide range of elemental concentration. The operating conditions of the discharge, i. e. low pressure and low power input, result in low kinetic temperatures of approximately 500–1000 K and emission lines with small physical widths. The spectral information obtained from the glow discharge source is, therefore, highly specific and selective. With constant excitation conditions, the spectral line intensities are proportional to the number densities of sputtered atoms in the plasma and hence the element concentration in the sample used. To maintain reproducible excitation conditions in the glow discharge source, the working conditions (e. g. argon pressure, dc-current or rf-power) are carefully controlled.

4.4.4 Quantification

Because of the complex nature of the discharge conditions, GD–OES is a comparative analytical method and standard reference materials must be used to establish a unique relationship between the measured line intensities and the elemental concentration. In quantitative bulk analysis, which has been developed to very high standards, calibration is performed with a set of calibration samples of composition similar to the unknown samples. Normally, a major element is used as reference and the internal standard method is applied. This approach is not generally applicable in depth-profile analysis, because the different layers encountered in a depth profile often comprise widely different types of material which means that a common reference element is not available.

The quantification algorithm most commonly used in dc GD–OES depth profiling is based on the concept of emission yield [4.184], R_{ik}, according to the observation that the emitted light per sputtered mass unit (i. e. emission yield) is an almost matrix-independent constant for each element, if the source is operated under constant excitation conditions. In this approach the observed line intensity, I_{ik}, is described by the concentration, c_i, of element, i, in the sample, j, and by the sputtering rate q_j:

$$I_{ik} = c_i q_j R_{ik} \tag{4.21}$$

The sputtered mass, δm_i, of element, i, during time increment, δt, is described by:

$$\delta m_i = I_{ik} \delta t / R_{ik} \tag{4.22}$$

The emission yield, R_{ik}, defined as the radiation of the spectral line, k, of an element, i, emitted per unit sputtered mass must be determined independently for each spectral line. The quantities q_j and R_{ik} are derived from a variety of different standard bulk samples with different sputtering rates. In practice, both sputtering rates and excitation probability are influenced by the working conditions of the discharge. Systematic variation of the discharge voltage, U_g, and current, I, leads to the empirical intensity expression [4.185]:

$$I_{ik} = K_{ik} c_i C_Q I^{A_k} f_k(U_g) \tag{4.23}$$

where K_{ik} is an atomic- and instrument-dependent constant characteristic of the spectral line, k, of element, i, A_k is a matrix-independent constant characteristic of the spectral line, k, $f_k(U_g)$ is a polynomial of degree 1–3, also characteristic of the spectral line k, and C_Q is a constant related to the probability of a sample atom being ejected during the sputter process. The most important reason for the success of the emission yield technique is that the total sputtered mass is easily determined by summing over all the elements present in each depth segment. The emission yield approach has proven to be extremely successful for a large and increasing number of applications and is currently implemented in commercial GD–OES depth-profiling instruments.

In contrast with the dc source, more variables are needed to describe the rf source, and most of these cannot be measured as accurately as necessary for analytical application. It has, however, been demonstrated that the concept of matrix-independent emission yields can continue to be used for quantitative depth-profile analysis with rf GD–OES, if the measurements are performed at constant discharge current and voltage and proper correction for variation of these two conditions are included in the quantification algorithm [4.186].

4.4.5 Depth Profiling

The primary information obtained in GD–OES depth profile measurements is the relative intensity, from the elemental detection channels, as a function of sputtering time. The intensity–time curves obtained for different elements can be converted into concentration–depth curves by applying eqs (4.21) to (4.23) and the sum normalization of all concentrations to 100%. The depth is determined from the sputtered mass, which is the quantity obtained by the emission-yield technique. The density of the composite material is calculated by an approach based on a weighted average of the density of pure elements. This method gives very accurate results for all types of metal alloy but tends to be less accurate for compounds which contains light and gaseous elements (oxides, nitrides, carbides, etc.). In general the accuracy, precision, and repeatability of quantitative GD–OES depth profile analyses are assured on

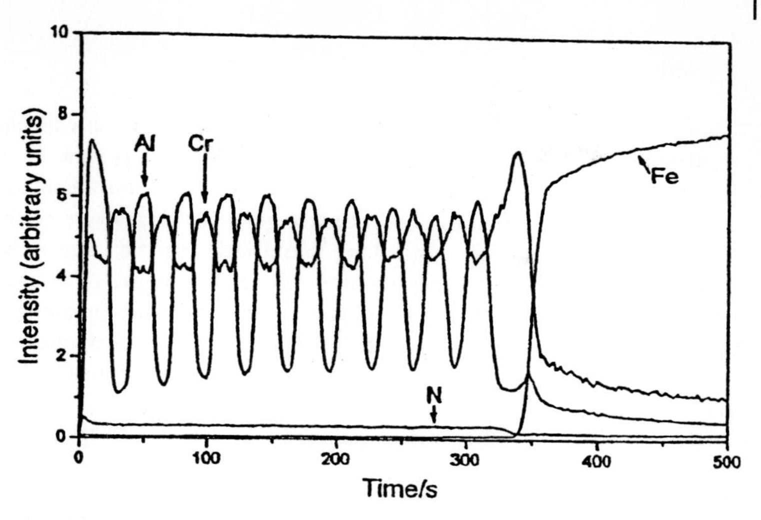

Fig. 4.37. Depth (temporal) profile obtained on a multilayer coating produced by plasma vapor deposition (PVD) using optimized rf-glow discharge conditions. Layer thickness: 30 × (300 nm CrN/300 nm CrAl)/100 nm Cr/steel [4.194].

the basis of inter-laboratory round-robin tests [4.187, 4.188], by use of a layered certified reference material [4.189], and evaluated by comparison with wet chemical analysis [4.188] or other surface techniques [4.190, 4.191].

The information depth achieved by use of the GD technique is determined, in principle, by the depth of penetration of the incident ions, which is in the range of a few nanometers at the relatively low energies employed in the discharge. The practical depth resolution is, however, almost larger and determined by several effects which are introduced by the sputter process (preferential sputtering, atomic mixing), the sample properties (surface roughness, polycrystalline structure), and most seriously by the non-uniform erosion of the sample material. Most of these effects are inherent also in other techniques which use sputtering for surface and depth-profile analysis. In practice, the depth resolution obtained on technical surfaces is roughly proportional to the sputtered depth, and usually deteriorates from the nanometer-range for near-surface layers to the micron-range at depths of several micrometers [4.192, 4.193]. Excellent depth resolution is realized on multilayer coatings, cf. Fig. 4.37, as produced by vacuum evaporation or by plasma vapor deposition (PVD) [4.194], especially if the curved sputtering crater bottom is taken into consideration in quantification by an iterative deconvolution technique [4.195]. The lateral resolution of the GD–OES technique, on the other hand, is restricted by the size of the sputtered area of the sample surface (usually 4–8 mm diameter) and is much larger than with other surface techniques. The lateral and depth resolution of GD–OES are, however, usually both adequate for rapid quantitative determination of the elemental composition of technical surfaces.

4.4.6
Applications

Glow discharges as used in GD–OES offer exciting analytical capabilities – short-term sample-to-sample reproducibility (typically 0.1–1%) for the determination of major and minor elements and detection limits (DL) for most elements in the 0.1–10 ppm range. These figures of merit make glow discharges very powerful atomization and excitation sources which have found wide analytical and technical application. Beside the field of spectrochemistry, where GD has gained renewed interest in bulk analysis [4.196–4.199], it is applied extensively as an ion source in glow discharge mass spectrometry (GDMS) [4.200, 4.201]. Those applications involving depth-profile analysis are the economically most important use of GD–OES today, and will probably become even more important in the future.

In comparison with other surface analytical techniques, GD–OES has the advantage of easy operation, rapid in-depth analysis, wide depth range, and sensitivity to all elements in the periodic table. For these reasons GD–OES depth profiling is of primary interest for industrial use, especially for the quality control of large-scale technical products. The numerous applications [4.202] cover the field of different coating technologies of metallic surfaces, the investigation of corrosion effects, and reactive diffusion processes in layer-bulk interfaces. Some typical applications of the GD–OES technique for depth-profile analysis of conducting and of non-conducting materials are summarized in Table 4.2.

4.4.6.1 dc GD sources

In the past, GD–OES with dc sources was mainly used for steel analysis, and the qualitative characterization of surfaces of cold rolled and hot rolled steels, to control enrichment or depletion of minors and traces during the process of manufacturing or the formation of oxide scales [4.203, 4.204]. Zinc-based coatings are currently of great industrial importance in building materials and materials for car manufacture. Because of the wide industrial application of hot-dip galvanizing and electroplating technology, the rapid and reliable determination of the thickness of the zinc coatings (typically ~5–30 µm) and the amounts of major elements in these layers [4.185, 4.205] is a great challenge to surface and depth-profiling methods. GD–OES has proven its capability in this field and is accepted as a standard method for quantitative depth-profile analysis of zinc-based metallic coatings [4.187]. Other typical applications of GD–OES include the analysis of surface-hardened steel after the process of nitriding or carburizing [4.206], and after different hard-coating processes, i. e. physical vapor deposition (PVD) of TiN layers [4.207], cf. Fig. 4.38.

In addition to the analysis of relatively thick metallic coatings and diffusion profiles, dc GD–OES has also been successfully applied to the analysis of thin protective layers (thickness typically <100 nm), e. g. phosphate and chromate layers on steel [4.185].

Tab. 4.2. Some typical applications of GD–OES depth-profile analysis.

Coating type	*Surface treatment*	*References*
Metallic coatings	Galvanized coatings on steel	4.205, 4.209
	Electroplated Zn coating on steel	4.187, 4.210, 4.216, 4.220
	Electroplated ZnNi coating on steel	4.187, 4.210
	Hot-dipped Zn coating on steel	4.187, 4.210
	AlZn alloy coating on steel	4.205, 4.209
Hard coatings	Nitriding, carburizing	4.206
	TiN, TiC (PVD) layers	4.207, 4.212, 4.221
Oxide scales	Oxide scale on alloyed steel	4.218
Polymer coatings	Painted automotive components	4.209, 4.215, 4.216
Ceramic coatings	SiC on Si_3N_4 ceramics	4.181
	SiO_2/MgO coating on steel	4.213
Thin layers	Anodic alumina films on aluminum	4.211
	Cr layers on a computer hard-disk	4.219
	Ni-P plated aluminum	4.222
	Cu/Cr-Ni superlattice material in silicon	4.213, 4.214
	Cr/Al multilayers on steel	4.194
	Ti/Al multilayers on aluminum alloy	4.189
Coated glass	Brass layer on glass	4.214

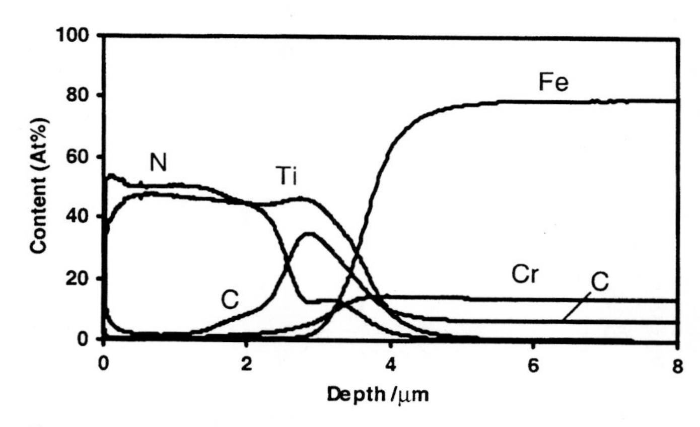

Fig. 4.38. Quantitative depth profile of TiN-coated steel [4.212].

4.4.6.2 rf GD sources

If the rf source is applied to the analysis of conducting bulk samples its figures of merit are very similar to those of the dc source [4.208]. This is also shown by comparative depth-profile analyses of commercial coatings an steel [4.209, 4.210]. The capability of the rf source is, however, unsurpassed in the analysis of poorly or non-conducting materials, e.g. anodic alumina films [4.211], chemical vapor deposition (CVD)-coated tool steels [4.212], composite materials such as ceramic coated steel [4.213], coated glass surfaces [4.214], and polymer coatings [4.209, 4.215, 4.216]. These coatings are used for automotive body parts and consist of a number of distinct polymer layers on a metallic substrate. The total thickness of the paint layers is typically more than 100 μm. An example of a quantitative depth profile on pre-painted metal-coated steel is shown as in Fig. 4.39.

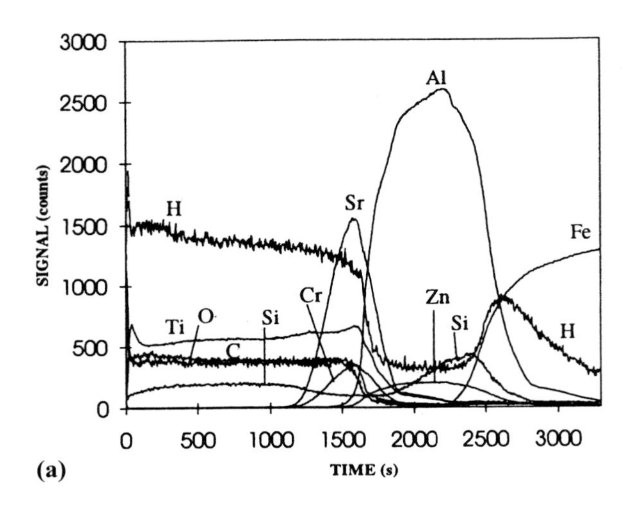

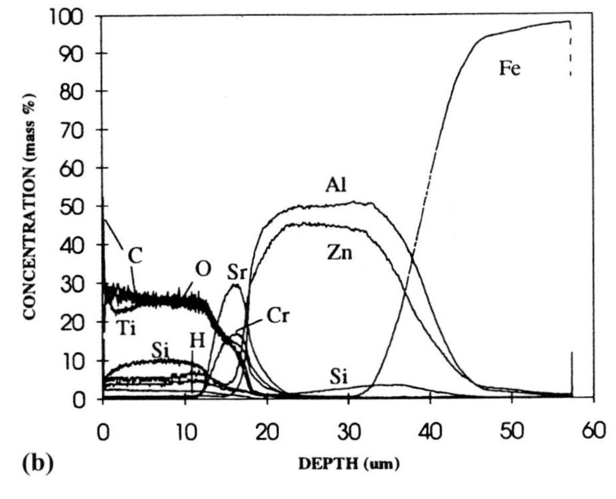

Fig. 4.39. Quantitative depth profile of a commercial pre-painted metal-coated steel sample: (a) as recorded; (b) quantitative [4.209].

In conclusion, GD–OES is a very versatile analytical technique which is still in a state of rapid technical development. In particular, the introduction of rf sources for non-conductive materials has opened up new areas of application. Further development of more advanced techniques, e.g. pulsed glow discharge operation combined with time-gated detection [4.217], is likely to improve the analytical capabilities of GD–OES in the near future.

4.5 Surface Analysis by Laser Ablation

Michail Bolshov

4.5.1 Introduction

Laser ablation (LA) is a popular and widespread sampling technique for direct solid-sample analysis. Several excellent monographs and review articles on the fundamentals of LA and its analytical applications have been published in the past three decades [4.223–4.225]. In contrast with most of the techniques used for surface analysis, LA is a "destructive" technique in which focused laser radiation is used to release material from a solid sample. The main goal of LA is analysis of the elemental composition of a solid sample. The material is removed by evaporation, or a kind of phase explosion, from the overheated spot of a sample. Evidently, under such conditions information on the structural composition of a sample surface and the nature of the binding of the components is lost. During last three decades most efforts were invested in the development of LA as a technique for bulk analysis. Typical values of the masses ablated in a single laser shot vary from several tens of nanograms to a few micrograms, depending on the type of a sample, which corresponds to the thickness of an ablated layer of several hundreds of nanometers to a few micrometers. These probing depths of LA are much larger than for most typical surface analytical techniques – e.g. SIMS, SNMS, TXRF etc. The first results from LA analysis of thin multicomponent and multilayer samples, with better depth resolution than that achieved hitherto, were reported only quite recently.

The advantages of LA are now well-known – no sample preparation is needed, conducting and non-conducting samples of arbitrary structure can be analyzed directly, spatial resolution up to a few microns can be obtained, high vacuum conditions are not required, rapid simultaneous multi-element analysis is possible, and it is possible to obtain complete analytical information with a single laser pulse. A brief overview of the potential and limitations of LA will be given in this chapter.

4.5.2
Instrumentation

4.5.2.1 Types of Laser

So far powerful lasers with picosecond to nanosecond pulse duration have usually been used for the ablation of material from a solid sample. The very first results from application of the lasers with femtosecond pulse duration were published only quite recently. The ablation thresholds vary within a pretty wide interval of laser fluences of 0.1–10 J cm^{-2}, depending on the type of a sample, the wavelength of the laser, and the pulse duration. Different advanced laser systems have been tested for LA:

(1) Nd:YAG laser (1064 nm) with its harmonics (532 nm, 354.7 nm, 266 nm, 213 nm); and
(2) excimer lasers XeCl (308 nm), KrF (248 nm), ArF (193 nm), F_2 (157 nm).

The main characteristics of some Nd:YAG and excimer lasers widely used for LA are listed in Tables 4.3 and 4.4.

Tab. 4.3. Some models of Nd:YAG laser.

Company	*Model*	*Pulse Energy [mJ]*		*Specification of the 4th Harmonic (266 nm)*
		at 1064 nm	*at 532 nm*	
Cetac	LSX-200	50	25	6 mJ, <4 ns, 20 Hz
Atos	MiniLase III-10	90	50	10 mJ, 5 ns, 10 Hz
Atos	Tempest 10	200	100	30 mJ, 4 ns, 10 Hz
Oriel	Brilliant W	360	180	40 mJ, 4 ns, 10 Hz/20 Hz
LaserGate	HYL-101	450	200	50 mJ, 5 ns, 20 Hz
Continuum	Surlite I-20			40 mJ, 5 ns, 20 Hz
Continuum	Surlite I-10			50 mJ, 5 ns, 10 Hz
Soliton	5011-COMP	450	200	50 mJ, 6 ns, 20 Hz

Tab. 4.4. Some models of excimer laser.

Company	*Model*	*Maximum repetition rate [Hz]*	*Pulse energy [mJ]*				
			F_2 157 nm	*ArF 193 nm*	*KrF 248 nm*	*XeCl 308 nm*	*XeF 351 nm*
Tui Laser	ExciStar		1.5	8	18	14	8
	S-200	200					
	S-500	400/500					
	Cera Tube mini	500		8	18	14	8
	Cera Tube midi	100		120	280	150	120
Lambda Physik	OPTex	200		13	22	10	10
	COMPex 102	20		200	300	200	150
	COMPex 110	110		200	300	200	150

In Table 4.3, the Cetac product LSX-200 is the specialized system for coupling with the ICP customer's system. It includes the laser, optical viewing system for exact positioning of the laser focus on a sample surface, and the sample cell mounted on the computer controlled XYZ translation stage. The system is also provided with the appropriate gas tubing for transport of the ablated material into an ICP-OES/MS.

A very important characteristic of laser radiation is the beam shape. So far most LA experiments have been performed with Gaussian laser beams. Lasers with uniform distribution of the beam cross-section have been used only recently to achieve high lateral and depth resolution. Specially designed beam homogenizers must be used for this purpose [4.226–4.228]. The Cetac LSX-200 system has a flat-top distribution of the laser beam.

Although, in principle, lasers with all wavelengths can be used for reproducible and accurate sampling, the experimental conditions, e.g. laser intensity and pulse length, buffer gas, and buffer gas pressure, delay and duration of the gate width for data acquisition must be optimized for specific excitation wavelength. There has recently been a trend toward UV-lasers because of greater absorption of the UV-radiation by most solid materials and less absorption of the UV-laser radiation by the plasma above a sample surface. Both enable more efficient coupling of laser energy to the solid sample. Because of this the lasers most widely used nowadays for LA are the 4th harmonic of Nd:YAG (266 nm) and ArF (197 nm), KrF (248 nm) and XeCl (308 nm) excimer lasers.

The disadvantage of lasers with nanosecond–picosecond pulse duration for depth profiling is the predominantly thermal character of the ablation process [4.229]. For metals the irradiated spot is melted and much of the material is evaporated from the melt. The melting of the sample causes modification and mixing of different layers followed by changes of phase composition during material evaporation (preferential volatilization) and bulk re-solidification [4.230]; this reduces the lateral and depth resolution of LA-based techniques.

If a laser pulse of sub-picosecond duration is used, deposition of the laser energy to the sample is so rapid that the thermal diffusion length is determined by the diffusion of hot electrons before they transfer the energy to the lattice of the solid sample. Less pronounced thermal diffusion provides better lateral and depth resolution and is the basis of successful application of femtosecond pulses in material processing and microstructuring [4.231, 4.232]. All-solid-state femtosecond lasers with a pulse duration of 100–200 fs and a pulse energy of approximately 1 mJ have recently become commercially available [4.233, 4.234].

4.5.2.2 Different Schemes of Laser Ablation

Different analytical techniques are used for detection of the elemental composition of the solid samples. The simplest is direct detection of emission from the plasma of the ablated material formed above a sample surface. This technique is generally referred to as LIBS or LIPS (*laser induced breakdown/plasma spectroscopy*). Strong continuous background radiation from the hot plasma plume does not enable detection of atomic and ionic lines of specific elements during the first few hundred nanoseconds of plasma evolution. One can achieve a reasonable signal-to-noise ratio for the measurement of atomic and ionic spectral line intensities by optimization of the experimental conditions – the type of laser and laser intensity, the type and pressure of buffer gas, and time delay and distance from the sample surface for data acquisition. Detection limits (DLs) in the mg g^{-1} to μg g^{-1} range have been achieved for direct analysis of metals, glasses, and ceramics by LIBS/LIPS techniques for variety of trace elements. Better DLs were achieved when LA was followed by detection of the ablated atoms or ions of a selected analyte by *laser atomic absorption spectroscopy* (LAAS) or by *laser-induced fluorescence* (LIF) *spectroscopy*. Although matrix effects are significant and detection limits are not as good as for more complex hyphenated techniques, LIBS has one major advantage – it enables rapid on line or remote analysis in an industrial environment, including characterization of solid samples in hazardous zones. Use of fibers for both excitation and emission collection enables the construction of a very robust system that allows on-line quality control.

The intrinsic drawback of LIBS is a short duration (less than a few hundreds microseconds) and strongly non-stationary conditions of a laser plume. Much higher sensitivity has been realized by transport of the ablated material into secondary atomic reservoirs such as a microwave-induced plasma (MIP) or an inductively coupled plasma (ICP). Owing to the much longer residence time of ablated atoms and ions in a stationary MIP (typically several ms compared with at most a hundred microseconds in a laser plume) and because of additional excitation of the radiating upper levels in the low pressure plasma, the line intensities of atoms and ions are greatly enhanced. Because of these factors the DLs of LA–MIP have been improved by one to two orders of magnitude compared with LIBS.

The sensitivity, accuracy, and precision of solid-sample analysis have been greatly improved by coupling LA with ICP–OES–MS. The ablated species are transported by means of a carrier gas (usually argon) into the plasma torch. Further atomization, excitation, and ionization of the ablated species in the stationary hot plasma result in a dramatic increase in the sensitivity of the detection of radiation (LA–ICP–OES) or of the detection of ions (LA–ICP–MS).

DLs in the sub-μg g^{-1} to ng g^{-1} range for different materials (metals, glasses, polymers) have been realized by LA–ICP–OES. Even higher sensitivity was achieved for LA–ICP–MS, because of the high efficiency of ion collection and detection. Under optimum ablation conditions (choice of gas, diameter of the crater, flow rate of the carrier gas) DLs in the ng g^{-1} to pg g^{-1} range are now routinely realized for most elements of the periodic table. For example the 3σ DLs listed in the LSX-200 prospectus vary from 10 ppt for Ho to approximately 14 ppb for Fe. The list of DLs includes 31 elements, 26 of which can be detected with DLs in sub-ppb range. As a result of its attractive characteristics LA–ICP–MS is currently being used for a large variety of applications, e.g. in-situ trace-element analysis of glasses, geological samples, metals, ceramics, polymers, and atmospheric particulate material. Fingerprinting (characterization of the trace-element composition) of diamonds, gold, glass, and steel by use of LA–ICP–MS was reported to be very useful for provenance determination and for forensic purposes. Today LA–ICP–MS is accepted as a most powerful technique for direct analysis of solid samples.

The analytical capabilities of LIBS and LA-MIP-OES were recently noticeably improved by use of an advanced detection scheme based on an Echelle spectrometer combined with a high-sensitivity ICCD (intensified charge-coupled device) detector. This combination enables simultaneous detection of a large spectral range from the ultraviolet to near-infrared in a single laser shot. It enables estimation of the temperature of a laser plume by constructing Boltzmann plots and correction for plasma temperature variations. The advantages of this technique are: complete sample analysis in a single laser shot, improved accuracy and precision, and the possibility of detection of sample inhomogeneities [4.235]. Lateral resolution of a few micrometers can be achieved with focused laser beams in the UV and visible spectral ranges.

4.5.3 Depth Profiling

Multilayer coatings of different composition and thickness are widely used in materials science and in the production of high-technology materials. The single- or multicomponent thin layers significantly improve important characteristics of the materials with, e.g., specific properties.

Improvement of the technology of such advanced materials requires appropriate methods for surface and depth-profile analysis in the nanometer to micrometer range. Most of the methods used for surface analysis are reviewed in this monograph. For example, *X-ray photoelectron spectroscopy* (XPS) and *secondary ion mass spectroscopy* (SIMS) are widely used for characterization of nanometer-thin layers. Both techniques cannot be directly applied for depth profiling of thicker layers (a few micrometers thick) and can hardly be applied for fast on-line process analysis. Another technique used for depth profiling is dc- or rf-glow discharge (GD) sputtering followed by detection of the sputtered material by *optical emission spectrometry* (GD–OES) or *mass spectrometry* (GD–MS) [4.236–4.238]. GD sputtering enables a low sputtering rate and depth resolution of approximately 10 nm. The limitations of GD are poor lateral resolution (at best a few millimeters), specific requirements on the form and dimensions of the sample, and the need for low-pressure conditions for sample sputtering. Laser ablation has been proven to be a reasonable alternative technique for direct solid sampling [4.239–4.244].

The potential of LA-based techniques for depth profiling of coated and multilayer samples have been exemplified in recent publications. The depth profiling of the zinc-coated steels by LIBS has been demonstrated [4.242]. An XeCl excimer laser with 28 ns pulse duration and variable pulse energy was used for ablation. The emission of the laser plume was monitored by use of a Czerny–Turner grating spectrometer with a CCD two-dimensional detector. The dependence of the intensities of the Zn and Fe lines on the number of laser shots applied to the same spot was measured and the depth profile of Zn coating was constructed by using the estimated ablation rate per laser shot. To obtain the true Zn–Fe profile the measured intensities of both analytes were normalized to the sum of the line intensities. The LIBS profile thus obtained correlated very well with the GD–OES profile of the same sample. Both profiles are shown in Fig. 4.40. The ablation rate of approximately 8 nm $shot^{-1}$

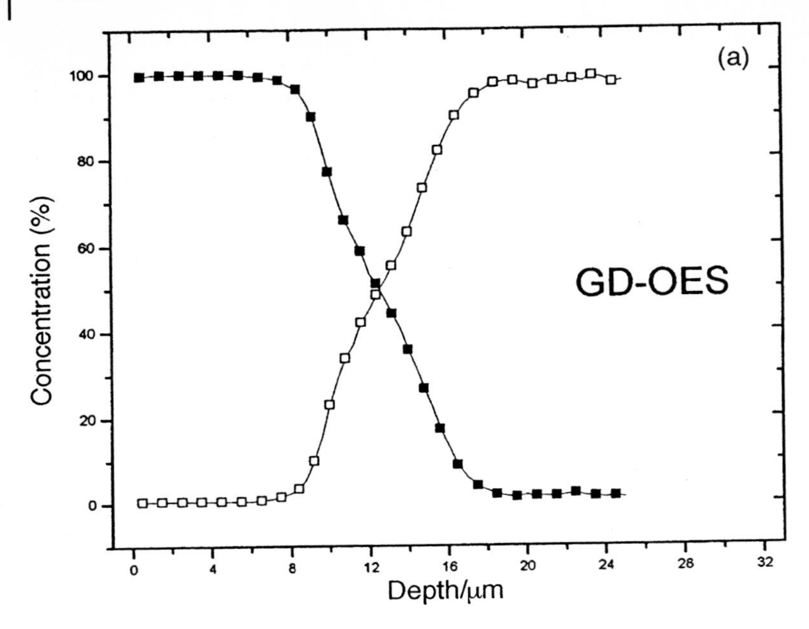

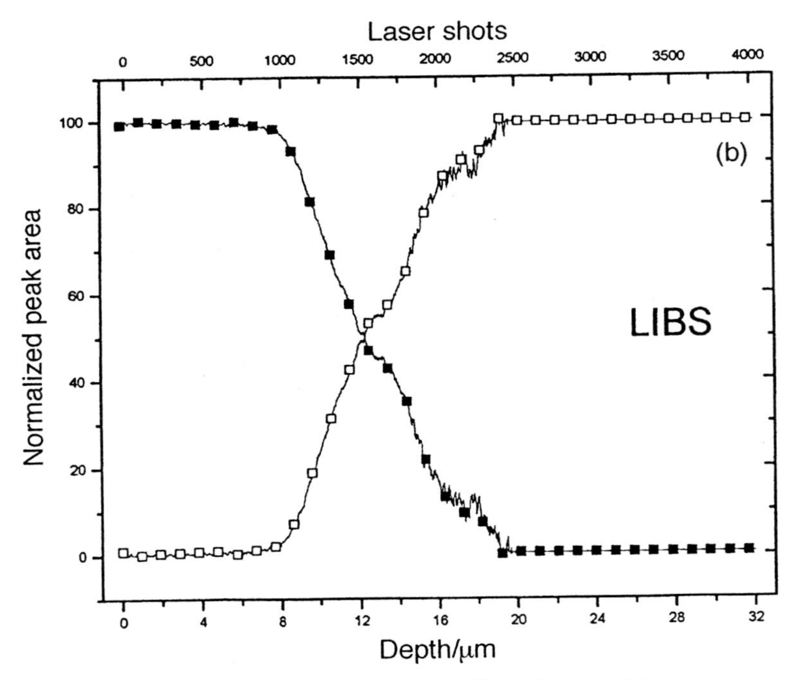

Fig. 4.40. (a) GD–OES and (b) LIBS emission profiles of Zn (■) and Fe (□) in a zinc-coated steel [4.242].

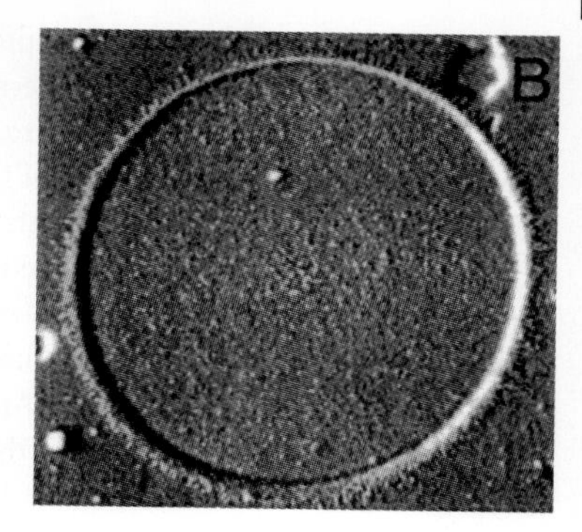

Fig. 4.41. The crater after two shots on TiN (120 μm crater diameter) [4.244].

was estimated on the basis of number of shots required to reach the Zn–Fe interface located 12 μm under the sample surface.

Depth-profile analysis of titanium-based coatings on steel samples has been performed by use of LA–ICP–MS [4.244]. A commercially available ArF excimer laser ablation system was used in combination with an ICP time-of-flight (TOF) MS. The laser pulse energy was varied in the 64–132 mJ range; the repetition rate was varied from 1 to 10 Hz. In these studies the laser beam was imaged by the optical system on to the sample surface to produce a crater 120 μm in diameter. The laser beam with flat-top intensity distribution was used to avoid material mixing from different layers. The flat morphology of the crater bottom obtained with the uniform laser beam readily apparent in Fig. 4.41. The lateral resolution of the system can be changed by realignment of the imaging optics. Craters approximately 10 μm wide with similar flat morphology could be obtained.

Different TiN-, TiC-, and TiAlN-based single layer coatings on steel alloyed with Cr, Ni, Mn, and WC were prepared by use of the cold vapor deposition technique. The thickness of the coatings varied from 2.7 to 6.4 μm.

The transient signals from $^{53}Cr^+$, $^{52}Cr^+$, and $^{182}W^+$ isotopes were used as the indicators of the steel substrate and ^{48}Ti and $^{48}Ti^{16}O^+$ were used as the indicators of the coating layer. Examples of the temporal profiles of the Ti and Cr isotopes for two samples with coatings of different thickness are shown in Fig. 4.42. The TOF tem-

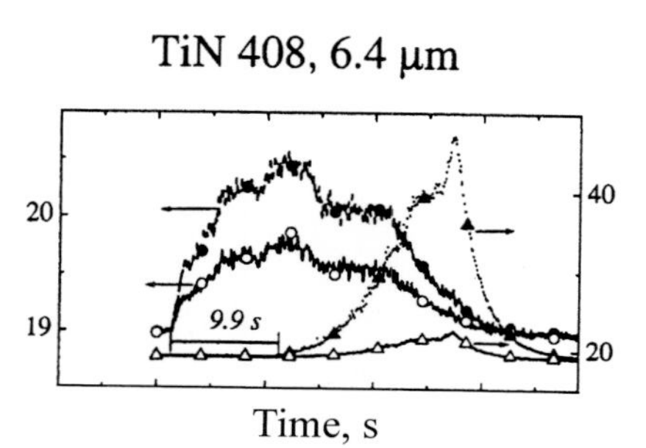

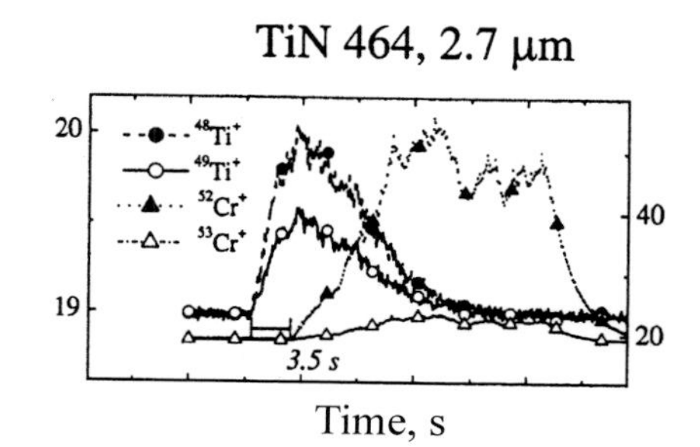

Fig. 4.42. Drill profiles through the two TiN coatings of 6.4 μm and 2.7 μm thickness (crater 120 μm, 100 mJ pulse energy, 3 Hz repetition rate). The titanium signal is given on the left y-axis, the chromium signal on the right y-axis [4.244].

poral profiles of Ti and matrix component Cr are well separated, which demonstrates the depth resolution of the LA system used. For the given laser fluence the drill time through the coating and the peak area of the TOF signals of the Ti isotopes were linearly proportional to the thickness of the coating.

Results from the first experiments on the application of the femtosecond laser for depth profiling have been published [4.245]. Two types of multi-component sample were investigated – multilayer Cu–Ag coatings on a Si substrate and TiN–TiAlN on an Fe substrate. The Cu–Ag multilayer samples were prepared as double and triple sandwiches of alternating Cu and Ag layers. The thickness of each copper and silver layer was approximately 600 nm. TiN/TiAlN samples comprised five TiN–TiAlN double layers on an iron substrate. The thickness of each TiN and TiAlN layer was equal – 280 nm. The nitrogen concentration was constant throughout the layers.

Two detection techniques were tested, LIBS for the Cu–Ag–Si samples and LA–TOF–MS for the TiN–TiAlN samples.

A commercial fs-laser (CPA-10; Clark-MXR, MI, USA) was used for ablation. The parameters used for the laser output pulses were: central wavelength 775 nm; pulse energy ~0.5 mJ; pulse duration 170–200 fs; and repetition rate from single pulse operation up to 10 Hz. In these experiments the laser with Gaussian beam profile was used because of the lack of commercial beam homogenizers for femtosecond lasers.

The depth profiling of the Cu–Ag–Si samples was performed with the laser beam focused on to the sample surface to the spot of approximately 30–40 μm. To obtain the best depth resolution the laser fluence was maintained near the 1 J cm^{-2} level, close to the threshold of the LIBS detection scheme. The intensity profiles of Cu and Ag emission lines are shown in Fig. 4.43. The individual layers of Cu and Ag were defi-

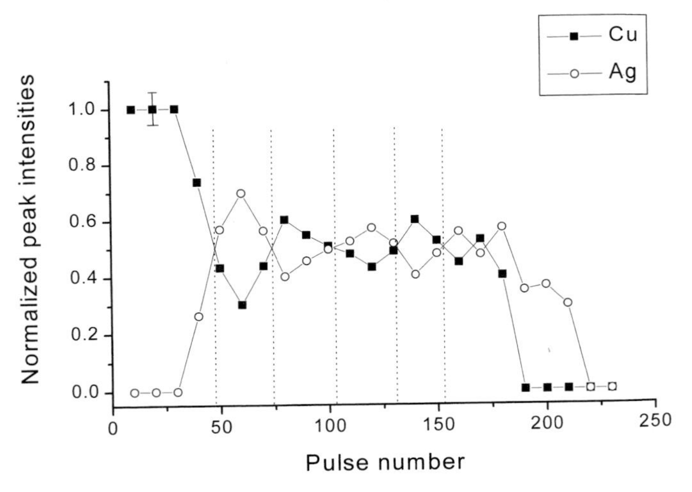

Fig. 4.43. Depth profile of a triple Cu-Ag sandwich (the laser fluence is about 1 J cm^{-2}; 10 pulses accumulation. The dashed lines indicate different layers [4.245].

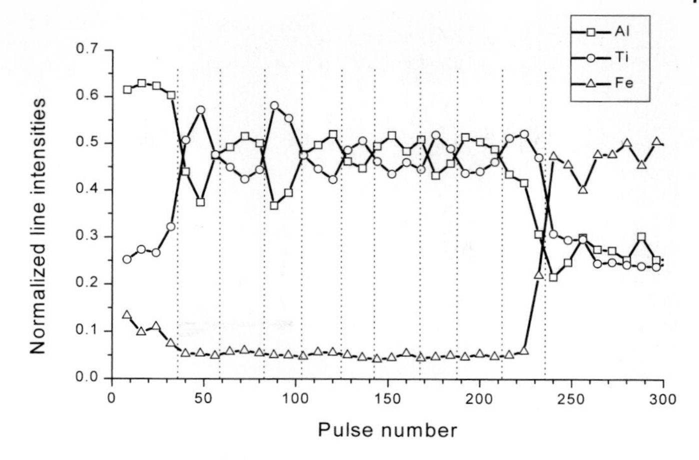

Fig. 4.44. Depth profile of TiN–TiAlN–Fe sample; normalization to the sum of all the major components is used (laser fluence 0.35 J cm^{-2}) [4.245].

nitely resolved, although the contrast of the normalized intensities in the successive layer is smoothed because of the bell-shaped profile of the laser beam. The estimated ablation rate for Cu–Ag–Si sandwiches varied between 15 and 30 nm per laser shot, depending on the beam fluence.

The "soft" ablation of the TiN–TiAlN samples by the low fluence laser beam was performed by use of LA–TOF–MS. Because of the greater sensitivity of this technique compared with direct LIBS the lower laser fluence of approximately 0.3–0.4 J cm^{-2} was used. One of the depth profiles, obtained by use of femtosecond LA–TOF–MS, is shown in Fig. 4.44. Each 280-nm-thick layer was ablated by approximately 20–25 pulses, which result in an average ablation rate of 11–14 nm pulse^{-1}. The ablation rate was low enough for resolution of all layers.

The crater surfaces obtained in the LA–TOF–MS experiment on the TiN–TiAlN–Fe sample were remarkably smooth and clearly demonstrated the Gaussian intensity distribution of the laser beam. Fig. 4.45 shows an SEM image of the crater after 100 laser pulses (fluence 0.35 J cm^{-2}). The crater is symmetrical and bell-shaped. There is no significant distortion of the single layers. Fig. 4.45 is an excellent demonstration of the potential of femtosecond laser ablation, if the laser beam had a flat-top, rather than Gaussian, intensity profile.

4.5.4
Conclusion

As already remarked in Sect. 4.5.1 (Introduction), LA was primarily designed as a technique for direct sampling in the bulk analysis of solid samples. The main advantages of LA are the possibility of ablating all types of solid material (metals, isolators, glasses, crystals, minerals ceramics, etc.), no special requirements on the

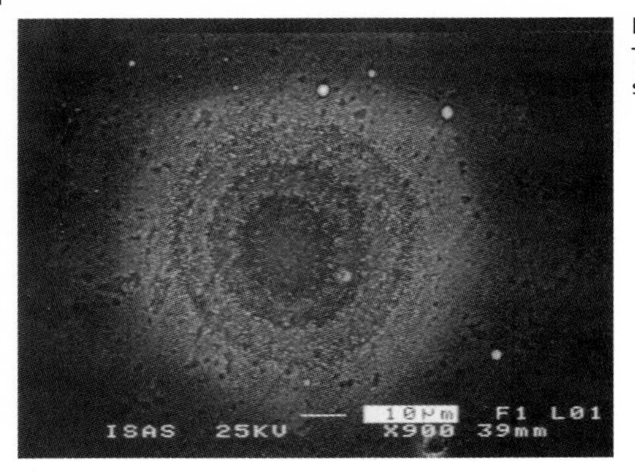

Fig. 4.45. SEM photograph of TiN–TiAlN–Fe sample after 100 laser shots (laser fluence 0.35 J cm^{-2}) [4.245].

form of the sample, high lateral resolution down to a few microns, and the possibility of remote sampling using a fiber optics. Different hyphenated techniques can be used for detection of the elemental composition of the ablated material. Among these the LIBS is the most flexible, technically the simplest, the cheapest, and enables moderate sensitivity with DL in the 10 to 100 ppm range. Much higher sensitivity can be realized by combination of LA with the ICP–OES–MS detector. Very low DL in ppb to ppt range have been achieved in the routine analysis of a wide variety of samples by LA–ICP–MS. LA–ICP–OES–MS instruments for direct analysis of solid samples are now commercially available. With this increased sensitivity, the amount of ablated material needed for analysis has been significantly reduced. This has transformed LA into an almost non-destructive technique – especially important for such applications as product certification and analysis of works of art and gemstones.

Initial results prove the high potential of LA-based hyphenated techniques for depth profiling of coatings and multilayer samples. These techniques can be used as complementary methods to other surface-analysis techniques. Probably the most reasonable application of laser ablation for depth profiling would be the range from a few tens of nanometers to a few tens of microns, a range which is difficult to analyze by other techniques, e.g. SIMS, SNMS, TXRF, GD–OES–MS, etc. The lateral and depth resolution of LA can both be improved by use of femtosecond lasers.

4.6 Ion Beam Spectrochemical Analysis (IBSCA)

Volker Rupertus

Ion beam spectrochemical analysis (IBSCA) is a sputtering-based surface analytical technique similar to SIMS/SNMS. In IBSCA the radiation emitted by excited sputtered secondary neutrals or ions is detected. IBSCA was developed parallel to SIMS in the nineteen-sixties and early nineteen-seventies [4.246, 4.247]. It is also known under the name SCANIIR (*surface composition by analysis of neutral and ion impact radiation*) or BLE (*beam induced light emission*).

For widespread use as a common analytical technique the detection of quantifiable signals is strictly necessary, and this is a principle problem that IBSCA shares with SIMS – application to the analysis of metals and semiconductors has shown that the photon yield is highly dependent on the surface oxygen content and oxygen partial pressure [4.248–4.251]. For oxidic samples, e. g. glass or glass ceramic, in contrast, the matrix-dependence usually proves to be negligible so IBSCA can be used for stoichiometric quantification [4.252, 4.253]. IBSCA is, therefore, mainly used as an analytical tool in combination with other methods, e. g. SIMS, for the analysis of highly insulating surfaces and thin films. Most applications deal with the depth profiling of multilayer structures on glass substrates, and with the characterization of the near-surface structures of glass samples which are often different from the bulk stoichiometry caused by interaction with the environment.

4.6.1 Principles

As is typical for a sputter-based erosion technique an ion beam of mostly positive noble gas ions (e. g. Ar^+) penetrates the surface. During the ion beam sputter processes at a non-elemental solid, many physical and chemical processes are initiated; these lead to a change in stoichiometry in the near-surface region by interaction of the primary ions with the atoms of the solid, then induced solid–solid atom interactions and, finally, surface erosion by emission of secondary particles [4.254, 4.255]. The ion beam-induced emission processes include emission of electrons and surface particles (atoms or molecules) in the charged, neutral, and, sometimes, excited states. Some theoretical aspects of the principle mechanism of producing excited sputtered particles are discussed in the literature; these include the "electron-transfer model" [4.256, 4.257], the "level-crossing model" [4.250], and the "LTE model" [4.258, 4.259]. The fluence of the sputtered particles is representative of the elemental composition of the near-surface region. Analysis of the emitted particle flux during the sputtering process enables calculation of quantitative concentration–depth profiles of the elemental components.

For the sputter erosion process energetic ions (0.5–10 keV) with current densities of 10–100 $\mu A\ cm^{-2}$ are commonly used. The lifetime of the excited states of the sputtered ions and neutrals is 10^{-13}–10^{-14} s, so the de-excitation processes occur up to 2 cm above the sample surface. The emitted photons can be detected in the visible range (250–900 nm) by use of transfer optics manufactured from fused silica. For the detection of UV lines, special optics and detection systems are required (lenses made of CaF_2, high-vacuum-pumped optical spectrometers, etc.). The radiation from atoms, ions, or molecules sputtered in an excited state, and, for glass, the emission of the luminescence excited within the solid, can be analyzed by means of optical multichannel analysis (OMA) or CCD cameras. Analysis of the sputtered species becomes possible by assigning peaks at measured wavelengths, λ, on the basis of wavelength-dependent listings of possible energetic transitions for all elements in spectral tables.

4.6.2
Instrumentation

For IBSCA analysis, standard HV or, better, UHV-equipment with turbomolecular pump and a residual gas pressure of less than 10^{-5} Pa is necessary. As is apparent from Fig. 4.46, the optical detection system, which consists of transfer optics, a spectrometer, and a lateral-sensitive detector, is often combined with a quadrupole mass spectrometer for analysis of secondary sputtered particles (ions or post-ionized neutrals).

Typical ion sources employ a noble gas (usually Ar). The ionization process works either by electron impact or within a plasma created by a discharge; the ions are then extracted from the region in which they are created. The ions are then accelerated and focused with two or more electrostatic lenses. These ion guns are normally operated to produce ions of 0.5–10 keV energy at currents between 1 and 10 μA (or, for a duoplasmatron, up to 20 μA). The chosen spot size varies between 100 μm and 5 mm in diameter.

In insulator analysis an electron gun is also necessary to compensate for the positive ion current at the sample surface. Two types of operation are typical.

(i) For lower electron-current densities a flood gun can be used to produce low-energy (1–20 eV) electrons directly; or

(ii) For higher electron-current densities, up to 300 μA cm^{-2}, standard electron guns are used. These emit electrons of 0.5–2 keV which are focused to a metal or carbon mask which covers the sample rim (Fig. 4.47). By proper choice of the primary electron energy, a secondary electron yield, $\delta > 1.5$, at perpendicular incidence can be achieved (e.g. δ_{max}(Ta) = 1.3 at E_P = 0.6 keV). Additional enlargement of the yield of secondary electrons by a factor of two can be achieved by use of bombardment angles of 50–60° normal to the sample surface [4.260]. By use of this procedure, low-energy

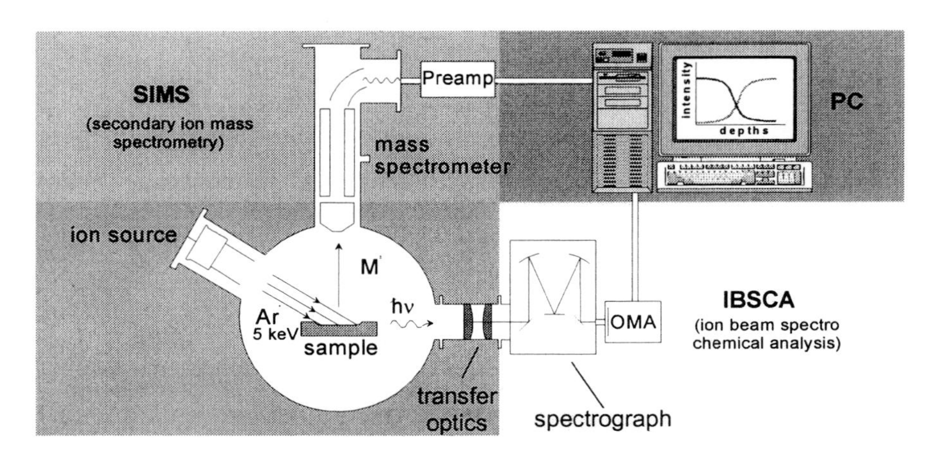

Fig. 4.46. Schematic diagram of IBSCA measurement equipment; this usually combined with a mass spectrometer (SIMS or SNMS).

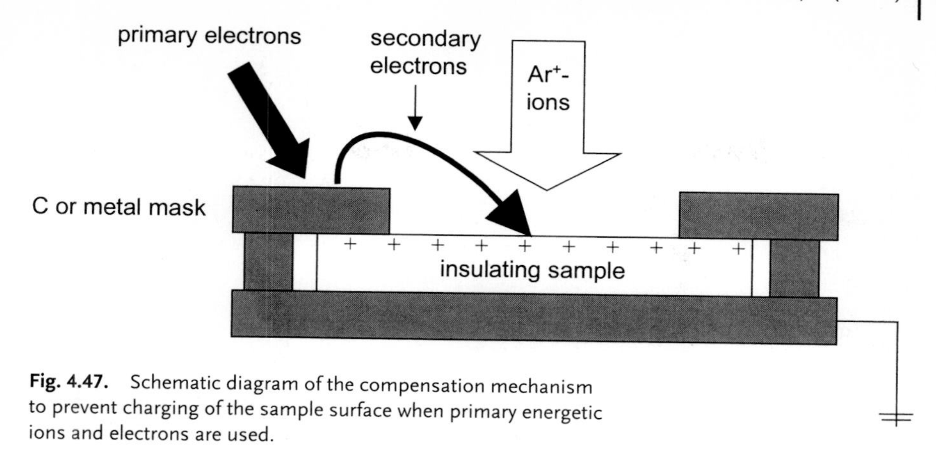

Fig. 4.47. Schematic diagram of the compensation mechanism to prevent charging of the sample surface when primary energetic ions and electrons are used.

secondary electrons are created with an energy distribution around the maximum of approximately 2 eV for Ta [4.261]. These low-energy electrons follow the attractive field forces at the sample surface directly, and compensate for the slight positive charging caused by ion impact. The penetration depth of the incoming secondary electrons is comparable with the penetration depth of the primary ions, so no local capacitor-like charging effects build up and, therefore, charging-induced migration/diffusion in the near-surface region is suppressed.

The photon detection system consists of transfer optics (a few lenses), made from fused silica for better transmission of down to 250 m. For the near-ultraviolet region, crystal materials such as CaF_2, etc., and a vacuum-pumped spectrometer, must be employed. The transfer optics focus the space volume in front of the penetrated sample surface to the entrance slits of a grating spectrometer typically in a Czerny–Turner arrangement (two mirrors and one grating). Depending on the grating used (120–2400 lines mm^{-1}), the resolution can be varied between 0.5 and 0.05 nm. The transmitted photons are registered with an optical multichannel analyzer or with a CCD-camera, so that the intensity can be detected with high lateral resolution. The detection system is computer-controlled so that the sputter time-dependent change of the spectra is stored and sputter depth profiles can subsequently be computed.

4.6.3
Spectral and Analytical Information

An IBSCA-spectrum (Fig. 4.48) consists of many peaks in the visible range (250–900 nm). Every peak can be related to an process of electron de-excitation of a sputtered particle from a higher to a lower state, for the more dominant peaks to the ground state. There are, in principle, two major types of peak family:
type I – photons emitted from excited sputtered secondary neutrals; and
type II – photons emitted from excited sputtered secondary ions (single charged).

The peaks can be identified from tables of spectra [4.262]. The variety of electronic transitions allowed results in many peaks in the visible range; these are sometimes

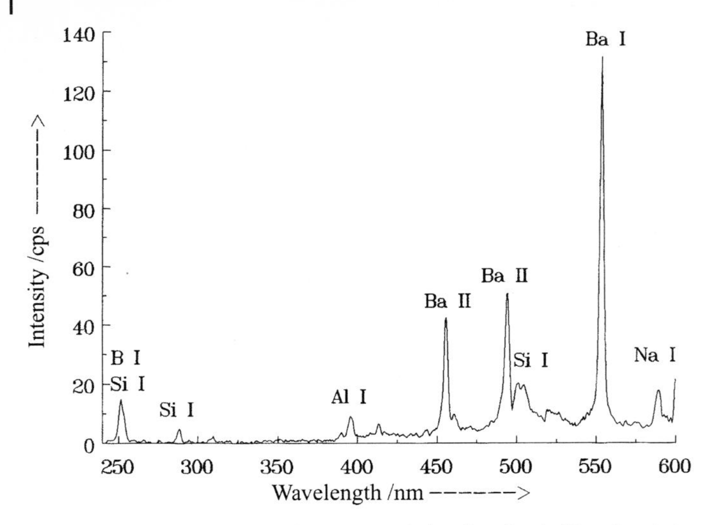

Fig. 4.48. IBSCA-spectrum of SK16 optical glass bombarded by 5 keV Ar^+.

energetically so close together that deconvolution of neighboring peaks is hampered. Proper choice of the lines to be used is, therefore, particularly important in quantitative interpretation of the analytical results from multicomponent samples.

A second source of failure is determination of the background of the spectra, which depends on the complete photon emission from the sample (solid-state luminescence). During the measurement, especially in depth profiling, the background level can change. Determination of the correct background is, therefore, important for quantitative analysis. Peak resolution depends on the grid size employed in the spectrometer, which also determines the detection interval (e.g. 250–600 nm). Better peak resolution correlates directly with a smaller detection interval and the grid size must, therefore, be optimized for each range of application. In Fig. 4.48 the main components of SK16 (Ba, B, Si, Al, Na) are visible. Obviously, more than one line for each element, excited neutral (label I) and ion (label II), is detectable in the detection range depicted. Only oxygen, the most abundant (~60 at%) of all the SK16 components, is not detectable because the main radiation lines are located in the vacuum-UV range. The relationship between recorded intensity and the actual surface concentration/chemical state of the surface is not as straightforward as in XPS or AES. In the same way as for the peak intensity in SIMS or SNMS, the height of the detected peaks depends to some extent on the chemical matrix of the sample surface.

4.6.4
Quantitative Analysis

The intensity I_{ki} (λ_A) of a spectral emission line, i.e. the radiative recombination of an electron of a species A from a higher energy level k to the lower level i, is characteristic of a sputtered element or molecule A and is calculated by use of the equation:

$$I_{ki}(\lambda_A) = I_P Y_A \eta_{ki} \alpha_{ki} / e_0 \tag{4.24}$$

where I_P is the primary ion current, λ_A is the wavelength of the transition from an excited state k to the ground state i, Y_A is the partial sputtering yield of the sputtered species A, η_{ki} is the detection factor of the optical system which quantitatively takes into account the aperture and the quantum efficiency of the optical detection system (electrons/quantum) for λ_A, α_{ki} is the excitation coefficient (or photon yield), which is defined as the ratio of the number of atoms of the element A showing this transition to the total number of the sputtered atoms, and e_0 is the electron charge.

The radiation factor α_{ki} is determined from:

$$\alpha_{ki} = \frac{N_k}{N_k + N_i} \varepsilon_{ki} P \tag{4.25}$$

where $N_k/(N_k + N_i)$ is the ratio of the excited sputtered secondary particles of species A, ε_{ki} is the probability of transition from level k to level i, and P is the probability that the de-excitation is a radiative process ($P \approx 1$ for insulating samples [4.250]).

Similar to other sputter-based techniques, a sensitivity factor can be determined:

$$D_A^{\lambda} = \eta_{ki} \alpha_{ki} \tag{4.26}$$

so that the intensity is given by:

$$I_{ki}(\lambda_A) = I_P Y_A D_A^{\lambda} \tag{4.27}$$

Taking atomic sputtering into account the proportion of the particles emitted as molecules is negligible and the partial sputtering yield for element A in sputter equilibrium can be determined by use of:

$$Y_A = c_A^b Y_{tot} \tag{4.28}$$

where c_A^b is the bulk concentration of element A and Y_{tot} is the total sputtering yield of the sample, where $Y_{tot} = \sum_A Y_A$.

Taking into account that $\sum_A c_A^b = 1$, the bulk concentration of element A is given by:

$$c_A^b = \frac{I_{ki}(\lambda_A)/D_A^{\lambda}}{\sum_j (I_{ki}(\lambda_j)/D_j^{\lambda})} \tag{4.29}$$

If relative sensitivity factors are used, reference measurement of standard samples is not necessary. The ratio of two different elemental concentrations in one sample is given by:

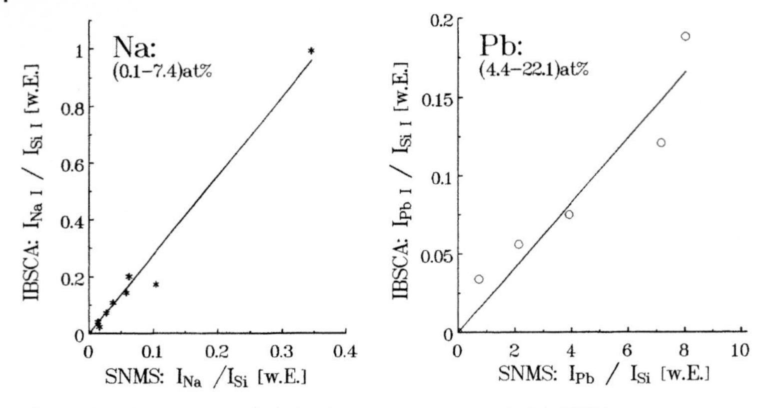

Fig. 4.49. Comparison of relative intensities measured with IBSCA and SNMS. Conditions: sputter equilibrium after bombardment with 5 keV Ar^{+}; the samples were oxidic glasses with different content of Na (0.1–7.4 at%) and Pb (4.4–22.1 at%).

$$\frac{c_A^b}{c_B^b} = \frac{I_A}{I_B}\frac{D_B}{D_A} \tag{4.30}$$

The ratio D_B/D_A is a so-called relative sensitivity factor D. This ratio is mostly determined by one element, e. g. the element for insulating samples, silicon, which is one of the main components of glasses. By use of the equation that the sum of the concentrations of all elements is equal to unity, the bulk concentrations can be determined directly from the measured intensities and the known D-factors, if all components of the sample are known. The linearity of the detected intensity and the flux of the sputtered neutrals in IBSCA and SNMS has been demonstrated for silicate glasses [4.253]. For SNMS the lower matrix dependence has been shown for a variety of samples [4.263]. Comparison of normalized SNMS and IBSCA signals for Na and Pb as prominent components of optical glasses shows that a fairly good linear dependence exists (Fig. 4.49).

From these results it can be postulated that for oxidic glasses a fixed proportion of sputtered secondary neutrals is emitted in an excited state. Such linearities can only be determined for similar matrices, which limits the use of D-factors to sample systems similar to the reference sample system used for the D-factor determination.

4.6.5
Applications

The depth profile mode of IBSCA is usually used for routine thin film analysis. A typical application is depicted in Fig. 4.50 – a fixed spectral range was monitored during the sputter erosion of the surface so that many IBSCA spectra were stored during sample analysis. This example shows how individual IBSCA spectra change during sputter erosion of an anti-reflecting coating on soda-lime glass (SiO_2–TiO_2–SiO_2/TiO_2-

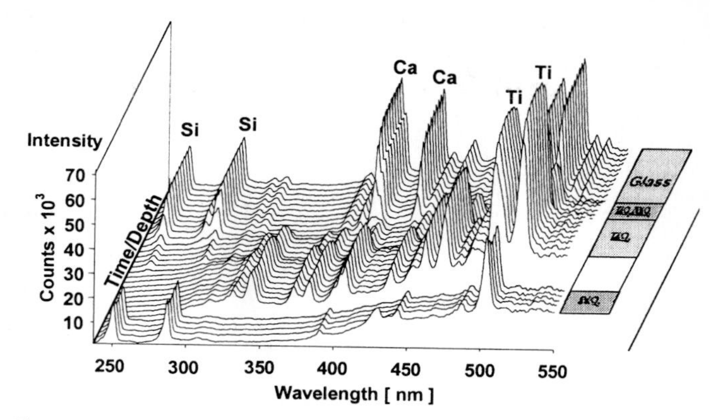

Fig. 4.50. IBSCA spectra of an antireflective coating on soda-lime glass (SiO_2–TiO_2–SiO_2/TiO_2-substrate); conditions: 5 keV Ar^+ bombardment.

substrate). The variation in the intensity of single peaks, e. g. those for Si or Ti, during the sputter erosion of the multilayer is apparent. Sputter depth profiles can be computed from such measurements (Fig. 4.51).

The lower matrix sensitivity compared with SIMS is apparent from Fig. 4.51. With both methods a TiO_2–SiO_2–TiO_2 multilayer coating on soda-lime glass was analyzed. A sol–gel technique was employed for the coating process, this was followed by a tempering process. During this tempering treatment, diffusion of alkaline substrate species material into the multilayer is initiated. Comparison of the positive SIMS profile with the IBSCA-depth profile reveals that characteristics are similar for the main components Ti and Si, but there are drastic differences between the shapes of the Na signals. For SIMS, intensity maxima at the SiO_2/TiO_2 interface and at the surface area are detected which are 20 % and 60 %, respectively, higher than for the bulk intensity of the substrate which is the Na source. Quite different behavior is apparent in the IBSCA profile (Fig. 4.51 b): The Na intensity also has maxima at the same positions, but only with local character. This Na profile reveals the diffusion process from the glass substrate to the surface and indicates that SiO_2 behaves as a diffusion barrier, so the interface from TiO_2 to SiO_2 is enriched in Na, because the diffusion coefficient of SiO_2 is lower than for TiO_2. The IBSCA–Na profile gives realistic information about the relative concentrations in comparison with the bulk of the glass substrate, which was supported by additional XPS-measurements. The enlargement of the Na signal for SIMS gives clear evidence of the influence of the chemical matrix. At the surface and at the single-layer interfaces, is a chemical environment which increases the probability of ionization. IBSCA is, therefore, a sensitive method which can be combined with SIMS for better interpretation of the multilayer structures of thin oxide films.

A second set of examples deals with the analysis of near-surface regions of glasses which normally have so-called altered or leached layers. The altered layer is found for soda-lime glasses and for many glasses used for optical applications. The chan-

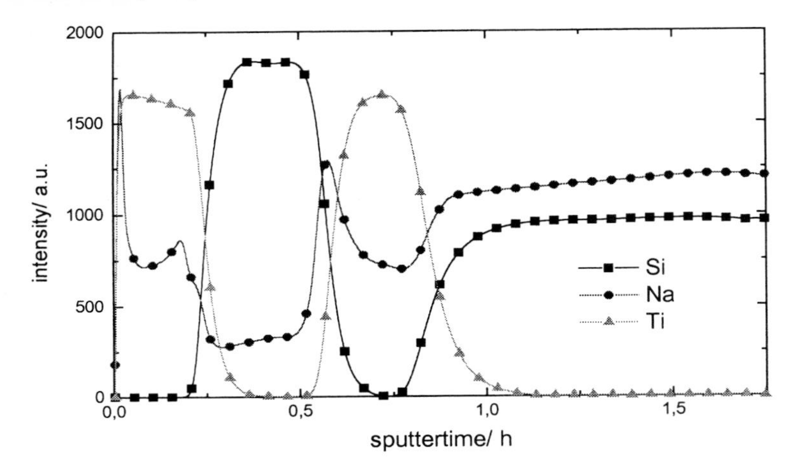

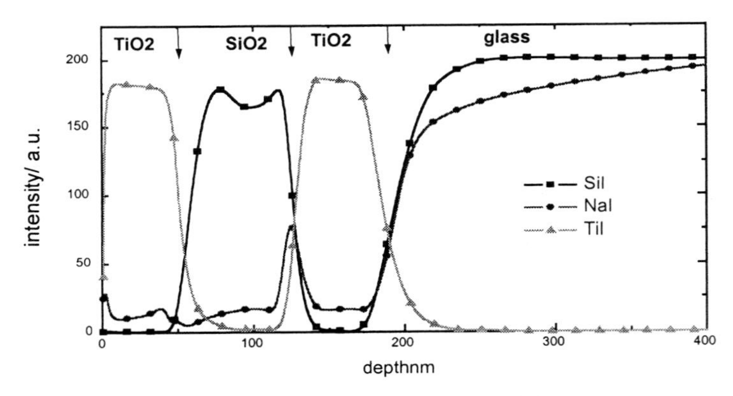

Fig. 4.51. SIMS (a) and IBSCA (b) depth profiles of an interference layer system used for automotive application (TiO_2–SiO_2–TiO_2–soda-lime glass); conditions: 5 keV Ar^+.

ged surface composition is caused by weathering effects and is indicative of ion exchange of alkaline species from the glass site and OH from the environment. This hydration effect leads to a change in the glassy network, with consequences for the physical properties of the altered layer in comparison with those of the bulk glass. The large influence on the chemical matrix is demonstrated in Fig. 4.52. The SIMS Si^+ signal behaves differently from the IBSCA Si I signal. This is indicative of the presence of H and/or OH in the uppermost region which increases the probability of ionization of Si^+ and leads to depletion of Li as a result of ion exchange (Li against H). The increase in the H content of that near-surface region was also detected by

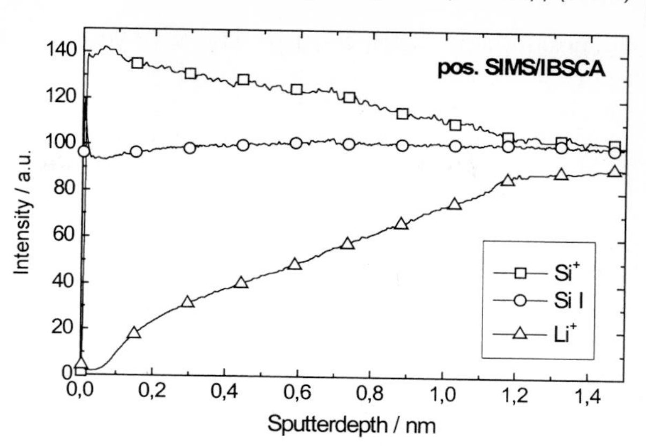

Fig. 4.52. SIMS and IBSCA depth profiles of the altered layer region of a lithium aluminosilicate (LAS) glass ceramic (conditions: 5keV Ar^+).

NRA [4.264] and supports the statement that SIMS and IBSCA react in quite a different manner to slight changes in the chemical matrix. This effect can be used as a indicator of increased levels of H and OH in the near-surface region of oxidic glasses when both methods are employed.

4.7 Reflection Absorption Infrared Spectroscopy (RAIRS)

Karsten Hinrichs

Reflection absorption infrared spectroscopy is a vibrational spectroscopic method frequently used in research in chemistry, physics, and biology [4.265–4.268]. The technique is well established for the identification and characterization of the chemical state and structure of molecules and thin films adsorbed on metallic and non-metallic surfaces. Vibrational spectra are used as characteristic fingerprints for adsorbate molecules, adsorption configurations, and structures. The method is known by different acronyms: RAIRS, FT-IRAS, IRRAS, and ERIRS (*external reflection infrared spectroscopy*); the different names refer to the same principle of measurement using infrared radiation and a reflection absorption geometry. This technique involves a single external reflection at a sample-covered substrate, whereby the sample molecules absorb radiation according to their vibrational frequencies and corresponding absorption bands occur in the spectrum of the reflected radiation. Substrates are chosen which do not produce disturbing absorption bands in the spectral range of interest.

4.7.1 Instrumentation

In the infrared spectral range in general Fourier transform (FT) interferometers are used. In comparison with dispersive spectrometers FTIR enables higher optical throughput and the multiplex advantage at equivalent high spectral resolution. In the near-infrared region a tungsten-filament lamp is conventionally employed as a source of radiation, whereas a "glowbar" is used in the mid-infrared region and a mercury-arc lamp in the far-infrared region. A synchrotron, a special radiation source characterized by high brilliance throughout the whole infrared region has recently been used for in-situ measurement during anodic oxidation of copper in aqueous solution environments [4.269]. RAIRS is often used as complementary method to HREELS (Sect. 2.3, EELS), which is also a standard method for investigation of dipole-active vibrations, but has a low spectral resolution (on semiconductor surfaces at approximately 20 cm^{-1}).

RAIRS is a non-destructive infrared technique with special versatility – it does not require the vacuum conditions essential for electron spectroscopic methods and is, therefore, in principle, applicable to the study of growth processes [4.270]. By use of a polarization modulation technique surfaces in a gas phase can be investigated. Higher surface sensitivity is achieved by modulation of the polarization between s and p. This method can also be used to discriminate between anisotropic near-surface absorption and isotropic absorption in the gas phase [4.271].

4.7.2
Principles

Conventionally RAIRS has been used for both qualitative and quantitative characterization of adsorbed molecules or films on mirror-like (metallic) substrates [4.265]. In the last decade the applicability of RAIRS to the quantitative analysis of adsorbates on non-metallic surfaces (e. g. semiconductors, glasses [4.267], and water [4.273]) has also been proven. The classical three-phase model for a thin isotropic adsorbate layer on a metallic surface was developed by Greenler [4.265, 4.272]. Calculations for the model have been extended to include description of anisotropic layers on dielectric substrates [4.274–4.276].

The dielectric properties of the substrate and the adsorbate determine the optimum conditions under which the RAIRS experiment should be performed. The excitation of vibrational adsorbate modes on metals is governed by a so-called "surface selection rule" [4.265]. The amplitude and phase of the reflected radiation depend on the direction of the electric field vector, which is composed of one component parallel to the reflection plane (p-polarized) and a second normal to the reflection plane (s-polarized). Considering Fresnel's equations, only the p-polarized radiation causes a significant electromagnetic field near the metallic surface for high incident angles of radiation near grazing incidence (at approximately 80°). In contrast the mean square electric field of radiation polarized in the surface plane is negligible because of the 180° phase shift after reflection at the metallic surface. Only vibrations with dipole components perpendicular to the surface can, therefore, be efficiently excited.

With p-polarized radiation and incident angles near grazing incidence an increase in sensitivity of approximately a factor of 25 can be achieved in comparison with transmission experiments [4.265]. This advantage is reduced to a factor of ~17 for a more realistic experimental situation in which the spread of incident angles is ca. ±5° at approximately 85°.

RAIRS spectra contain absorption band structures related to electronic transitions and vibrations of the bulk, the surface, or adsorbed molecules. In reflectance spectroscopy the absorbance is usually determined by calculating $-\log(R_S/R_O)$, where R_S represents the reflectance from the adsorbate-covered substrate and R_O is the reflectance from the bare substrate. For thin films with strong dipole oscillators, the Berreman effect, which can lead to an additional feature in the reflectance spectrum, must also be considered (Sect. 4.9; Ellipsometry). The frequencies, intensities, full widths at half maximum, and band line-shapes in the absorption spectrum yield information about adsorption states, chemical environment, ordering effects, and vibrational coupling.

For films on non-metallic substrates (semiconductors, dielectrics) the situation is much more complex. In contrast with metallic surfaces both parallel and perpendicular vibrational components of the adsorbate can be detected. The sign and intensity of RAIRS-bands depend heavily on the angle of incidence, on the polarization of the radiation, and on the orientation of vibrational transition moments [4.267].

4.7.3
Applications

RAIRS is routinely used for the analysis of chemically modified surfaces – surface systems in electrochemistry [4.277], polymer research [4.266, 4.278], catalysis [4.265, 4.271], self-assembling monolayers [4.267, 4.268], and protein adsorption [4.268, 4.279] have been investigated.

Beside studies of adsorbates on metal and semiconductor surfaces, considerable interest has been shown in the structural characterization of monolayers at the air–water interface [4.273, 4.280]. New experiments at the gas–water interface have been performed in biochemical research in the characterization of membrane proteins [4.279] or the structures of other biological molecules [4.281]. Experimental and simulated RAIRS spectra of single monolayers of PBG (poly-γ-benzyl-l-glutamate) and the synthetic peptide $K(LK)_7$ at the air–water interface are depicted in Fig. 4.53 [4.281]. In the simulations the different secondary structures of both molecules, and anisotropic optical constants, were taken into account. The spectra of PBG contain significant features resulting from the amide I, amide II, and ester bands in the α-helix structure. For $K(LK)_7$ the simulations reveal molecules lying in antiparallel β-sheets on the water surface. The frequency of the amide I mode (1622 cm^{-1}) provides information about interactions between the antiparallel β-sheets and the substrate. Knowledge of such optical data is of great interest in the determination of the main secondary structures (α-helix, β-sheets) found in proteins and polypeptides [4.281].

Another use of RAIRS is in the determination of the average tilt angles of molecules on a surface. In comparison with calculations of relative intensities of absorption bands in Ref. [4.267] an average tilt angle of the hydrocarbon chains of approximately 10° towards the surface normal was concluded. Fig. 4.54 shows the relevant infrared reflection spectra from an octadecylsiloxane (ODS) monolayer on a silicon substrate [4.267] for s- and p-polarized radiation at different incident angles. The absorption bands are assigned to fundamental symmetric and asymmetric CH stretching vibra-

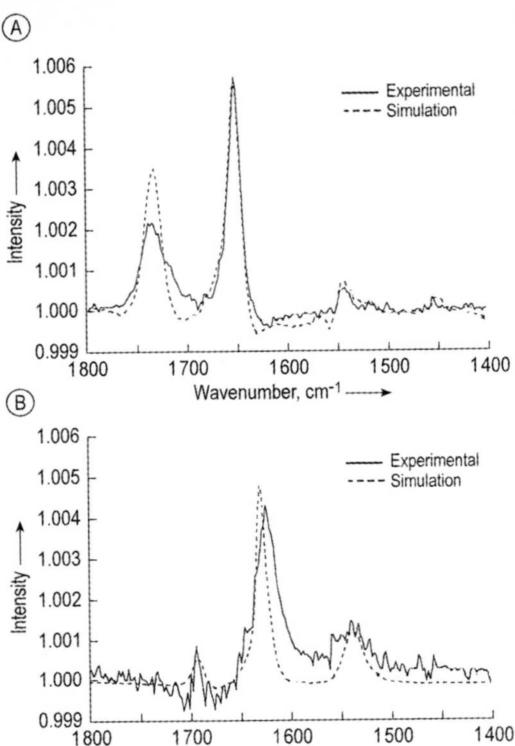

Fig. 4.53. Experimental and simulated PM (polarization modulated) IRRAS spectra of single monolayers of (A) PBG and (B) $K(LK)_7$ at the air–water interface. The surface pressure was 20 mN m^{-1} [4.281].

tions. The spectra change significantly as a function of radiation incidence angle and the direction of polarization. For p-polarization the sign of the RAIRS bands changes by crossing the Brewster angle at 73°. The spectra also provide information about the bonding between the adsorbate and substrate. In addition to the structure and orientation of molecules, reaction kinetics, e.g. of alkoxides on Cu(111) surfaces [4.282], can also be investigated.

The line shapes and the intensities of reflection spectra from thin films are usually different from those of transmission spectra. Calculated and measured spectra for isotropic PMMA (poly(methyl methacrylate)) films of different thickness on glassy carbon are shown in Fig. 4.55. The spectra obtained with incident light with s- and p-polarization are completely different for the three films. In each the band shapes were distorted compared with those from conventional transmission experiments, which measure the material dispersed in a KBr matrix [4.266]. The observed absorbance maxima are approximately 10 cm^{-1} higher in energy than the maximum of 1731 cm^{-1} in conventional transmission experiments [4.266]. It was concluded that band-shape distortions result from “optical effects”, which means all effects which cause changes in the recorded spectrum except those resulting from surface perturbations and changes in structure or chemical bonding. Optical effects are taken into account by optical theory [4.276, 4.283].

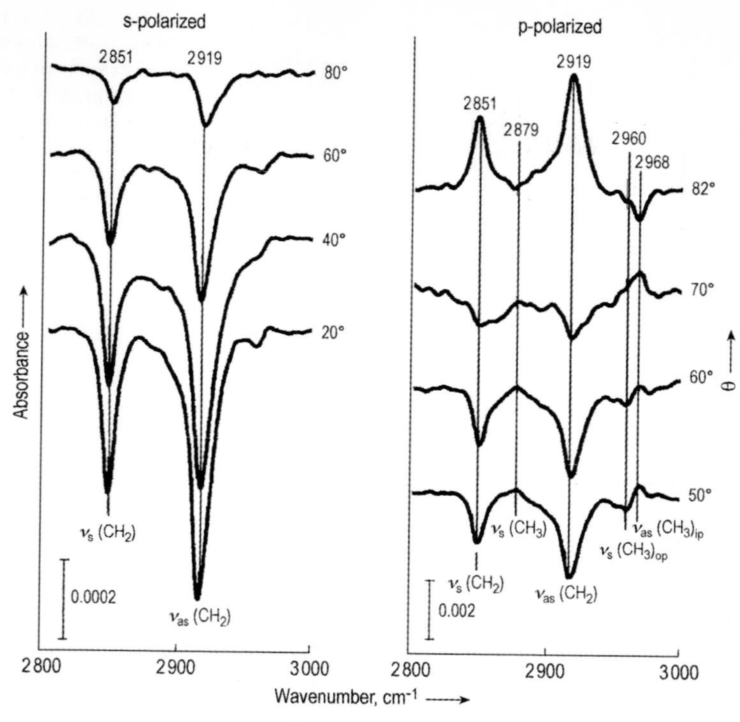

Fig. 4.54. IR reflection spectra from an ODS monolayer on silicon for s- and p-polarized radiation at different incident angles θ. Symmetric (s), asymmetric (as), in plane (ip), and out-off plane (op) CH_2 and CH_3 vibrations are marked [4.267].

To complete this treatment it is necessary to discuss another important surface infrared spectroscopy, *infrared absorption spectroscopy* (IRAS). In contrast with RAIRS this technique works with internal attenuated-total-reflection (ATR) configurations. For example, monolayers of small molecules on semiconductors can be detected when the detection sensitivity is increased by employing multiple internal reflections [4.284, 4.285]. Edges of samples must be beveled, to couple the radiation in and out. This method has, for example, been applied to an investigation of the morphology of H-terminated Si surfaces after etching [4.284] and an investigation of the surface chemistry of organometallic chemical vapor deposition, the technique of choice for fabricating III/V compound semiconductor devices [4.286].

In conclusion RAIRS, which affords high spectral resolution, is a very versatile non-destructive optical technique which does not depend on a vacuum environment. Vibrational spectra also serve as characteristic fingerprints for adsorbate molecules, adsorption configurations, and structures on metallic and dielectric substrates. Extension to include dielectric substrates opened new fields of application in polymer and biochemical research.

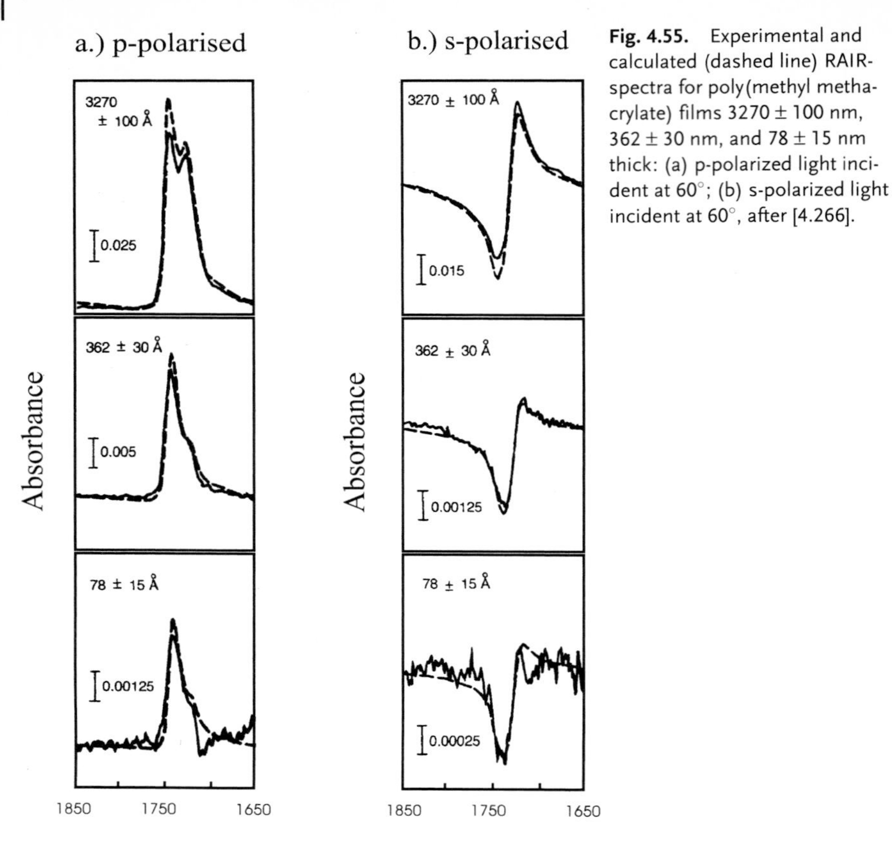

Fig. 4.55. Experimental and calculated (dashed line) RAIR-spectra for poly(methyl methacrylate) films 3270 ± 100 nm, 362 ± 30 nm, and 78 ± 15 nm thick: (a) p-polarized light incident at 60°; (b) s-polarized light incident at 60°, after [4.266].

4.8
Surface Raman Spectroscopy

Wieland Hill

4.8.1
Principles

Raman scattering [4.287] is one of the phenomena that occur when electromagnetic radiation interacts with a molecule or a crystal. As a result of the inelastic scattering process, photons lose energy, which excites molecular or lattice vibrations (Stokes scattering), or can even gain energy (anti-Stokes scattering) when the transition starts from an excited vibrational state and the final state is of lower energy (e. g. the ground state). When a sample is irradiated by monochromatic light, the scattered light contains a spectrum of wavelengths in which the intensity peaks are shifted from the excitation wavelength by energy-equivalent amounts corresponding to that for the excitation of the molecular vibration modes or crystal phonons. These vibrational spectra are highly specific for the material studied and can be used for unambiguous identification of substances.

Because the Raman cross-section of molecules is usually low, intense light sources and low-noise detectors must be used, and high sensitivities – as required for surface analysis – are difficult to achieve. Different approaches, singly and in combination, enable the detection of Raman spectroscopy bands from surfaces.

(1) Ultrasensitive Equipment: In recent years all components of Raman equipment (laser, sampling optics, filtering, monochromator, and detector) have been clearly improved. This has led to an enormous increase in sensitivity and has enabled direct observation of adsorbed molecules with carefully optimized instruments without the need for further enhancement or resonance effects.
(2) Large Specific Surface Area: Porous materials can have a large proportion of surface atoms – their surface area within a typical sampling volume of 10^3 μm^3 can reach 10^5 μm^2, which is approximately 10^3 larger than for a smooth surface crossing the same volume. These effects lead to clearly increased Raman intensities of surface species and also to improved intensity ratios of surface and bulk Raman bands.
(3) Resonant Excitation: Excitation by a laser, which is resonant with an electronic transition of the material under investigation, can increase the Raman cross-section by approximately 10^2. The transitions and thus the resonance wavelengths are specific for the substances. Resonance excitation thus leads to selectivity that can be useful for suppressing bulk bands, but can also complicate the detection of mixtures of substance with different absorption spectra.
(4) Surface-enhancement: Electromagnetic and chemical effects can enhance the Raman intensities from substances in close proximity to appropriate metal surfaces by several orders of magnitude.
(5) Other Considerations: Another issue in Raman spectroscopy is the superposition of the weak Raman spectrum by the more intense fluorescence of the substances or their impurities. This problem can be overcome by the use of longer excitation wavelengths, which cannot excite electronic transitions. Although this concept works well for many samples, near-infrared excitation wavelengths result in lower sensitivity in comparison with visible excitation, because of the noise of detectors working in this range and because of the λ^{-4} dependence of the Raman scattering intensity.

Raman spectroscopy does not need vacuum and is, in principle, applicable to all optically accessible samples – even inclusions in glass or minerals, substances in aqueous solutions, or substances packaged in transparent glass or plastics can be analyzed. Small particles down to diameters below 1 μm can also be characterized by proper focussing of the exciting laser beam. Because transmission is not required for back-scattering measurements, strongly scattering materials can also be investigated.

4.8.2
Surface-Enhanced Raman Scattering (SERS)

The application of Raman spectroscopy in surface analysis is limited by the low scattering cross-section of the molecules and the correspondingly low intensities of their Raman bands. This situation can be substantially improved by use of so-called surface-enhanced Raman scattering (SERS) [4.288, 4.289]: The Raman scattering of molecules adsorbed by rough metal surfaces or by small metal particles (Fig. 4.56) can be increased by six, or more, orders of magnitude by electromagnetic and chemical effects. This large enhancement enables easy analysis of molecular monolayers at appropriate surfaces – even 1% of a monolayer has been observed and the detection of single molecules on optimum metal particles was recently reported [4.290].

The first SERS experiments were performed with electrochemically roughened electrodes and metal colloids, and many other types of suitable SERS substrates are known – e. g. metal island films, metal films over nanoparticles (see Fig. 4.58, below) or rough substrates, gratings, and sputter-deposited metal particles.

SERS is usually restricted to specially prepared "active" surfaces. A broad range of other surfaces also becomes accessible to SERS spectroscopy after special preparation techniques:

(1) Metal island films form spontaneously during slow evaporation of the metal on to supports with low adhesion. The metal islands can produce the surface-enhancement required for Raman characterization of the surface of the support.
(2) Appropriate metal particles can be sputter-deposited on almost any support which can be exposed to vacuum.
(3) Ultrathin metallic, semiconductor, insulator, or organic overlayers can be deposited on SERS-active metal surfaces.

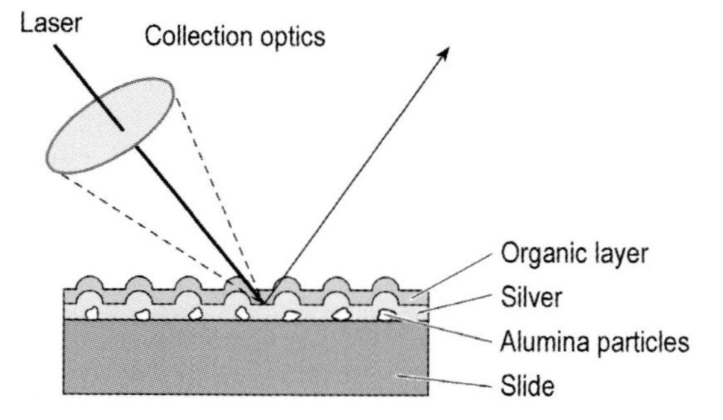

Fig. 4.56. Schematic diagram of a SERS-active substrate and the measurement arrangement. Alumina nanoparticles are deposited on a glass surface and produce the required roughness. A thin silver layer is evaporated on to the nanoparticles and serves for the enhancement. Organic molecules adsorbed on the silver surface can be detected by irradiation with a laser and collecting the Raman scattered light.

For overlayer thicknesses of a few atomic or molecular layers, the supporting metal can produce surface-enhanced fields at the surface of the overlayer. Then, composition and structure of the overlayer surface can be analyzed by SERS spectroscopy [4.291].

4.8.3
Instrumentation

The scattering effect was first observed in 1928 by C.V. Raman, who used sunlight and complementary filters [4.287]. Later, mercury vapor lamps were used for illumination of samples. Current Raman spectrometers use lasers as monochromatic light sources. Different types of laser can be used, depending on the wavelength, sensitivity, and spectral resolution required. Gas lasers such as Ar^+ and HeNe lasers have been quite common because of their narrow, well-defined spectral lines, stability, and beam quality. High-intensity Ar^+ ion lasers, in particular, are expensive to maintain, because they consume much electrical power and cooling water. Diode lasers are more convenient. Powerful diode lasers for Raman spectroscopy usually emit red wavelengths which are suitable for avoiding fluorescence and fit the sensitivity range of silicon detectors. Diode lasers must be frequency stabilized, because their emission frequency is not defined by narrow transitions. Spontaneous emission background and low-intensity side-bands must, furthermore, be filtered out when diode lasers are used for high-sensitivity Raman measurements. Solid-state lasers emitting in the near-infrared, e. g. Nd:YAG lasers are used in combination with Fourier-transform spectrometers.

Sample optics focus the laser on to the sample and collect the Raman scattered light. In the widespread-back-scattering or 180° arrangement a single objective serves for both, focusing the laser and collecting the scattered light. This makes measurements easy, because only the distance between sample and objective has to be adjusted. Because transparency of the whole sample is not needed with a back-scattering arrangement, Raman spectroscopy is well suited for in-situ measurements at interfaces between gases, liquids, and solids with at least one transparent phase.

Elastically scattered laser light must be removed in front of the spectrometer, because it is several orders of magnitude more intense than the Raman scattered radiation and would otherwise produce stray radiation at the detector. Nowadays, holographic notch filters are usually used for this purpose, because of their high transmission of required wavelengths and strong suppression of unwanted wavelengths. Premonochromators are preferable for measurements at low wavenumbers and for use of varying laser wavelengths.

Spectral analysis of the Raman scattered light is achieved by use of dispersive or Fourier-transform (FT) spectrometers. Dispersive spectrometers decompose the radiation according to its wavelength by means of a grating monochromator (Fig. 4.57). FT spectrometers calculate the spectrum from an interference pattern (interferogram) produced by an interferometer with a variable path-length within one of the interferometer arms. Monochromators are preferable in the visible spectral range. Compact and robust monochromators without moving parts can be used in rough environments, whereas larger high-resolution devices are mainly used in research.

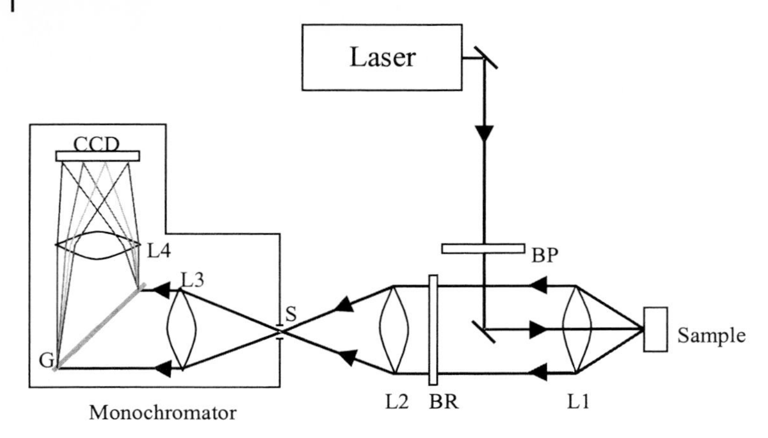

Fig. 4.57. Example of sensitive Raman equipment. The band pass filter, BP, cleans the laser radiation. The high NA objective lens L1 focuses the laser on the sample and collects the Raman scattered radiation within a large solid angle. The band-rejection filter, BR, blocks elastically scattered radiation. Imaging of the sample on the entrance slit, S, of the monochromator reduces stray radiation. Within the monochromator, the Raman scattered radiation is dispersed by a phase grating and finally focused on the CCD multichannel detector by lens L4.

In combination with array detectors, imaging monochromators can use the multichannel advantage and thus enable large signal-to-noise (S/N) ratios for short measurement times. FT spectrometers also afford high S/N ratios because of their throughput advantage. The noise of their near-infrared detectors is, however, considerably larger than that of detectors operating in the visible region. Major advantages of FT spectrometers are strong suppression of fluorescence, as a result of their use of long-wavelength excitation, and variable spectral resolution.

Charge-coupled device (CCD) detectors are preferred in the visible spectral range, because of their low dark current, high quantum efficiency, and multichannel capability. They must be cooled thermoelectrically or by use of liquid nitrogen for high-sensitivity applications, e. g. surface spectroscopy. Because the sensitivity of CCD detectors decreases considerably at wavelengths >1 μm, owing to the band-gap of the silicon material, other semiconductor detectors with larger dark current, e. g. germanium or InGaAs, must be used for near-infrared Raman spectroscopy.

Different accessories are available for Raman spectroscopy. Confocal microscopes afford lateral resolution of approximately 1 μm and depth resolution down to approximately 2 μm. The depth resolution is especially important for surface and interface spectroscopy, because it helps to eliminate Raman intensities from the bulk phases. Fiber optics can be used to guide the exciting laser light to the sample and the Raman scattered light to the spectrometer, thus making it easier to change samples and eliminating the hazards of freely propagating laser light. Special sample cells have been constructed for measurement at high or low temperatures and at high pressures, and for laser multi-pass through gaseous samples.

4.8.4
Spectral Information

For large molecules, at least, Raman spectra contain numerous bands which cannot always completely be assigned to particular vibrational modes. The large number of bands can, however, when measured with appropriate spectral resolution, enable unambiguous identification of substances by comparing the spectral pattern ("fingerprint") with those of reference spectra, if they are available.

Several bands in characteristic spectral regions can definitely be assigned to molecular or crystal vibrations, especially if information on the chemical composition of the sample is available. For example, these bands enable the identification of aromatic or aliphatic hydrogen, carbonyls, thiols, double or triple bonds, heterocycles and other groups or skeletal components. Vibrations of weakly polar and even symmetrically bonded atoms usually result in intense Raman bands because Raman activity is connected with changes in molecular polarizability during the vibration. This is a major difference from infrared spectroscopy, in which bands from vibrations causing changes of the dipole moment dominate the spectrum. Raman bands of carbon–carbon bonds are easily observable. Strong vibrational coupling along carbon chains and symmetry selection rules cause these bands to be sensitive markers of substitution at aromatic rings, and for conformation of the carbon backbone. Structural isomers usually have very different spectra and can, therefore, be easily distinguished.

Polarization effects are another feature of Raman spectroscopy that improves the assignment of bands and enables the determination of molecular orientation. Analysis of the polarized and non-polarized bands of isotropic phases enables determination of the symmetry of the respective vibrations. For aligned molecules in crystals or at surfaces it is possible to measure the dependence of up to six independent Raman spectra on the polarization and direction of propagation of incident and scattered light relative to the molecular or crystal axes.

4.8.5
Quantification

In Raman spectroscopy the intensity of scattered radiation depends not only on the polarizability and concentration of the analyte molecules, but also on the optical properties of the sample and the adjustment of the instrument. Absolute Raman intensities are not, therefore, inherently a very accurate measure of concentration. These intensities are, of course, useful for quantification under well-defined experimental conditions and for well characterized samples; otherwise relative intensities should be used instead. Raman bands of the major component, the solvent, or another component of known concentration can be used as internal standards. For isotropic phases, intensity ratios of Raman bands of the analyte and the reference compound depend linearly on the concentration ratio over a wide concentration range and are, therefore, very well-suited for quantification. Changes of temperature and the refractive index of the sample can, however, influence Raman intensities, and the band positions can be shifted by different solvation at higher concentrations or in different solvent mixtures. Although these effect can be negligible, they must be considered if high accuracy is required. Chemometric computations might be required for quantification of complex mixtures. They work much better than, e.g., in near-infrared spectroscopy, because of the usually narrow and well-resolved Raman bands.

Quantification at surfaces is more difficult, because the Raman intensities depend not only on the surface concentration but also on the orientation of the Raman scatterers and the, usually unknown, refractive index of the surface layer. If noticeable changes of orientation and refractive index can be excluded, the Raman intensities are roughly proportional to the surface concentration, and intensity ratios with a reference substance at the surface give quite accurate concentration data.

4.8.6 Applications

4.8.6.1 Unenhanced Raman Spectroscopy at Smooth Surfaces

Recent developments in Raman equipment has led to a considerable increase in sensitivity. This has enabled the monitoring of reactions of organic monolayers on glassy carbon [4.292] and diamond surfaces and analysis of the structure of Langmuir–Blodgett monolayers without any enhancement effects. Although this unenhanced surface-Raman spectroscopy is expected to be applicable to a variety of technically or scientifically important surfaces and interfaces, it nevertheless requires careful optimization of the apparatus, data treatment, and sample preparation.

Nitrophenyl groups covalently bonded to classy carbon and graphite surfaces have been detected and characterized by unenhanced Raman spectroscopy in combination with voltammetry and XPS [4.292]. Difference spectra from glassy carbon with and without nitrophenyl modification contained several Raman bands from the nitrophenyl group with a comparatively large signal-to-noise ratio (Fig. 4.58). Electrochemical modification of the adsorbed monolayer was observed spectrally, because this led to clear changes in the Raman spectrum.

Utilization of resonance effects can facilitate unenhanced Raman measurement of surfaces and make the technique more versatile. For instance, a fluorescein derivative and another dye were used as resonantly Raman scattering labels for hydroxyl and carbonyl groups on glassy carbon surfaces. The labels were covalently bonded to the surface, their fluorescence was quenched by the carbon surface, and their resonance Raman spectra could be observed at surface coverages of approximately 1%. These labels enabled assess to changes in surface coverage by C–OH and C=O with acidic or alkaline pretreatment [4.293].

Raman spectroscopy has also been used to determine the structure of $Fe(CN)_6^{3-}$ and its reaction products in chromate conversion coatings on an aluminum aircraft alloy, and to monitor coating growth rates and their dependence on coating bath composition. The Raman spectra of the coatings contained bands of Berlin green, a Fe^{3+}–CN–Fe^{3+} polymer, and $Fe(CN)_6^{3-}$ physisorbed on $Cr(OH)_3$. Results from different coating baths indicated a redox-mediation action for $Fe(CN)_6^{3-/4-}$ as the mechanism of acceleration of coating formation [4.294].

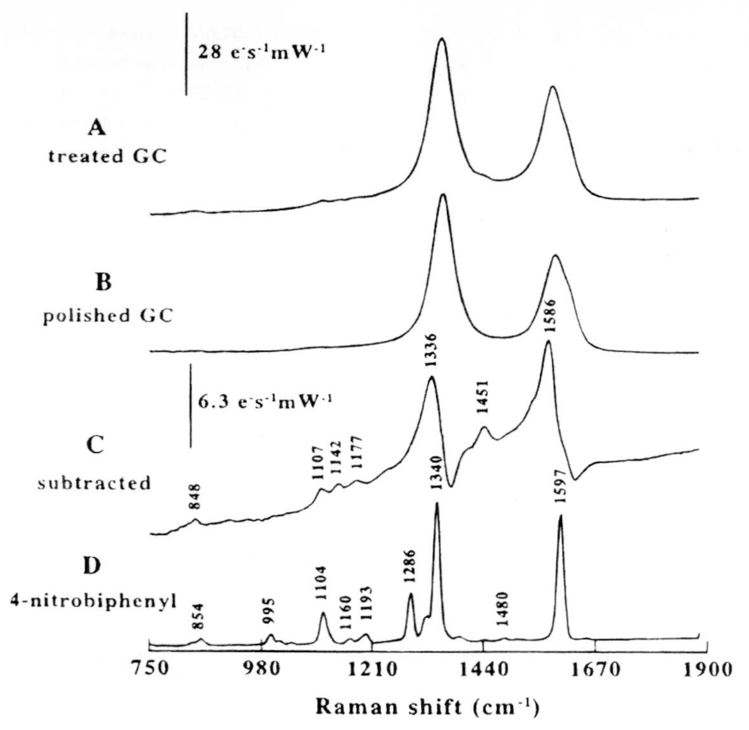

Fig. 4.58. Raman spectra of nitrophenyl modified (A) and untreated (B) glassy carbon, the difference between spectra A and B (C), and the reference spectrum of solid 4-nitrobiphenyl (D). The strongest surface bands at approximately 1336 and 1586 cm^{-1} might be affected by subtraction of closely neighboring, much stronger bands of the carbon bulk. Five further marked surface bands are clearly visible in the difference spectrum C [4.292].

4.8.6.2 **Porous Materials**

The surfaces of porous materials, e.g. catalysts, molecular sieves, or adsorbents, are much more readily accessible than smooth surfaces to Raman spectroscopy, because larger amounts of adsorbed substance can be placed within the laser focus, thus contributing to the scattering process.

Raman spectroscopy has provided information on catalytically active transition metal oxide species (e.g. V, Nb, Cr, Mo, W, and Re) present on the surface of different oxide supports (e.g. alumina, titania, zirconia, niobia, and silica). The structures of the surface metal oxide species were reflected in the terminal M=O and bridging M–O–M vibrations. The location of the surface metal oxide species on the oxide supports was determined by monitoring the specific surface hydroxyls of the support that were being titrated. The surface coverage of the metal oxide species on the oxide supports could be quantitatively obtained, because at monolayer coverage all the reactive surface hydroxyls were titrated and additional metal oxide resulted in the formation of crystalline metal oxide particles. The nature of surface Lewis and Brønsted acid sites in supported metal oxide catalysts has been determined by adsorbing probe molecules such as pyridine. Information on the behavior of the surface metal oxide species during catalytic reactions has been provided by in-situ characterization studies [4.295]. Investigations of synergy effects were performed on ^{18}O-exchange and reoxidation of SbO_x–O_3 physical mixtures. Peroxide species were detected on working methane-coupling catalysts at temperatures up to 1070 K, and were identified as centers for the activation of methane [4.296].

The diffusion, location and interactions of guests in zeolite frameworks has been studied by in-situ Raman spectroscopy and Raman microscopy. For example, the location and orientation of crown ethers used as templates in the synthesis of faujasite polymorphs has been studied in the framework they helped to form [4.297]. Polarized Raman spectra of *p*-nitroaniline molecules adsorbed in the channels of $AlPO_4$-5 molecular sieves revealed their physical state and orientation – molecules within the channels formed either a phase of head-to-tail chains similar to that in the solid crystalline substance, with a characteristic ω_3 band at 1282 cm^{-1}, or a second phase, which is characterized by a similarly strong band around 1295 cm^{-1}. This second phase consisted of weakly interacting molecules in a pseudo-quinonoid state similar to that of molten *p*-nitroaniline [4.298].

4.8.6.3 SERS

Because silver, gold and copper electrodes are easily activated for SERS by roughening by use of reduction–oxidation cycles, SERS has been widely applied in electrochemistry to monitor the adsorption, orientation, and reactions of molecules at those electrodes in-situ. Special cells for SERS spectroelectrochemistry have been manufactured from chemically resistant materials and with a working electrode accessible to the laser radiation. The versatility of such a cell has been demonstrated in electrochemical reactions of corrosive, moisture-sensitive materials such as oxyhalide electrolytes [4.299].

Langmuir–Blodgett films (LB) and self-assembled monolayers (SAM) deposited on metal surfaces have been studied by SERS spectroscopy in several investigations. For example, mono- and bilayers of phospholipids and cholesterol deposited on a rutile prism with a silver coating have been analyzed in contact with water. The study showed that in these models of biological membranes the second layer modified the fluidity of the first monolayer, and revealed the conformation of the polar head close to the silver [4.300].

Coating of metal SERS substrates with organic receptor layers makes SERS suitable for highly sensitive chemo-optical sensors for organic trace analysis in water and air [4.301, 4.302]. The combination of selective adsorption by the receptor with measurement of molecule-specific SERS spectra can reduce cross-sensitivities considerably. Inclusion complexation with immobilized calix[4]arene receptors led to clearly decreased detection limits for aromatic compounds without groups that can become attached to metals. SERS bands of the receptor were used as internal standards for the surface concentration of the aromatic compounds. The intensity ratios of the aromatic compound and receptor bands gave a measure of the concentration of the solution of the aromatic compounds with a dynamic range of two orders of magnitude.

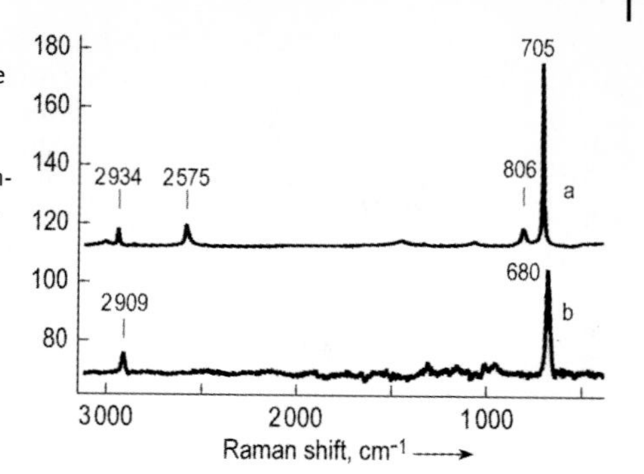

Fig. 4.59. Raman spectrum of methyl mercaptan (a) and SERS spectrum of methyl mercaptide (b) formed by adsorption of the mercaptan on a silver surface. The surface reaction is proven by the disappearance of the S–H stretching and bending bands at 2575 cm^{-1} and 806 cm^{-1}, respectively. The Raman shift of the C–S stretching band at approximately 700 cm^{-1} is reduced during adsorption by withdrawal of electron density from the C–S, because of bonding to the silver. The symmetric methyl stretching appears above 2900 cm^{-1} [4.303].

The enthalpy of the adsorption of chlorobenzene was determined from the temperature-dependence of the SERS intensities [4.301].

SERS substrates with bare metal surfaces irreversibly adsorb thioorganics (Fig. 4.59) and other compounds and can thus serve for the detection and identification of very low gas or solution concentrations of these substances [4.303]. SERS is especially well suited for the analysis of traces of gases, because it combines measurement of surface concentration with extremely high sensitivity. A monolayer in a typical focus of a laser with a diameter of 10 μm has a mass in the range of 10 femtograms! even smaller amounts of substance are easily detectable, because 1% of a monolayer in a region 1-μm in diameter results in SERS of sufficient intensity.

SERS has also been applied as a sensitive, molecule-specific detection method in chromatography, e.g. thin layer, liquid, and gas chromatography. SERS-active colloids were deposited on the thin layer plates or mixed continuously with the liquid mobile phases. After adsorption of the analytes, characteristic spectra of the fractions were obtained and enabled unambiguous identification of very small amounts of substance.

The label-free detection of biomolecules is another promising field of application for SERS spectroscopy. Tiniest amounts of these molecules can be adsorbed by specific interactions with receptors immobilized on SERS-active surfaces. They can then be identified by their spectra, or specific interactions can be distinguished from unspecific interactions by monitoring characteristic changes in the conformation sensitive SERS spectra of the receptors.

4.8.6.4 **Near-Field Raman Spectroscopy**

Raman scattering excited by near-fields close to either a small aperture of a tapered optical fiber, a single small field-enhancing particle, or an accordingly prepared tip of a scanning probe microscope enables analysis with a spatial resolution beyond the diffraction limit. This brand-new technique enables not only characterization of lateral structures with nanometer dimensions, but also measurement of very thin layers without strong background from the bulk. Initial investigations demonstrated the detection of dye and C_{60} molecules on glass and metal surfaces [4.304–4.306] (Fig. 4.60). Further optimization of field-enhancing probes is expected to result in sensitivity sufficient for the detection of non-resonantly scattering molecules on a wide variety of surfaces.

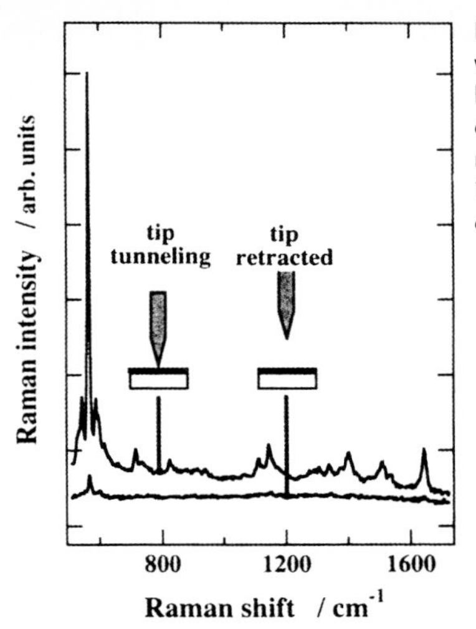

Fig. 4.60. Comparison of resonance Raman spectra with and without tip enhancement for 0.5 monolayers of brilliant cresyl blue on a smooth gold film. The tip increased the total Raman intensity by a factor of approximately 15, when positioned at a tunneling distance of 1 nm. Several other bands were made visible as a result of the tip enhancement [4.306].

4.8.7
Non-linear Optical Spectroscopy

Unlike linear optical effects such as absorption, reflection, and scattering, second order non-linear optical effects are inherently specific for surfaces and interfaces. These effects, namely *second harmonic generation* (SHG) and *sum frequency generation* (SFG), are dipole-forbidden in the bulk of centrosymmetric media. In the investigation of isotropic phases such as liquids, gases, and amorphous solids, in particular, signals arise exclusively from the surface or interface region, where the symmetry is disrupted. Non-linear optics are applicable in-situ without the need for a vacuum, and the time response is rapid.

Surface SHG [4.307] produces frequency-doubled radiation from a single pulsed laser beam. Intensity, polarization dependence, and rotational anisotropy of the SHG provide information about the surface concentration and orientation of adsorbed molecules and on the symmetry of surface structures. SHG has been successfully used for analysis of adsorption kinetics and ordering effects at surfaces and interfaces, reconstruction of solid surfaces and other surface phase transitions, and potential-induced phenomena at electrode surfaces. For example, orientation measurements were used to probe the intermolecular structure at air–methanol, air–water, and alkane–water interfaces and within mono- and multilayer molecular films. Time-resolved investigations have revealed the orientational dynamics at liquid–liquid, liquid–solid, liquid–air, and air–solid interfaces [4.307].

SFG [4.309, 4.310] uses visible and infrared lasers for generation of their sum frequency. Tuning the infrared laser in a certain spectral range enables monitoring of molecular vibrations of adsorbed molecules with surface selectivity. SFG includes the capabilities of SHG and can, in addition, be used to identify molecules and their structure on the surface by analyzing the vibration modes. It has been used to observe surfactants at liquid surfaces and interfaces and the ordering of interfacial water molecules. Despite the power of SFG in surface investigations, applications are still limited by the shortage of powerful tunable mid-infrared lasers and by the experimental expense.

4.9 UV–Vis–IR Ellipsometry (ELL)

Bernd Gruska and Arthur Röseler

4.9.1 Principles

Ellipsometry is a method of measuring the film thickness, refractive index, and extinction coefficient of single films, layer stacks, and substrate materials with very high sensitivity. Rough surfaces, interfaces, material gradients and mixtures of different materials can be analyzed.

Film thicknesses between 0.1 nm and 100 μm can be measured, depending on the spectral range used for the analysis and the homogeneity of the thicker films. Thicknesses <1 μm can be determined with a sensitivity better than 0.01 nm. Thicknesses in the micron-range can be analyzed with sensitivity typically better than 1 nm.

The refractive index of a film or a substrate material can be measured with a sensitivity better than 5×10^{-4}, the best available for non-invasive optical measurement methods, especially for thin films. The extinction coefficient can be measured with almost the same sensitivity, which corresponds to a lower limit of 10–100 cm^{-1} for the absorption coefficient of the material.

The number of measurable layers of a stack is limited only by the optical contrast between the different layers. In practice stacks of ten layers and more can be analyzed by ellipsometry. Further advantages of ellipsometry compared with other metrological methods are the non-invasive and non-destructive character of the optical method, the low energy entry into the sample, the direct measurement of the dielectric function of materials, and the possibility of making the measurement in any kind of optical transparent environment.

The principal of measurement is shown schematically in Fig. 4.61. Linear polarized light is reflected from a sample surface which must be flat and sufficiently reflecting. The state of polarization of the incident light is changed, by reflection, into ellipti-

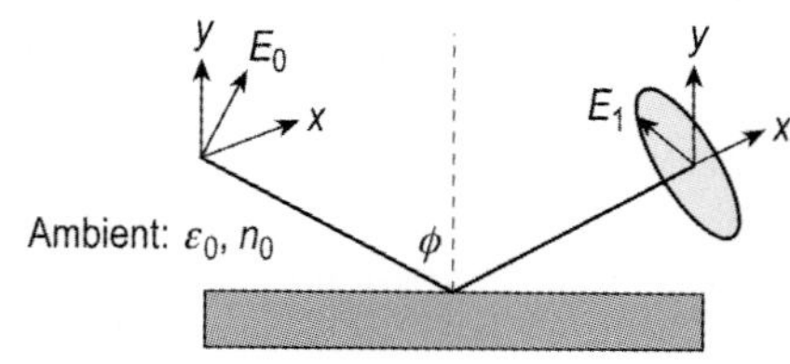

Fig. 4.61. Principle of measurement in ellipsometry.

cally polarized light. The properties measured by ellipsometry are tan_, the amplitude ratio of the resolved components of the electric field vector of the reflected light parallel to and perpendicular to the plane of incidence, and cos_, the phase difference between the two components. Unambiguous determination of the phase angle, Δ, over the whole data range between 0° and 360° is possible using a retarder to direct circular polarized light as incident light in a second measurement – $\tan\Psi$ and $\sin\Delta$ are determined.

Information about the properties of the sample are contained in the complex ratio, ρ, of the Fresnel coefficients of reflection of the parallel (r_P) and perpendicular (r_S) incident plane polarized electrical field vectors.

$$\rho = \frac{r_P}{r_S} \tag{4.31}$$

The fundamental equation of ellipsometry [4.311]:

$$\rho = \tan \Psi e^{i}\Delta \tag{4.32}$$

describes the connection between the measured quantities Ψ and Δ and the sample properties contained in the coefficients of reflection, and hence in ρ. In principal this complex equation can be solved analytically for pure substrates only. Each other more complex structure requires optical modeling of the sample and fitting of the calculated ratio, ρ, to the measured quantities Ψ and Δ. Ψ and Δ are always correct quantities assuming the ellipsometer is working correctly. Film thickness, refractive index, extinction coefficient and other properties are calculated quantities based on a model. The choice of the correct optical model for the sample is a fundamental assumption for correct values of all the properties calculated.

The thickness of a film influences the interference of light waves reflected from the front and back of the film, and hence the reflectance. The thickness of an absorbing film can, therefore, be measured only as long as there is still a contribution of from the back of the film to the reflectance of the sample. Typical measurable thicknesses of metallic layers are < 50 nm.

Ellipsometric measurements depends on the incident angle, ϕ, because of the angular dependence of the coefficients of reflection r_S and r_P. Large differences between both quantities are found for angles $>50^\circ$, where r_P has a local minimum at the so-called Brewster angle. This angle is between 50 and 60° for transparent materials and $>70^\circ$ for almost all absorbing materials. Ellipsometric measurements close to the Brewster angle of the substrate are especially sensitive for very thin layers on top of the substrate. Multiple angle measurements are suitable for confirming optical models of sample structure.

Two measured ellipsometric angles Ψ and Δ at a fixed wavelength and a fixed angle of incidence enable calculation of a maximum of two other properties, e. g. the film thickness and refractive index of a transparent layer. Multiple angle measurements increase the number of measured quantities and hence the number of properties which can be determined for a specific sample, although even under these conditions the number of measurable properties will typically be no larger than 4–5. Wavelength-dependent ellipsometric measurements (spectroscopic ellipsometry) further increase the number of measurable properties, and hence the complexity of the analyzable sample structure. Spectroscopic ellipsometry enables measurement of the dispersion of the refractive index and the extinction coefficient of materials, and the analysis of more complex sample structures (e.g. multilayer stacks, interfaces, material gradients, material compositions) on the basis of parameterized dielectric functions which drastically reduce the number of unknown properties compared with the number of measured ellipsometric angles.

Spectroscopic ellipsometry is sensitive to the dielectric functions of the different materials used in a layer stack. But it is not a compositional analytical technique. Combination with one of the compositional techniques, e.g. AES or XPS and with XTEM, to furnish information about the vertical structure, can provide valuable additional information enabling creation of a suitable optical model for an unknown complex sample structure.

4.9.2 Instrumentation

The basic configuration of each ellipsometer is very simple: light source, polarizer, sample, polarizer, detector. Advanced configurations use a phase shift device (retarder) before or after the sample. There is a large number of possibilities in which a specific ellipsometer can be realized. The main differences consist in the way the light is polarized and the way the state of polarization of the reflected light is detected. The light source can be monochromatic (lasers), white light (xenon arc lamp, deuterium lamp, halogen lamp) with a monochromator, or silicon carbide rods (“*glowbars*” – used mainly for infrared ellipsometry).

Photomultiplier tubes (vacuum UV), silicon photodiodes (UV–Vis), Ge- or InGaAs photodiodes (near IR), MCT (HgCdTe), or DTGS detectors (mid-MIR) are used as broadband detectors. They must be operated with a monochromator if a white light source is used. Fast spectroscopic ellipsometers use photometers and diode arrays as detectors. This enables simultaneous detection of multiple wavelengths, rather than sequential detection. State-of-the-art ellipsometry in the infrared spectral range (approximately 0.8–100 μm wavelength) uses modulated white light from an FTIR instrument and broadband detectors to measure the reflected light. Intensity spectra are generated by Fourier transformation of the detected signal. Ellipsometric measurements are performed at a fixed angle of incidence. Each ellipsometer configuration can be operated in a discrete wavelength or spectroscopic (i.e. variable angle of incidence) manner.

Null ellipsometers are among the oldest configurations. In these a revolving polarizer and a revolving compensator are used to change the state of polarization of the incident light and a revolving polarizer (analyzer) is used to analyze the reflected light. The angular position of the optical elements is changed until the intensity of the light on the detector is zero. This configuration is one of the most accurate but also the slowest, even if automated.

Rotating element ellipsometers have a continuously rotating element (10–40 Hz frequency of revolution) which enables automated acquisition and faster measurement. Most of the instruments used nowadays have this configuration. The polarizer (rotating polarizer ellipsometer, RPE) or analyzer (rotating analyzer ellipsometer, RAE) can be used as the rotating element. The detector signal is measured as function of time and is then Fourier-analyzed to obtain the ellipsometric angles Ψ and Δ. The RPE configuration has the advantage that any detector (photomultiplier tube, monochromator) polarization sensitivity does not influence the measurement. The state of polarization of the detected light is fixed by the fixed analyzer position. The disadvantage is the sensitivity to residual polarized light of the light beam incident on the rotating polarizer. The RAE configuration has the advantage of being insensitive to any residual polarization of the light source (laser, white light source with monochromator), but errors can occur as a result of the polarization sensitivity of the detector (photomultiplier tube). If polarization-insensitive Si photodiodes are used as detectors this configuration is most often used for discrete wavelength ellipsometers. Single wavelength ellipsometers are typically equipped with a quarter-wave plate (compensator) which can be moved into the light beam. The linearly polarized light is changed into circularly polarized (in general elliptically polarized) light after passing through the quarter-wave plate. The measured ellipsometric angle, Δ, is thereby shifted by 90°; it can be determined in the whole data range of 0–360°, and the large uncertainty of Δ is removed when the value of Δ is close to 0° or 180°.

Spectroscopic ellipsometers that measure Δ in the whole data range require an achromatic compensator with a phase shift close to 90° over a large spectral range. The compensator can be located between polarizer and sample or between sample and analyzer.

Step scan polarizer/analyzer (SSP/SSA) and continuously rotating element configurations are used for diode-array- or FTIR-based ellipsometers. With the SSP or SSA modes complete intensity spectra are measured at several angle positions of the revolving polarizer/analyzer. The acquisition time varies between seconds (UV-Vis) and minutes (IR), depending on the intensity of the incident beam and the reflectance of the sample. The measurements are rapid compared with the monochromator based configurations and the positioning of the revolving element is highly exact. The continuously rotating element configuration is used for very fast spectroscopic ellipsometers measuring complete ellipsometric spectra in less than 100 ms. For such instruments the intensity at each wavelength (pixel of the diode array) is accumulated over a 45° revolution angle of the rotating element.

A few instruments have a continuously rotating compensator and fixed polarizer and analyzer. The advantage of this configuration is insensitivity to residual polarization of the light source or to polarization sensitivity of the detector. The disadvantage is the large effect of a small misalignment of the rotating achromatic compensator on the performance of the instrument.

Polarization modulation ellipsometers use a photo-elastic modulator to modulate the state of polarization of the incident beam. Polarizer and analyzer are fixed during the measurement. Fourier analysis of the time dependent signal gives the ellipsometric parameters Ψ and Δ. The advantages of this configuration are the very high measuring speed (typical modulation frequency 50 kHz) for discrete wavelengths and the lack of moving parts. The ellipsometric angle, Δ, can be measured over the whole range of 0–360°. Disadvantages are the frequent calibration of the instrument owing to the high temperature-dependence of the polarization modulator and the limited measurement speed for complete spectra.

The infrared ellipsometer is a combination of a Fourier-transform spectrometer (FTS) with a photometric ellipsometer. One of the two polarizers (the analyzer) is moved step by step in four or more azimuths, because the spectrum must be constant during the scan of the FTS. From these spectra, the $\tan\Psi$ and $\cos\Delta$ spectra are calculated. In this instance only Δ is determined in the range 0–180°, with severely reduced accuracy in the neighborhood of 0° and 180°. This problem can be overcome by using a retarder (compensator) with a phase shift of approximately 90° for a second measurement –$\cos\Delta$ and $\sin\Delta$ are thereby measured independently with the full Δ information [4.315].

In a second kind of infrared ellipsometer a dynamic retarder, consisting of a photoelastic modulator (PEM), replaces the static one. The PEM produces a sinusoidal phase shift of approximately 40 kHz and supplies the detector exit with signals of the ground frequency and the second harmonic. From these two frequencies and two settings of the polarizer and PEM the ellipsometric spectra are determined [4.316]. This ellipsometer system is mainly used for rapid and relative measurements.

4.9.3 Applications

4.9.3.1 UV–Vis–NIR Spectral Range

Spectroscopic ellipsometry in the UV–Vis–NIR is mainly used for the characterization of new materials (e.g. photoresist, polymers, low-*k* dielectrics, semiconductors, composed material), the analysis of multilayer stacks (e.g. low-*k* and AR layer stacks on architecture glass, flat panel applications, silicon IC technology, optoelectronic devices, photonic devices), qualification of a large number of deposition processes for production and R&D applications, and in-situ monitoring of growth and deposition processes. Spectral range and speed of the measurement are key features which strongly force the use of spectroscopic ellipsometers for R&D and for production surveillance. The availability of spectroscopic ellipsometers in the spectral range 140–2.3 μm enables many applications in metrology.

Ellipsometry in the vacuum UV (< 190 nm) enables the analysis of materials for the next generation lithography (photoresist, AR coatings) at the latest exposure wavelengths (157 nm and 193 nm). The short wavelengths increase the sensitivity of ellipsometric measurements of ultra thin films (< 10 nm). New prospects are expected for the analysis of thin metallic and dielectric layers.

In the UV most of the materials of interest, e.g. Si, polysilicon, SiGe, GaAs, and other semiconductor materials, are strongly absorbing; this enables surface-sensitive measurements, Surface roughness, native oxide covering, material composition, and structural properties can be analyzed.

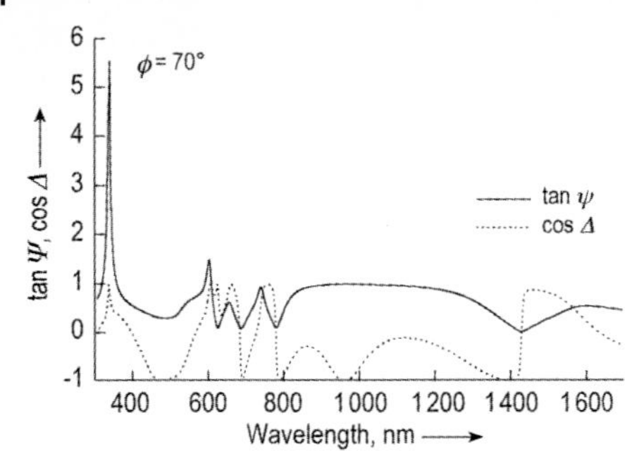

Fig. 4.62. Measured ellipsometric spectra of an a-Si/SiN multilayer stack (seven layers) on GaAs.

The Vis-NIR spectral range is mainly used to measure the thickness of single films and layer stacks. Many materials (dielectrics, semiconductors, polymers) are transparent in this spectral range. The analysis of complex layer stacks often requires a knowledge of the dielectric function of each material in the stack or parameterized dispersion models to reduce the number of unknown parameters. The Cauchy formula is often used for transparent layers, the Drude–Lorentz formula works well for metals and infrared active materials, the Leng formula [4.312] can be used for semiconductor materials, and the Tauc-Lorentz formula [4.313] works best for amorphous materials. The general Gaussian-broadened polynomial superposition (GBPS) parametric dispersion model [4.314] can be broadly applied to most materials, including crystalline and amorphous semiconductors, metals, and organic compounds. Fig. 4.62 shows measured ellipsometric spectra of an a-Si/Si_xN_y multilayer deposited on GaAs. Four layers of Si_xN_y and three layers of a-Si were deposited alternately. The dispersion of each layer material was measured on single films. Film thicknesses of 78 nm for a-Si and 145 nm for Si_xN_y were determined for the different layers of the layer stack, by use of appropriate parameterized dispersion models for each material. Other typical multilayer applications are dielectric stacks on silicon, spectral selective mirrors stacks, or Bragg reflectors for laser devices.

Ellipsometric measurements in the NIR are often used to determine the thickness and refractive index of thicker dielectric films and semiconductor stacks based on InP and GaAs, which are transparent in this spectral region. For highly sensitive ellipsometric measurement of film thickness and refractive index the wavelength and film thickness used should not differ much. Fig. 4.63 shows ellipsometric spectra obtained from a 2.5 μm thick SiO_2 layer on silicon measured between the wavelengths 350 nm and 2.3 μm. The higher sensitivity in the near-infrared spectral range is obvious. The growing number of applications of 1.55 μm semiconductor lasers has resulted in the use of near-infrared ellipsometers for control and monitoring of AR coatings on laser facets.

Laterally inhomogeneous films and patterned structures of microelectronic and optoelectronic applications require small measuring spots. Today's measurements in 50 μm × 50 μm areas are standard for μ-spot spectroscopic ellipsometers used in fablines. Areas more than ten times smaller can be analyzed by use of discrete-wavelength ellipsometers equipped with laser-light sources.

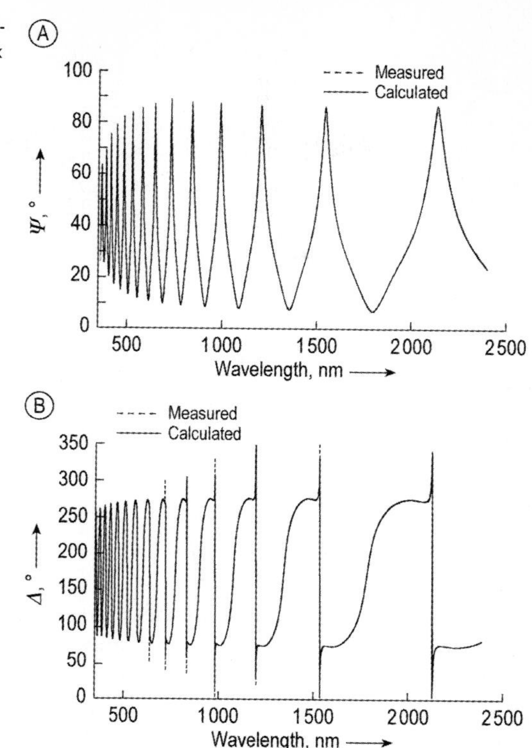

Fig. 4.63. Measured and calculated ellipsometric spectra from a 2.5 µm thick SiO_2 layer on Si.

4.9.3.2 **Infrared Ellipsometry**

Infrared ellipsometry is typically performed in the mid-infrared range of 400 to 5000 cm^{-1}, but also in the near- and far-infrared. The resonances of molecular vibrations or phonons in the solid state generate typical features in the $\tan\Psi$ and Δ spectra in the form of relative minima or maxima and dispersion-like structures. For the isotropic bulk calculation of optical constants – refractive index n and extinction coefficient k – is straightforward. For all other applications (thin films and anisotropic materials) iteration procedures are used. In ellipsometry only angles are measured. The results are also absolute values, obtained without the use of a standard.

Two types of resonance must be distinguished in the infrared, that of strong and weak oscillators. The strong oscillator is characterized by a region in which the refractive index is <1 for wave numbers higher than the oscillator frequency; this is usually observed for inorganic compounds with reststrahlen bands (n, k spectrum of quartz glass, see Fig. 4.67 b, below). Weak oscillators, typically organic compounds, always have refractive indexes >1 (Fig. 4.64 a). The optical constants of these spectra are calculated by use of relationships derived for metal optics [4.317]. From ellipsometrically determined optical constants other experimental configurations can be calculated. Figure 4.64 b shows the transmission spectrum calculated from the spectra in Fig. 4.64 a; this is now is suitable for direct comparison with library spectra.

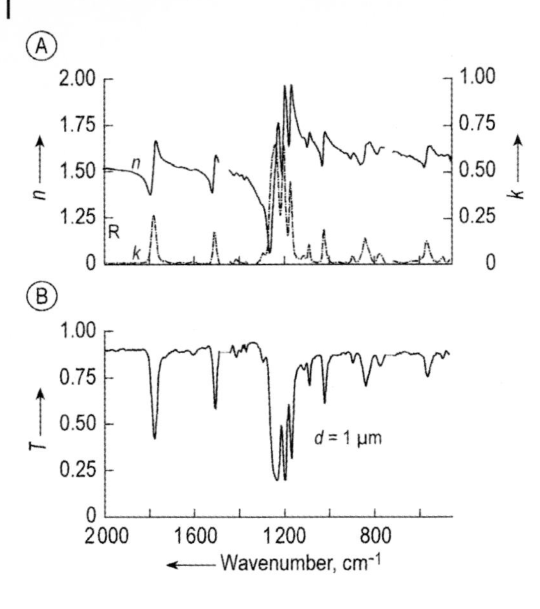

Fig. 4.64. (a) Bulk spectrum of polycarbonate *n* and *k*; (b) with *n*, *k* calculated transmission spectrum for a thickness of 1 µm.

The spectra from strong oscillators have special features which are different from those from metallic and dielectric substrates. Different structures in $\tan\Psi$ and Δ are observed on a metallic substrate, dependent on the thickness of the film (Fig. 4.65). For very thin films up to approximately 100 nm the Berreman effect is found near the position of $n = k$ and $n < 1$ with a shift to higher wavenumbers in relation to the oscillator frequency. This effect decreases with increasing thickness ($d \geq$ approx. 100 nm) and is replaced by excitation of a surface wave at the boundary of the dielectric film and metal. The oscillator frequency (TO mode) can now also be observed. On metallic substrates for thin films ($d \leq$ approx. 2 µm) only the z-component of the electric field is relevant. With thin films on a dielectric substrate the oscillator frequency and the Berreman effect are always observed simultaneously, because in these circumstances all three components of the electric field are possible (Fig. 4.66).

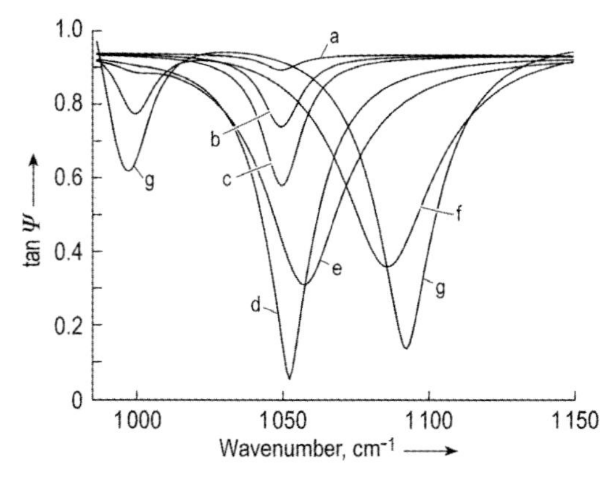

Fig. 4.65. Different spectral features of $\tan\Psi$ for a strong model oscillator at 1000 cm^{-1} on a metal substrate. The TO mode (1000 cm^{-1}), Berreman effect (1050 cm^{-1}), and excitation of a surface wave (1090 cm^{-1}) are seen for different thicknesses – 1, 5, 10, 50, 100, 500, and 1000 nm.

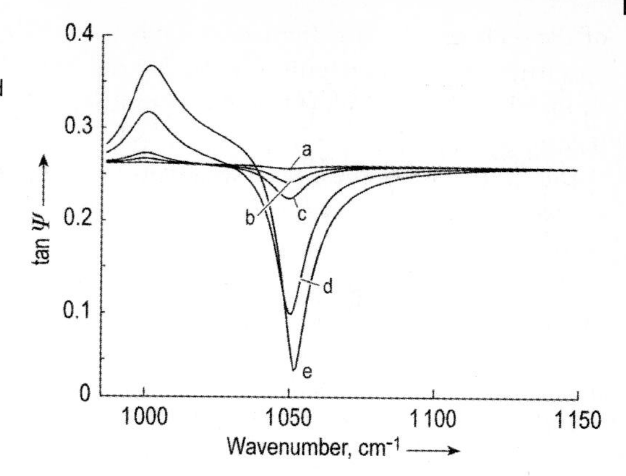

Fig. 4.66. tanΨ spectra for the same oscillator as in Fig. 4.65 for a silicon substrate –thicknesses 1, 5, 10, 50, and 100 nm.

The relative slopes of tanΨ (maximum or minimum), Δ, and reflectivity for the parallel component depend on the position relative to the Brewster angle (greater or lower). Thicker films on both types of substrate lead to dominant interferences.

Double and buried layers are readily detected by infrared ellipsometry if the film material is characterized by reststrahlen bands. The response is observed at the position

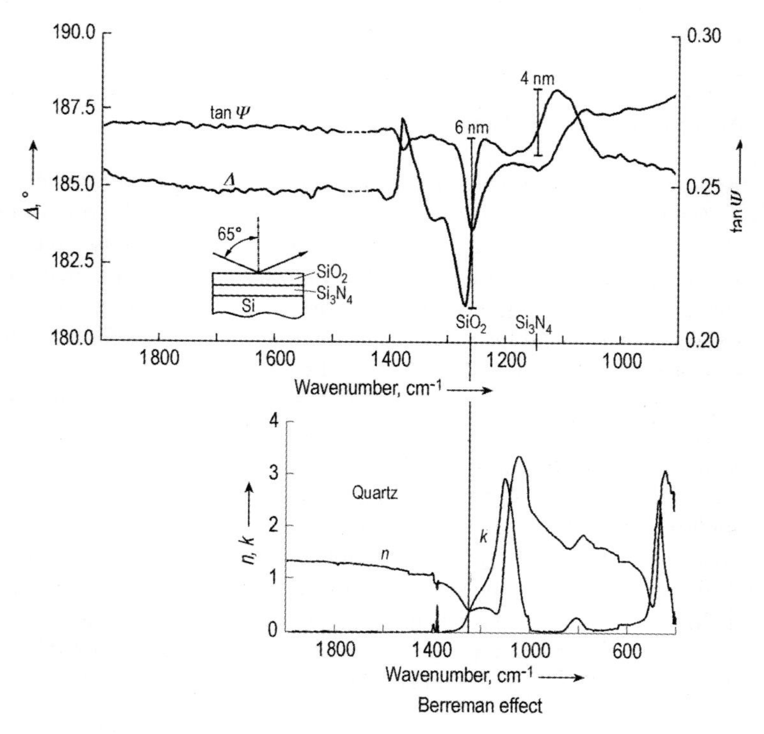

Fig. 4.67. (a) tanΨ and δ spectrum of a double layer of SiO_2 (6 nm) and Si_3N_4 (4 nm) on a silicon substrate, (b) SiO_2 spectrum for comparison.

of the Berreman effect (Figs 4.67 a and b for SiO_2 and Si_3N_4 layers, and the SiO_2 spectrum for comparison) and also by the TO mode for greater thicknesses. The thinnest film was observed by the H-terminated silicon surface with a thickness of 0.2 nm [4.318].

Determination of the optical constants and the thickness is affected by the problem of calculating three results from two ellipsometric values. This problem can be solved by use of the oscillator fit in a suitable wavenumber range or by using the fact that ranges free from absorption always occur in the infrared. In these circumstances the thickness and the refractive index outside the resonances can be determined – by the algorithm of Reinberg [4.317], for example. With this result only two data have to be calculated.

It is usually possible to investigate very thin films (up to the subnanometer range) by use of infrared wavelengths, which are much greater than the thickness of the film (a factor of 10000) because of the interference optics of the strong oscillator (Berreman effect).

4.10 Other Photon-detecting Techniques

John C. Rivière

4.10.1 Appearance Potential Methods

Appearance potential methods all depend on detecting the threshold of ionization of a shallow core level and the fine structure near the threshold; they differ only in the way in which detection is performed. In all of these methods the primary electron energy is ramped upward from near zero to whatever is appropriate for the sample material, while the primary current to the sample is kept constant. As the incident energy is increased, it passes through successive thresholds for ionization of core levels of atoms in the surface. An ionized core level, as discussed earlier, can recombine by emission either of a characteristic X-ray photon or of an Auger electron.

4.10.1.1 Soft X-ray Appearance Potential Spectroscopy (SXAPS)

In SXAPS the X-ray photons emitted by the sample are detected, normally by letting them strike a photosensitive surface from which photoelectrons are collected, but also – with the advent of X-ray detectors of increased sensitivity – by direct detection. Above the X-ray emission threshold from a particular core level the excitation probability is a function of the densities of unoccupied electronic states. Because two electrons are involved, incident and the excited, the shape of the spectral structure is proportional to the self-convolution of the unoccupied state densities.

4.10.1.2 **Disappearance Potential Spectroscopy (DAPS)**

Crossing an ionization threshold means that electrons are lost from the primary beam as a result of ionization of a core hole. Thus if the reflected current of electrons at the primary energy, more usually termed the elastically reflected current, is monitored as a function of energy, a sharp decrease should be observed as a threshold is crossed. This is the principle of operation of DAPS. It is, in a sense, the inverse of AEAPS, and, indeed, if spectra from the two techniques from the same surface are compared, they can be seen to be mirror images. Background problems occur in DAPS also.

The principal advantages of AEAPS and DAPS over SXAPS is that they can be operated at much lower primary electron currents, thus causing less disturbance to any adsorbed species.

4.10.2
Inverse Photoemission Spectroscopy (IPES) and Bremsstrahlung Isochromat Spectroscopy (BIS)

Irradiation of a surface with electrons leads to emission not only of X-ray photons of energies characteristic of the material, but also of a continuous background of photons called bremsstrahlung radiation. If a detector is set to detect only those background photons of a particular energy, and the primary electron energy is ramped upward from zero, the variations in photon flux should mirror variations in the densities of unoccupied electron states. The process is called inverse photoemission, because it is clearly the opposite of ordinary photoemission, as observed in XPS and UPS. Although the name *inverse photoemission spectroscopy* (IPES) should apply to all forms of the inverse photoemission technique, it is, in fact, confined by usage to the form in which the energy at which the photons are detected is in the UV region, typically with $h\nu = 9.7$ eV. If, on the other hand, a crystal monochromator normally used to provide monochromatized X-rays for XPS is used in reverse, with a detector placed where the electron source is in Fig. 2.3, detection is then at the X-ray energy $h\nu = 1486.6$ eV, and the technique is called *bremsstrahlung isochromat spectroscopy* (BIS).

Because IPES maps the densities of unoccupied states, it is related to other techniques that do the same (e.g. STS and SXAPS). When used in conjunction with a technique that maps the densities of occupied surface states, e.g. UPS or ELS, a continuous spectrum of state density from occupied to unoccupied can be obtained. Just as in UPS, in which angular resolution enables elucidation of the three-dimensional occupied band structure, so in IPES angular resolution enables mapping of the three-dimensional unoccupied band structure. This version is called KRIPES (i.e. *K-resolved* IPES).

5
Scanning Probe Microscopy

Gernot Friedbacher

The invention of the *scanning tunneling microscope* (STM) by Binnig and Rohrer in 1981 [5.1, 5.2], has triggered a tremendous development of *scanning probe microscopy* (SPM) techniques [5.3]. The basic principle all scanning probe microscopes have in common, is shown in Fig. 5.1. A sharp probe (e. g. tip, optical fiber, pipette) is raster-scanned across a sample surface by means of piezoelectric translators, while a certain signal is recorded by the probe for every single image point. Since the size of the probe's apex and the distance between probe and sample surface can be smaller than the wavelength of the analytical signal (e. g. electrons, electromagnetic radiation) information can be obtained in the so called near-field and thus, the resolution is no longer diffraction limited as given by the wavelength of the signal. Since the measured signal constitutes local information of the surface below the probe, the term local probe technique can be found frequently. The most important perspective of this concept is to use the local signal for monitoring the distance between probe and surface. Thus, topographical information – in principle down to atomic resolution – can be obtained in real space. Moreover, the possibility of positioning sharp probes above a surface with Angstrom and sub-Angstrom precision, opens up the exciting potential of accessing local spectroscopic information even confined to spots as small as a single atom.

Although a large number of scanning probe microscopy techniques utilizing various signal generation mechanisms has emerged in recent years, *atomic force microscopy* and *scanning tunneling microscopy* are the most important techniques with a wide range of applications in science and technology.

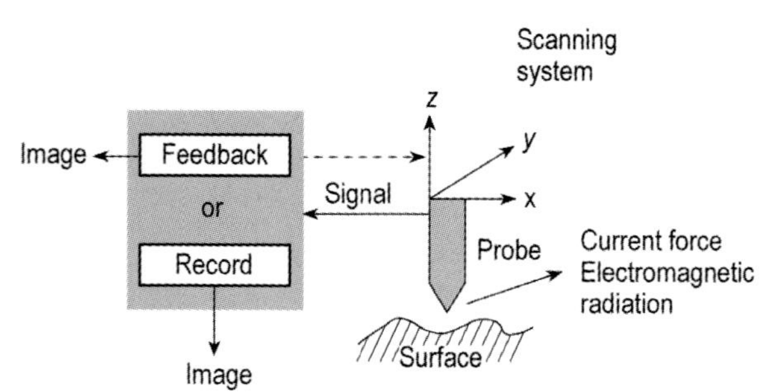

Fig. 5.1. General scheme of scanning probe microscopy.

5.1
Atomic Force Microscopy (AFM)

5.1.1
Principles

Atomic force microscopy (AFM), invented in 1985 by Binnig, Quate, and Gerber [5.4, 5.5], belongs to the group of *scanning force microscopy* (SFM) techniques, a group of techniques which are based on the measurement of different forces (e. g. attractive, repulsive, magnetic, electrostatic, van der Waals) between a sharp tip and the sample surface. Imaging is accomplished by measuring the interaction force via deflection of a soft cantilever while raster-scanning the tip across the surface (Fig. 5.2). Although various types of forces are encountered when a tip approaches the sample surface, signal generation in AFM is essentially based on interatomic repulsive forces which are of extreme short range nature (Fig. 5.3). In the ideal case one can imagine the tip to be terminated by a single atom, which means that direct contact of the tip with the sample surface is confined to an extremely small area. As a consequence there is always an interatomic repulsive force at this small contact area owing to penetration of the electronic shells of the tip and substrate atoms. Since the interatomic repulsive force is influenced by the total electron density around an atom, and not just by particular states of energy like in *scanning tunneling microscopy*, this force can be used to map the topography of the surface down to atomic dimensions. In addition to this short-range force, also long-range forces (e. g. Coulomb forces between charges, dipole–dipole interactions, polarization forces, van der Waals dispersion forces, capillary forces, due to adsorbate films between tip and substrate), which can be attractive or repulsive are encountered (Fig. 5.3). Although both types of forces contribute to the total force acting on the probing tip, only the varying repulsive interatomic force enables high resolution imaging of surfaces, because it is of extreme short range nature. Since long range attractive forces pull the tip towards the sample surface, they give rise to an increase in the local repulsive force which can deteriorate the experimental conditions. This can

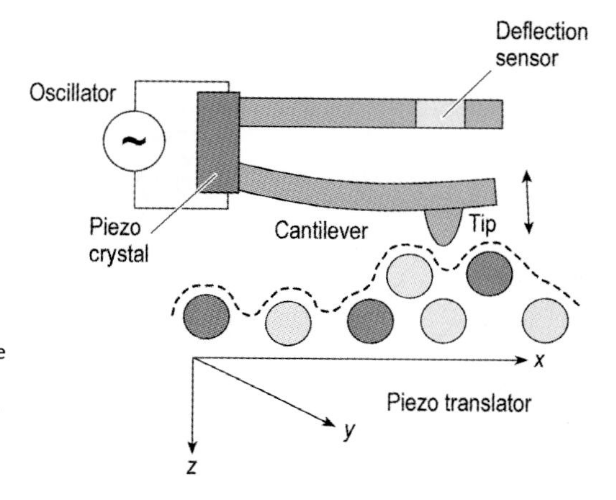

Fig. 5.2. Principle of AFM. The sample symbolized by the circles is scanned by means of a piezoelectric translator. The piezo crystal and the oscillator is only needed for tapping mode operation.

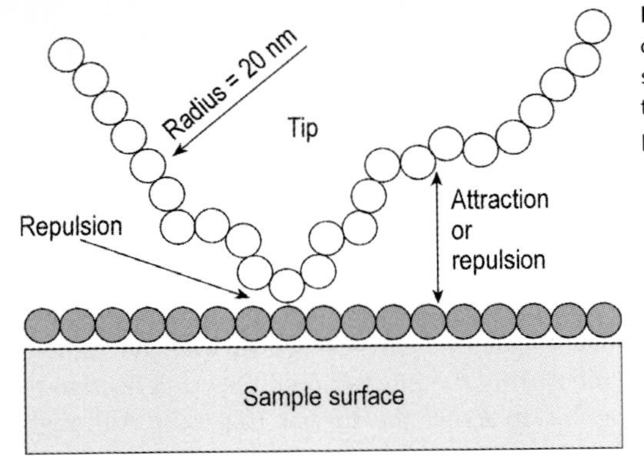

Fig. 5.3. Schematic view of forces encountered when the tip touches the sample surface. Bright circles symbolize tip atoms, dark circles symbolize sample atoms.

lead, for instance, to fallacious results, because of sample deformation or in the worst case the sample can be destroyed by the tip scratching across the surface. Thus, it is important to minimize those long range forces (e.g. by imaging under liquids) to achieve very low repulsive forces (nano-Newton and less) in the contact area between tip and sample. This is especially important when looking at soft materials, which can be deformed or destroyed easily by the load of the tip.

The interaction between tip and sample can be described by force–distance curves. The force–distance curve in Fig. 5.4 shows how the force changes, when the sample surface approaches the tip. At large separations there is no interaction and the observed force is zero (straight line between 1 and 2, if we assume that there are no electrostatic charging forces). At position 2 the tip jumps into contact, because of attractive van der Waals interaction. As the sample is further moved towards the tip, the total force acting on the cantilever becomes repulsive. When the sample is retracted again, the force is reduced along the line from position 3 to 4. Below the zero force line in the diagram the net force acting on the cantilever becomes attractive, because the tip is held at the surface, because of adhesion. In 4 the adhesion force and the cantilever load are just balanced and the tip flips off the surface when further retracting the sample. For AFM measurements the force can be set along the curve be-

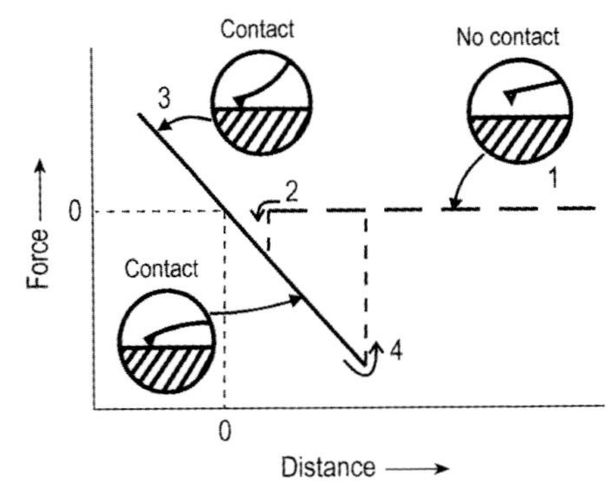

Fig. 5.4. Force–distance curve depicting the interaction of the AFM tip with the sample surface. The operation range of the AFM is between 3 and 4. For explanation refer to the text.

tween position 3 and 4, preferably close to 4, in order to minimize the contact force. The value of the pull-off force can be reduced significantly by imaging under liquids, because of elimination of capillary condensation forces, which pull the tip towards the sample.

One option to image surfaces under more gentle conditions is the tapping mode AFM [5.6, 5.7]. In this mode of operation the cantilever with the tip is driven near its resonance frequency by means of a piezo oscillator (Fig. 5.2). Thus, only intermittent contact between tip and sample occurs. In tapping mode AFM topographic information is retrieved from the amplitude signal of the oscillating cantilever. The advantage of this technique is that, because of intermittent contact, lateral shear forces can be eliminated. Thus, scratching of soft samples or removal of loosely bound surface features (e. g. particles, thin films) can be avoided.

AFMs can be operated in two different modes, the constant force mode and the constant height mode. In the constant force mode the cantilever deflection – and thus the force – is kept constant by readjusting the sample in vertical direction following the topographic features on the surface. In this mode comparatively large and rough sample areas can be imaged without destroying the tip and/or sample surface ("tip crash"), however, the scan rates must be kept comparatively low, to enable the feedback system to respond to height changes. In the constant height mode the vertical position of the sample is kept constant and the varying deflection of the cantilever is recorded. In this mode higher scan rates can be achieved which is advantageous for eliminating thermal drifts in high resolution imaging, however, large scan sizes should be avoided, because "tip crashes" are possible.

5.1.2
Instrumentation

From the general principles described in the previous section the following basic components of an AFM can be identified:

(1) sharp tip mounted on a soft cantilever
(2) detection system for measuring the deflection of the cantilever
(3) piezoelectric translator to move the probe relative to the sample
(4) feedback system to keep the deflection constant by height readjustment of the probe
(5) imaging system to convert the single data points into an image

The tips and the cantilevers have to fulfill a number of requirements. The tip should be very sharp with a small radius of curvature and a high aspect ratio, to be able to trace fine details on the surface. The cantilever must be softer than the bonds between the atoms of the sample, to achieve a measurable deflection without destructive displacement of surface atoms. Moreover, the resonance frequency should be high, to minimize noise, because of coupling with low frequency noise (e. g. vibrations of buildings) from the environment. Low spring constants and high resonance frequencies can only be achieved by making the cantilevers very small. Today cantilevers with force constants of less than 0.1 N m^{-1} and resonance frequencies up to several 100 kHz can be produced by lithographic multi-step processes common in mi-

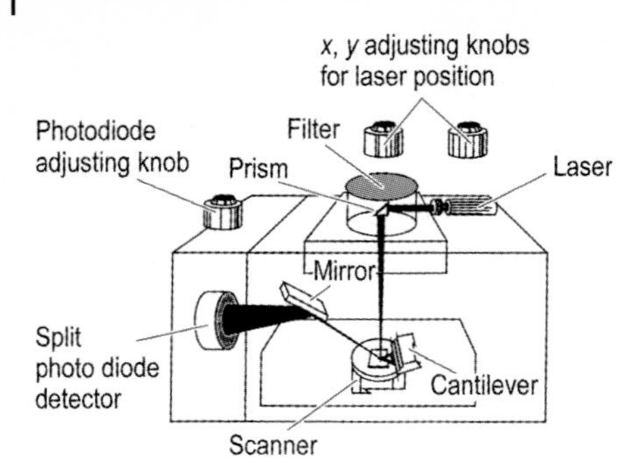

Fig. 5.5. Schematic view of the deflection sensing system as used in the NanoScope III AFM (Digital Instruments, Santa Barbara, CA, USA). The deflection of the cantilever is amplified by a laser beam focused on the rear of the cantilever and reflected towards a split photodiode detector.

croelectronic technology. Such micro-fabricated cantilevers [5.8, 5.9] enable imaging with forces typically in the nano-Newton and sub-nano-Newton range. These forces are about 10 000 to 100 000 times lower than the force of gravity introduced by a fly (1 mg) sitting on a surface. Commercially available cantilevers are usually made of silicon nitride or single crystalline silicon. The length of such cantilevers is typically 100 to 200 μm, the total size of the integrated tip is of the order of 10 μm and the radius of curvature at the apex can be as small as 5 to 10 nm.

The force acting on the tip is sensed by measuring the deflection of the cantilever. A large number of options (e.g. tunneling current, interferometry, capacitance, laser diode feedback) [5.5] for measuring this deflection has been described, however, the most important one is the optical lever technique [5.10] depicted in Fig. 5.5. In this technique a laser beam is focused and positioned on the rear of the cantilever. The reflected laser beam is directed towards a double segment photodiode. As the cantilever deflection – and thus the mirror plane for the laser beam – changes, the position of the laser beam on the photodiode – and thus the difference signal between the two segments – changes, too. Therefore, the difference signal is a sensitive measure for the cantilever deflection with a resolution of 0.01 nm.

Usually, in AFM the position of the tip is fixed and the sample is raster-scanned. After manual course approach with fine-thread screws, motion of the sample is performed with a piezo translator made of piezo ceramics like e.g. lead zirconate titanate (PZT), which can be either a piezo tripod or a single tube scanner. Single tube scanners are more difficult to calibrate, but they can be built more rigid and are thus less sensitive towards vibrational perturbations.

In contrast to many other surface analytical techniques, like e.g. *scanning electron microscopy*, AFM does not require vacuum. Therefore, it can be operated under ambient conditions which enables direct observation of processes at solid–gas and solid–liquid interfaces. The latter can be accomplished by means of a liquid cell which is schematically shown in Fig. 5.6. The cell is formed by the sample at the bottom, a glass cover – holding the cantilever – at the top, and a silicone o-ring seal between. Studies with such a liquid cell can also be performed under potential control which opens up valuable opportunities for electrochemistry [5.11, 5.12]. Moreover, imaging under liquids opens up the possibility to protect sensitive surfaces by in-situ preparation and imaging under an inert fluid [5.13].

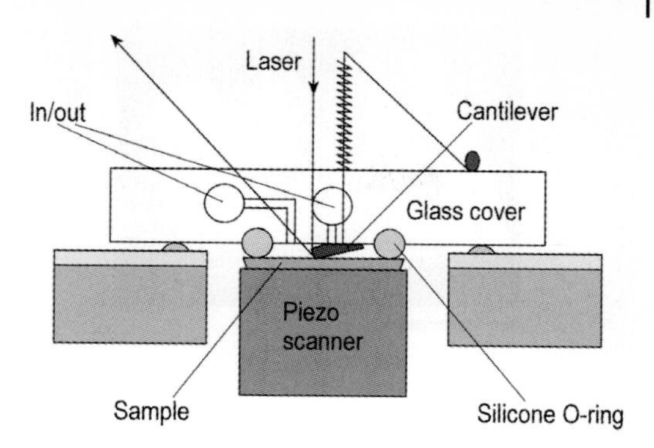

Fig. 5.6. Cross-sectional view of a liquid cell for in-situ AFM measurements of surface processes.

5.1.3
Applications

The great analytical potential of AFM is based on a number of properties summarized below. Atomic resolution can be achieved, but imaging can also be performed on areas larger than 100 μm in square. This capability enables acquisition of overview images, which can be used to zoom in details with high resolution without changing the sample or the instrumental setup, which would make it impossible to find the identical sample position again. Figure 5.7A shows as an example an image of a PVD gold film on silicon, which gives a good overview about size and distribution of the single crystallites, which are typically 100 nm in diameter [5.14]. The rms-roughness (rms stands for root-mean-square and reflects the standard deviation of

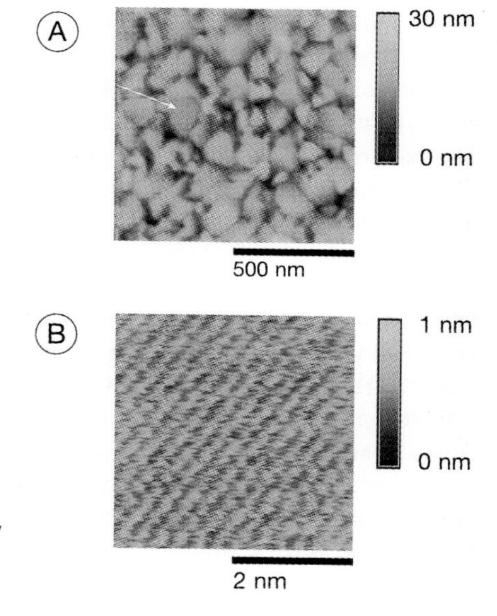

Fig. 5.7. (A) AFM image of a PVD gold film on silicon. Image size: 1 μm × 1 μm, depth scale: 30 nm from black to white. (B) Atomic resolution image of the crystallite marked by the arrow in A. Image size: 4 nm × 4 nm, depth scale: 1 nm from black to white.

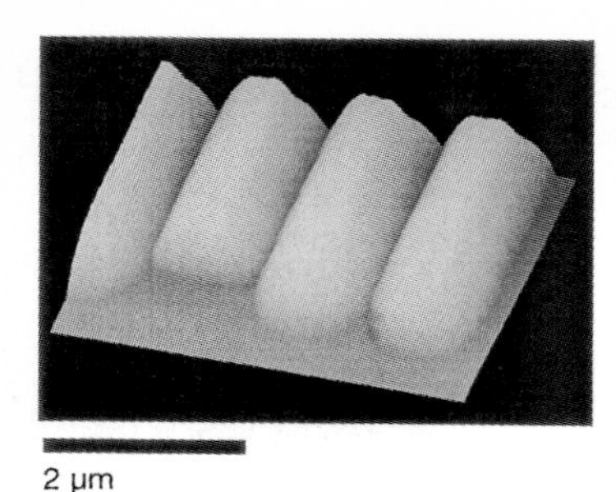

B
250 nm
-250 nm

Fig. 5.8. (A) AFM image of a photoresist layer on silicon structured by ion lithography. (B) Cross-sectional profile through the AFM image shown in (A).

all height values within a considered area) determined from that image is 3 nm. From XRD measurements it is known that the surface of the gold film is mainly a (111) surface. Atomic resolution images of the position marked with the arrow in Fig. 5.7A revealed a gold (111) surface for that specific crystallite (Fig. 5.7B). Although the corrugation in one lattice direction is more prominent, which can be caused by convolution with an asymmetric tip, a hexagonal symmetry can be observed and the spacings are in good agreement with the expected values (0.29 nm).

A further important advantage of AFM is the fact that images contain direct depth information which makes the technique a valuable metrological tool. Fig. 5.8A shows as an example an AFM image of a structured photoresist on a silicon surface. The pattern observed has been produced by ion lithography. Fig. 5.8B depicts a cross-sectional profile through the image shown in Fig. 5.8A. This example nicely demonstrates that the AFM data can be conveniently analyzed by the software, to measure the height of surface features directly.

Besides conducting samples also insulators can be imaged readily with AFM without the need to coat them with a conductive film. This means that sample preparation is facilitated in many cases and that artifacts introduced by the coating process can be avoided. Therefore, a wide variety of organic [5.15, 5.16], biological [5.17, 5.18] but also inorganic insulators [5.19] can be studied in a straightforward manner. Fig. 5.9 shows as an example the topography image of an organic filtration membrane. The pores observed in this Lavsan (polyethylene terephthalate) membrane have been produced by alpha particle bombardment and their diameter is approximately 200 nm. Between the pores a uniform surface with an rms-roughness of 7 nm can be observed.

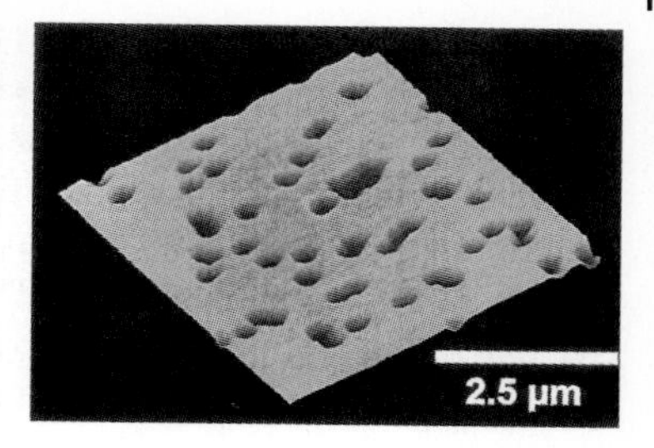

Fig. 5.9. AFM image of a Lavsan filtration membrane. Image size: 5 µm × 5 µm, depth scale: 500 nm from black to white.

An important potential of AFM for surface chemistry is its capability to perform in-situ measurements under liquids and in air. This opens up the opportunity to directly investigate surface processes like corrosion, crystal growth, thin film deposition, or electrochemical processes at electrode surfaces [5.11–5.13, 5.20, 5.21]. As an example for in-situ AFM imaging, Fig. 5.10 shows a series of AFM images depicting the deposition of an octadecylsiloxane (ODS) monolayer on silicon. Such self-assembled monolayers (SAM) can be obtained through deposition of octadecyltrichlorosilane from toluene [5.22]. It can be seen that large islands with diameters of typically 5 µm are deposited on the surface right from the beginning of the growth experiment. This indicates that, under the given conditions, such ordered structures already exist in the deposition solution. Once the uncovered surface area does not supply sufficient space for deposition of further larger islands anymore, continuing film formation can only proceed by growth of already deposited islands through attachment of smaller units, a process which is comparatively slow. During course of time the growing islands coalesce and it can take up to 1 h or more until a full coverage of the surface is reached. From such AFM images also surface coverages can also be evaluated

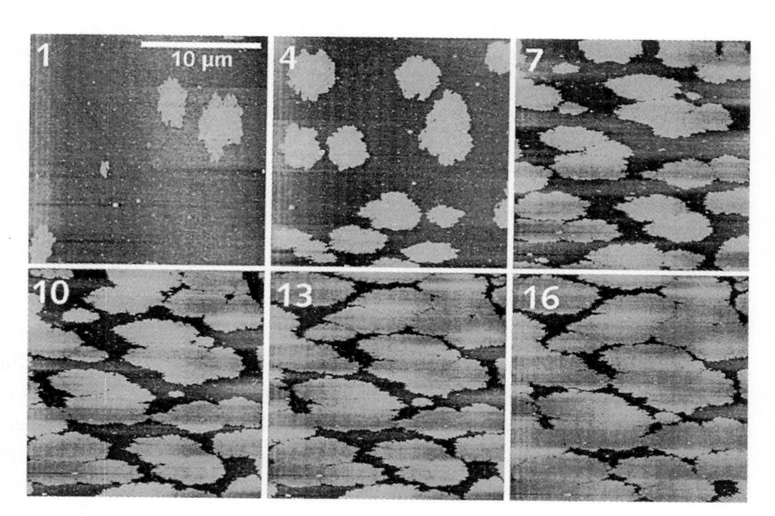

Fig. 5.10. Series of sequential in-situ AFM images of the growth of ODS on silicon. The numbers indicate the adsorption time in minutes.

on a quantitative basis. Since AFM is able to reveal the structure of sub-monolayer islands in such deposition experiments with sub-micrometer resolution, it has significantly contributed to the understanding of the respective growth mechanisms, which is important for controlling film growth through variation of deposition parameters like the water content of the precursor solution or the temperature [5.22].

Concluding this section, it should be mentioned that one shortcoming of AFM is the fact that chemical information or material specific properties cannot be obtained from AFM images directly. Several AFM-related techniques have, however, been introduced which enable access to information in addition to the topography. *Friction force microscopy* (FFM) or *lateral force microscopy* (LFM) measures the friction between tip and sample surface. This is of particular interest for research in the field of tribology [5.23, 5.24]. The principle of the FFM has also been used in a development called *chemical force microscopy* (CFM) [5.25]. In this method the tip has been chemically modified to make it sensitive towards different chemical surface compositions. In this way image contrast could be achieved, because of variations in adhesion (and thus also in friction) between tip and sample. In *Young's modulus microscopy* (YMM) [5.26] differences in the surface elasticity can be accessed through measuring of the response of the tip on the vertical modulation of the sample while the tip is in contact with the surface. The adhesion between tip and sample surface can also be visualized by force–distance curve measurements which have already been addressed in 5.1.1. (Fig. 5.4) [5.27–5.29]. *Electric force microscopy* (EFM) utilizes electrostatic forces or force gradients between a conductive tip and a sample for signal generation. In this way information on surface charges [5.30, 5.31], topography, capacitance (dielectric constant) and potential (see Ref. [5.5] and references cited therein) can be obtained. In *magnetic force microscopy* (MFM) [5.32] magnetostatic forces are measured by interaction of magnetic domains on a surface with a magnetized tip. MFM is mainly interesting as a tool for applications in magnetic storage device manufacturing [5.33] enabling imaging of magnetic patterns with a resolution of approximately 50 nm.

5.2 Scanning Tunneling Microscopy (STM)

5.2.1 Principles

In the scanning tunneling microscopy (STM) [5.1, 5.2] the probe is a sharp metal tip scanned across a conducting surface at distances of the order of typically 1 nm (Fig. 5.11). A bias voltage of typically a few millivolts is applied between the tip and the sample leading to a tunneling current of the order of a few nanoamperes. Figure 5.12 shows what happens when a voltage is applied between two metal electrodes separated by a very small distance.

If a voltage V is applied then the Fermi levels E_F are shifted against each other by an energy $e \times V$, where e is the electrostatic charge of an electron. Because of the energy

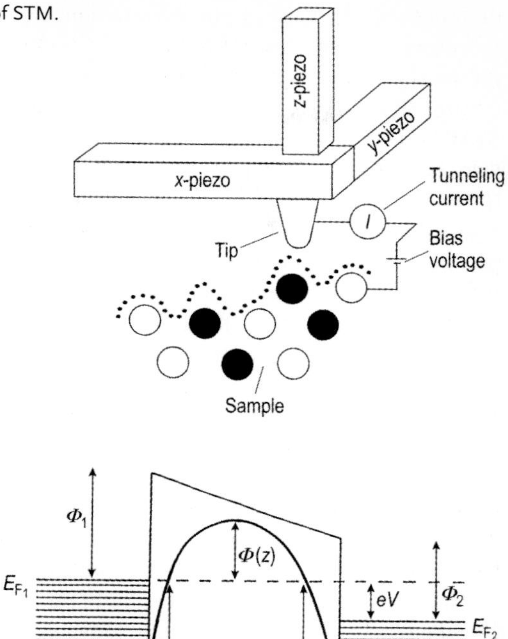

Fig. 5.11. Principle of STM.

Fig. 5.12. Energy level diagram of a tunneling junction [5.34].

barrier between the two metals, which is in the order of the metals' work functions Φ, at small voltages a conventional current cannot flow. However, because the electrons can be described by wave functions decaying exponentially across the energy barrier, the probability to find an electron on the opposite side of the barrier is not zero. This phenomenon is known as the quantum mechanical tunneling effect. As a consequence, electrons can flow from occupied states on the negative electrode to unoccupied states on the positive electrode, whereby at small voltages they do not change their energy, which is described as elastic tunneling. The dependence of the tunneling current I_T on the applied voltage V_T and the tip–surface separation s is given by

$$I_T \approx V_T \exp\left(-A\Phi^{1/2} s\right) \tag{5.1}$$

where A is a constant given by $A = 2\left((2m_e)^{1/2}\right)/h$ (m_e is the mass of an electron and h is Planck's constant). Because of this exponential relationship, the tunneling current is an extremely sensitive measure to control the separation. For example, a variation of the separation by 0.1 nm changes the tunneling current by a factor of approximately 10. This means that in principle variations in the separation of 0.001 nm can be monitored by keeping the current constant within a few percent. However, the tunneling current does not only depend on the tip–surface separation, but it is also influenced by the electronic structure of the surface. This is important when interpreting STM images on the atomic scale, because in that case, in contrast to the AFM, in general the image does no longer simply show the topography of the surface, but rather a contour map of the local density of states near the Fermi level ($LDOS_{\mathrm{EF}}$) as long as small voltages in the millivolt range are applied. For larger voltages up to a few volts the density of other electronic states can be imaged. At this point it should be mentioned that the relations stated above are only an approximation assuming that the tip and the applied voltage does not influence the electronic structure of the surface. Furthermore, for an exact description three-dimensional tunneling must be considered, to take into account the complex spatial structure of the electronic states (wave functions) involved [5.35].

There are two ways of operating an STM, the constant current mode and the constant height mode. In the constant current mode the tip is scanned across the sample surface at a fixed bias voltage and the tunneling current is kept constant by readjusting the tip in vertical direction following the topographic features (or, on the atomic level, the local density of states) on the surface. In this mode comparatively large and rough sample areas can be imaged without destroying tip and/or sample surface ("tip crash"), however, the scan rates must be kept comparatively low, to enable the feedback system to respond to height changes. For images taken in constant current mode at a certain bias voltage sometimes the term constant current topograph (CCT) is used. In the constant height mode the vertical position of the tip is kept constant and the varying tunneling current is recorded. In this mode higher scan rates can be achieved which is advantageous for eliminating thermal drifts in high resolution imaging, however, large scan sizes should be avoided, because "tip crashes" are possible.

5.2.2 Instrumentation

Figure 5.13 shows a schematic overview of an STM system consisting of the following basic components:
(1) sharp metal tip
(2) piezoelectric translator to move the tip relative to the sample
(3) control electronics for applying the bias voltage and measuring the tunneling current
(4) feedback system to keep the tunneling current constant by height readjustment of the tip
(5) imaging system to convert the single data points into an image

The tips used for STM experiments should be sharp and stable. Chemical stability can be achieved by using a noble metal. Mechanical rigidity can be reached by short wires. Alloys of Pt and Ir are frequently used for fabrication of STM tips. They can be produced in a surprisingly simple way just by cutting a metal wire with conventional cutting tools. Because of their high chemical stability, such Pt/Ir tips are well

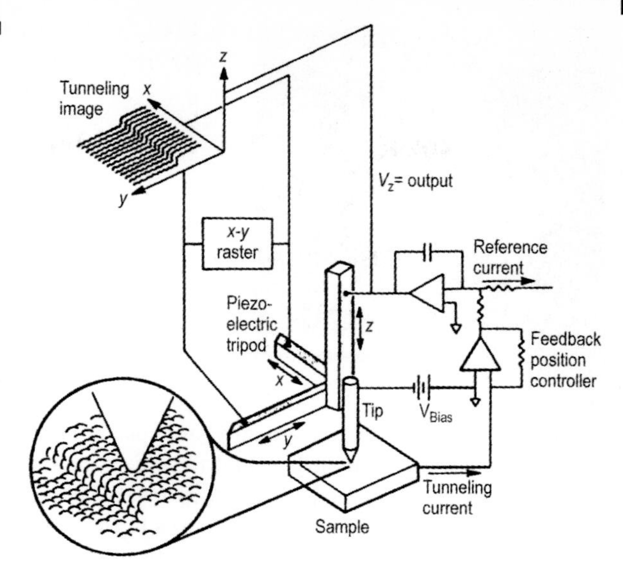

Fig. 5.13. Scheme of an STM system [5.36].

suited for atomic resolution experiments on flat samples. Because of their low aspect ratio near the apex, however, they fail to trace steep features and narrow trenches. Therefore, electrolytically etched tungsten tips with higher aspect ratios are sometimes used as an alternative, although they are less stable against oxidation.

Usually, in STM the position of the sample is fixed and the tip is raster-scanned. Like in AFM, after manual course approach with fine-thread screws, motion of the tip is performed with a piezo translator made of piezo ceramics like e.g. lead zirconate titanate (PZT), which can again be either a piezo tripod or a single tube scanner.

Although physical studies of the electronic structure of surfaces have to be performed under UHV conditions to guarantee clean uncontaminated samples, the technique does not require vacuum for its operation. Thus, in-situ observation of processes at solid–gas and solid–liquid interfaces is possible as well. This has been utilized, for instance, to directly observe corrosion and electrode processes with atomic resolution [5.2, 5.37].

5.2.3 Lateral and Spectroscopic Information

STM images can only be interpreted in terms of surface topography for surface structures with dimensions well above the atomic scale. In general, images taken in constant current mode deliver contour maps of the local density of states. If the polarity of the sample is negative, then states in the valence band are imaged. For a positive polarity of the sample the distribution of electronic states in the conduction band can be recorded. In case different chemical species are present on the surface, the image contrast is further influenced by the varying effective barrier height (work function) at different positions. As an example Fig. 5.14 shows an STM image of a si-

Fig. 5.14. STM image of silicon covered with 1/3 of a monolayer of silver [5.38].

licon (111) surface with 1/3 of a silver monolayer on top of it [5.38]. The high contrast between the silicon surface and the silver islands does not represent the real height of the islands, but it appears exaggerated, because of the lower work function in the silver regions. Such local differences in the effective barrier height can be directly imaged by modulating the tip in vertical direction and recording dI/ds, which according to Eq. (5.1) is proportional to $\Phi^{1/2}xI$. STM can also yield spectroscopic information by recording dI/dU curves at fixed positions, which is called scanning tunneling spectroscopy (STS). This opens up the opportunity to perform electron spectroscopy with a resolution down to one single atom. An other option for obtaining spectroscopic information is simultaneous recording of STM images at various bias voltages. This can be accomplished by performing point spectroscopy at every image point or by modulating the bias voltage while scanning. In the literature simultaneous recording of images at various bias voltages has been called current imaging tunneling spectroscopy (CITS). As one example, the discrimination between different chemical species (Ga and As atoms on a GaAs (110) surface) in an atomically resolved STM image by simultaneously recording of two images at reversed bias voltage, the work of Feenstra et al. [5.39] should be cited.

5.2.4
Applications

Similar to the AFM, one advantage of STM is the fact that a wide range of scan sizes from more than 100 μm down to the atomic level can be covered in one experiment. Because the information is obtained in real space, local defects (e.g. mono-atomic defects, steps, dislocations) can be investigated. This clearly is a great advantage compared to diffraction methods relying on extended periodic structures, and thus showing averaged information. Moreover, the possibility to obtain spectroscopic information makes the STM a very valuable tool for studying surface processes on the atomic scale. An instructive example highlighting several figures of merit of STM has been introduced by the group of Avouris [5.40–5.42], which will be described in the following. Figure 5.15 shows an STM image of the unoccupied states of a Si (111)-7 × 7 surface exposed to 0.2 L O_2 at a temperature of 300 K. Besides the charac-

Fig. 5.15. STM image of a Si (111)-7×7 surface exposed to 0.2 L of O_2 at 300 K. The sample bias voltage was 2 V. Dark and bright sites generated by oxygen exposure are marked with A and B, respectively [5.41].

teristic pattern of the 7×7 reconstruction of the Si (111) surface, pronounced dark and bright spots can be observed as a result of exposure to oxygen. In order to clarify the mechanistic origin of these spots, tunneling spectra at various atomic sites in the 7×7 unit cell (Fig. 5.16) have been recorded and compared with UPS spectra and theoretical results from tight-binding slab calculations for various adsorption models of oxygen. First of all, Fig. 5.16 shows that the different sites can be clearly distinguished by the respective tunneling spectra, whereas photoemission spectra only deliver averaged information on the electronic structure of the surface. The spectra labeled with A, B, and C have been obtained on sites which are not changed by adsorption of oxygen. The spectra labeled with D, E, and F have been obtained on oxygen-induced dark, bright, and perturbed (gray) adatom sites, respectively. Comparison of these spectra with theoretical calculations and a thorough evaluation including UPS spectra for different temperatures and oxygen pressures lead to the conclusion that the bright spots (E) are silicon atoms which have maintained their dangling bond but are perturbed by a neighboring oxygen atom inserted in the back bond of the adatom. The dark spots have been interpreted to be oxygen atoms saturating the dangling bond of a silicon atom which has another oxygen inserted in its back bond. Although this is a rather early work in the field of scanning tunneling spectroscopy, it is still an excellent example for atom-resolved investigation of surface chemistry with STM.

STM has, however, not only been applied to metals and semiconductors, but also to a wide variety of organic and biological systems, like thin films on conducting substrates [5.15, 5.16] as well as protein and DNA molecules [5.43–5.46].

Concluding this section, two interesting variants of the STM should be addressed. The spin-polarized STM (SPSTM), which works with a ferromagnetic tip, can be used to probe surface magnetism with high resolution [5.47, 5.48]. Other modifications of the STM involve electromagnetic radiation, whereby two basic concepts can

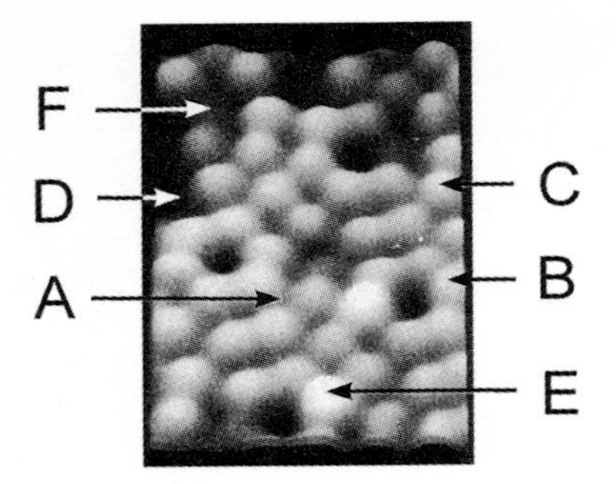

Fig. 5.16. Top: STM image of a region of an O_2-exposed Si (111)-7 × 7 surface. The sites labeled with A, B, and C have not been changed by exposure to oxygen. D, E, and F correspond to oxygen-induced dark, bright, and perturbed (gray) sites, respectively. Bottom: Tunneling spectra corresponding to the different sites [5.41].

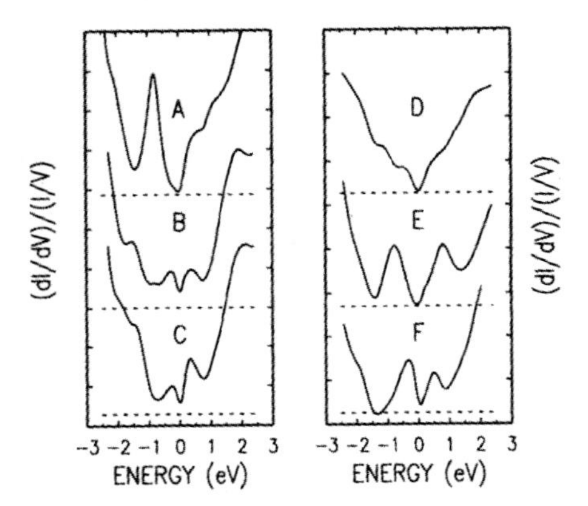

be distinguished: measuring of photons generated in the tunneling gap by inelastic tunneling effects [5.49, 5.50], which is called *photon emission scanning tunneling microscopy* (PESTM), and generation of electrical currents by irradiation of the tunneling gap [5.51–5.54], which is know as *photon assisted scanning tunneling microscopy* (PASTM).

6
Summary and Comparison of Techniques

Important surface and thin film analytical techniques are summarized and compared in Table 6.1.

Tab. 6.1. Typical data for the more important techniques of surface and thin-film analysis.

Technique	*Primary probe*	*Elemental range*	*Type of information*	*Depth of information*	*Lateral resolution*	*Sensitivity (at. %)*	*Ease of quantification*	*Insulator analysis*	*Destructive*	*UHV environment*
XPS	X-ray photons (Al or Mg Kα)	All except H, He	Elemental, chemical	4–8 Monolayers	100 μm (standard), 3 μm (imaging)	0.1	Good	Yes	Sometimes	Yes
UPS	UV photons (He I, He II)	(Adsorbed molecules)	Chemical (valence band)	2–10 Monolayers	1 mm	1	n.a.	Yes	Sometimes	Yes
AES	e^- (3–10 keV)	All except H, He	Elemental, (chemical)	2–6 Monolayers	20 nm	0.2	Moderate	No	Often	Yes
SAM	e^- (20–50 keV)	All except H, He	Elemental, (chemical)	2–6 Monolayers	5 nm	0.2	Moderate	No	Often	Yes
EELS	e^- (60–1000 keV)	All (H, He only hardly)	Elemental, chemical	(None)	1 nm	0.1–1	Moderate	Yes	Yes	No
EFTEM	e^- (60–1000 keV)	All except H, He	Elemental, chemical	(None)	≤1 nm	0.1–1	Moderate	Yes	Yes	No
LEED	e^-	All	Crystallo-graphic	~5 Mono-layers	0.01 nm	n.a.	n.a.	No	No	Yes
SSIMS	Ar^+, Ga^+	All	Chemical	1–2 Monolayers	50 nm	10^{-6}–10^{-2}	Poor	(Yes)**	(Yes)	Yes
DSIMS	Ar^+, Ga^+	All	Chemical	1–2 Monolayers	20 nm (imaging) 50 μm (depth profile)	10^{-7}–10^{-1}	Poor*	(Yes)**	Yes	Yes

SNMS e⁻-beam	Ar^+, Cs^+	All	Elemental	1–2 Monolayers	5 µm	10^{-2}–1	Moderate*	Yes (with rf)	Yes	Yes
SNMS plasma	Ar^+, Kr^+	All	Elemental	1–2 Monolayers	1 mm	10^{-3}–10^{-1}	Moderate*	Yes	Yes	Yes
SNMS laser	Ar^+, Ga^+	All	Chemical, elemental	1–2 Monolayers	50 nm	10^{-7}–1	Moderate*	Yes	Yes	Yes
RBS	H^+, He^+	High Z on low Z	Elemental, depth distribution	1 Mono-layer–2 µm	5 µm–1 mm	10^{-6}	Good	Yes	No	No
LEIS	He^+, Ne^+	All except H, He	Elemental	1–2 Monolayers	~1 mm	10^{-4}	Good	Yes	Not with He^+	Yes
ERDA	He^+–Au^+	Low Z	Elemental, depth distribution	< µm	1 mm	1–10^{-4}	Good	Yes	Yes	No
NRA	H^+, D^+, N^+	Low Z	Isotopic, depth distribution	< µm	10–100 µm	1–10^{-4}	Good	Yes	Yes	No
TXRF	X-ray photons	Z > 14	Elemental	1 Monolayer	(None)	10^{-4}–10^{-1}	Yes	Yes	No	No
EDXS	e^- (few keV–1000 keV)	All Z ≥ 5	Elemental, (chemical)	(None)	~5 nm	0.1–1	Yes	Yes	Yes	No
GAXRD	X-ray photons	All	Crystalline structure	Monolayers to microns	~5 mm	n.a	n.a.	Yes	No	No
GXRR	X-ray photons	All	Thickness, density, roughness	0–400 nm	~1 cm	n.a.	n.a.	Yes	No	No

Tab. 6.1. (continued)

Technique	*Primary probe*	*Elemental range*	*Type of information*	*Depth of information*	*Lateral resolution*	*Sensitivity (at. %)*	*Ease of quantification*	*Insulator analysis*	*Destructive*	*UHV environment*
Refl-EXAFS	X-ray photons	All	Local atomic environment	~2 nm	~1 cm	10^{-2}–10^{-4}	n.a.	Yes	No	No
GD-OES	A^+ (Ar^+)	All	Elemental, depth profile	10 nm	None	10^{-6}	Moderate	Yes	Yes	No
L-OES	UV–Vis photons	All	Elemental, (depth profile)	~0.1–1 µm	~10 µm	10^{-3}–10^{-2}	Moderate	Yes	Yes	No
IBSCA	Ar^+ (2–20 keV)	All	Elemental	0.3 nm	100 µm	10^{-4}	Moderate	Yes	Yes	Yes
RAIRS	IR photons	(Adsorbed molecules)	Chemical (vibrational)	1–2 Monolayers	~1 mm	10^{-4}	n.a.	No	No	No
SERS	Laser photons	Restricted	Chemical (vibrational)	10–100 nm	~0.5 mm	10^{-6}	Moderate	Yes	No	Yes
ELL	UV/VIS/IR photons	Index of reflection	Chemical	1 nm	~1 mm	1 of adsorbant	Poor	(Yes)	No	No
AFM	Mechanical force	All	Topographic	1 Monolayer	1 Atom	(1 Atom)	n.a.	Yes	No	No
STM STS	High field	All	Topographic, chemical	1 Monolayer	1 Atom	(1 Atom)	n.a.	No	Possibly	Yes

n.a. = not applicable
* good with matrix-near standards
** with e^--charge compensation

7
Surface and Thin Film Analytical Equipment Suppliers

Many companies now supply surface and thin film analytical equipment; although a few provide several techniques, most specialize in just one or two. (Only suppliers of the equipment which is discussed in more detail are given in the list below. The authors cannot guarantee completeness.)

Supplier	***Technique***			
	Electron detection	***Ion detection***	***Photon detection***	***Scanning probe microscopy***
Gammadata Scienta P.O. Box 15120 750 15 Uppsala Sweden www.gammadata.se	XPS			
Jeol USA, Inc. 11 Dearborn Road Peabody MA 01960 USA www.jeol.com	XPS, AES EELS, EFTEM	SIMS	EDXS, WDXS	
LEO Electron Microscopy Ltd. Clifton Road Cambridge, CB1 3QH United Kingdom www.leo-em.co.uk	EELS, EFTEM			
Gatan Inc. 5933 Coronado Lane Pleasonton CA 94588 USA www.gatan.com	EELS, EFTEM			
Kratos Analytical Wharfside Trafford Wharf Road Manchester, M17 1GP United Kingdom www.kratos.com	XPS, AES		GAXRD-GXRR	

Supplier	***Technique***			
	Electron detection	***Ion detection***	***Photon detection***	***Scanning probe microscopy***
Omicron Vakuumphysik GmbH Idsteiner Straße 78 65232 Taunusstein Germany www.omicron-instruments.com	XPS, AES, SAM, UPS, LEED	ISS		STM, AFM
Physical Electronics, Inc. 6509 Flying Cloud Drive Eden Prairie MN 55344 USA www.phi.com	XPS, AES, SAM, UPS	SIMS		
Specs GmbH Voltastraße 5 13355 Berlin Germany www.specs.de	XPS, AES, UPS, LEED	SIMS, SNMS, ISS		
Staib Instruments GmbH Obere Hauptstraße 45 85354 Freising Germany www.staib-instruments.com	AES, RHEED			
Tectra GmbH Reuterweg 65 60323 Frankfurt a. M. Germany www.tectra.de	LEED			
VG Scientific The Birches Industrial Estate Imberhorne Lane East Grinstead West Sussex RH19 1UB United Kingdom www.vgscientific.com	XPS, AES, SAM, UPS, HREELS, LEED	SIMS, SNMS, GDMS, APFIM, ISS	BIS	STM, AFM
VSW Ltd. Unit 4, Heather Close Lyme Green Business Park Macclesfield, Cheshire SK11 0LR www.vsw.co.uk	XPS, AES, UPS, EELS	ISS	BIS	
Atomika Instruments Bruckmannring 40 85764 Oberschleißheim Germany www.atomika.com		SIMS	TXRF	

Supplier	Technique			
	Electron detection	Ion detection	Photon detection	Scanning probe microscopy
Cameca France 103 Boulevard Saint Denis 92403 Courbevoie Cedex France www.cameca.fr		SIMS		
Hiden Analytical Ltd. 420 Europa Boulevard Warrington WA5 5UN United Kingdom www.hiden.co.uk		SIMS		
National Electrostatics Corp. 7540 Graber Road P.O. Box 620310 Middleton WI 53562–0310 USA www.pelletron.com		RBS		
Charles Evans and Associates 810, Kifer Road Sunnyvale CA 94086 USA www.cea.com		RBS		
EDAX USA 91 McKee Drive Mahwah NJ 07430 USA www.edax.com			EDXS, XRF	
Thermo NORAN 2551 W. Beltline Highway Middleton WI 53562–2697 USA www.noran.com			EDXS, WDXS XRF	
Princeton Gamma-Tech (PGT) Inc. C/N 863 Princeton NJ 08542–0863 USA www.pgt.com			EDXS, WDXS	
RÖNTEC GmbH Germany Schwarzschildstr. 12 12489 Berlin Germany www.roentec.de			EDXS, TXRF	

Supplier	Technique			
	Electron detection	*Ion detection*	*Photon detection*	*Scanning probe microscopy*
Ital Structures Via M. Miscone 11/d – Z.I. Baltera 38066 Riva del Garda Italy www.italstructures.com			TXRF	
Philips Analytical Lelyweg 1 7602 EA Almelo The Netherlands www.analytical.philips.com			TXRF, ELL GAXRD-GXRR	
Rigaku/USA, Inc. 9009 New Trails Drive The Woodlands, TX 77381 USA www.rigaku.com			TXRF GAXRD-GXRR	
Jobin Yvon S.A. 16–18, rue du Canal 91165 Longjumeau Cedex France www.jyinc.com			GD-OES, RAIRS, ELL	
LECO Instrumente GmbH Benzstraße 5 b 85551 Kirchheim bei München Germany www.leco.com			GD-OES	
SPECTRUMA Spectro Analytical Instruments GmbH Boschstraße 10 47533 Kleve Germany www.spectruma.de			GD-OES	
ABB Bomem inc. 585 Charest East, Suite 300 Québec, QC G1K 9H4 Canada www.bomem.com			RAIRS	
Beaglehole Instruments Ltd. 127 Kelburn Parade Wellington New Zealand www.beaglehole.com			ELL	

Supplier	Technique			
	Electron detection	Ion detection	Photon detection	Scanning probe microscopy
J.A. Woollam Co. 650 J Street Suite 39 Lincoln NE 68508 USA www.jawoollam.com			ELL	
NFT-Nanofilm Technologie GmbH Anna-Vandenhoeck-Ring 5 37081 Göttingen Germany www.nanofilm.de			ELL	
Sentech Gesellschaft für Sensortechnik GmbH Bahnstraße 3 82131 Stockdorf Germany www.sentech.com			ELL	
Sopra, Inc. 26, rue Pierre Joigneaux 92270 Bois-Colombes France www.sopra-sa.com			ELL	
Waterloo Digital Electronics Division of WDE Inc. 279 Weber St. N. Waterloo, Ontario N2J 3H8 Canada www.onramp.tuscaloosa.al.us			ELL	
Bruker Optics Inc. 19 Fortune Dr. Manning Park Billerica MA 01821–3991 USA www.bruker.com			RAIRS, Raman	
ChemIcon Inc. 7301 Penn Avenue Pittsburgh Pennsylvania 15208 USA www.chemimage.com			RAIRS, Raman	

Supplier	Technique			
	Electron detection	*Ion detection*	*Photon detection*	*Scanning probe microscopy*
Chromex 2705-B Pan American Fwy NE Albuquerque NM 87107 USA www.chromexinc.com			Raman	
Kaiser Optical Systems Inc. Box 983 371 Parkland Plaza Ann Arbor MI 48106–0983 USA www.kosi.com			Raman	
Nicolet Instrument Corporation 5225 Verona RD Bldg 5 Madison WI 53711–4495 USA www.nicolet.com			RAIRS, Raman	
Ocean Optics Europe Soerense Zand 4 a 6961 LL Eerbeek The Netherlands www.topsensors.nl			Raman	
Perkin Elmer Instruments 761 Main Avenue Norwalk CT 06859 USA www.instruments.perkinelmer.com			Rairs, Raman	
Renishaw plc New Mills Wotton-under-Edge Gloucestershire GL12 8JR United Kingdom www.renishaw.com			Raman	
Bede Scientific Instruments Ltd, Bowburn South Industial Estate Unit 13D, Bowburn, County Durham DH6 5AD, UK www.bede.co.uk			GAXRD- GXRR	

Supplier	*Technique*			
	Electron detection	*Ion detection*	*Photon detection*	*Scanning probe microscopy*
Bourevestnik, Inc. Malookchtinsky pr., 68 St. Petersburg, 195112, Russia www.bourevestnik.com			GAXRD-GXRR	
BRUKER AXS, Inc. 5465 East Cheryl Parkway Madison WI 53711–5373 USA www.bruker-axs.com			GAXRD-GXRR	
Crystal Logic Inc. 10573 W. Pico Blvd. PMB 106 Los Angeles, CA 90064–2348 USA www.xtallogic.com/index.html			GAXRD-GXRR	
INEL Z.A., C.D. 405 45410 Artenay France www.valcofim.fr/inel			GAXRD-GXRR	
Nonius B.V. Röntgenweg 1, 2624 BD Delft The Netherlands www.nonius.com			GAXRD-GXRR	
Rich. Seifert and Co. SEIFERT FPM Ahrensburg Germany www.roentgenseifert.com			GAXRD-GXRR	
STOE and CIE GmbH Hilpertstr. 10 64295 Darmstadt Germany www.stoe.com			GAXRD-GXRR	

Supplier	***Technique***			
	Electron detection	***Ion detection***	***Photon detection***	***Scanning probe microscopy***
Burleigh Instruments Inc. Burleigh Park Fishers NY 14453 USA www.burleigh.com				STM, AFM
Digital Instruments GmbH Janderstraße 9 68199 Mannheim Germany www.digital-instruments.com				STM, AFM
LK Technologies Inc. 3910 Roll Avenue Bloomington IN 47403 USA www.lktech.com				STM
ThermoMicroscopes 1171 Borregas Avenue Sunnyvale CA 94089 USA www.thermomicro.com				AFM, STM

References

References to Chapter 1

1-1 J. C. Rivière: *Surface Analytical Techniques,* Oxford University Press, Oxford **1990**.

1-2 M. P. Seah, D. Briggs in: D. Briggs, M. P. Seah (eds.): *Practical Surface Analysis,* 2nd ed., vol. 1, John Wiley and Sons, Chichester **1990**, p. 1–18.

1-3 D. M. Hercules, S. H. Hercules, *J. Chem. Educ.* **61 (1984)** 402–409.

1-4 M. J. Higatsberger, *Adv. Electron. Electron. Phys.* **56 (1981)** 291–358.

1-5 H. W. Werner, *Mikrochim. Acta Suppl.* **8 (1979)** 25–50.

1-6 A. E. Morgan, H. W. Werner, *Phys. Scr.* **18 (1978)** 451–463.

1-7 J. Tousset, *Le Vide* **33 (1978)** 201–211.

1-8 G. B. Larrabee, *Scanning Electron Micros.* part I **1977**, 639–650.

1-9 A. Benninghoven, *Appl. Phys.* **1 (1973)** 3–16.

References to Chapter 2

2-1 K. Siegbahn, C. Nordling, A. Fahlman, R. Nordberg, K. Hamrin, J. Hedman, G. Johansson, T. Bergmark, S.-E. Karlsson, I. Lindgreen, B. Lindberg: *ESCA: Atomic, Molecular, and Solid State Structure Studied by Means of Electron Spectroscopy.* Almqvist and Wiksells, Uppsala **1967**.

2-2 P. Auger, *J. Phys. Radium* **6 (1925)** 205–209.

2-3 M. P. Seah, W. A. Dench, *Surf. Interface Anal.* **1 (1979)** 2–11.

2-4 D. Briggs, J. C. Rivière in: D. Briggs, M. P. Seah (eds.): *Practical Surface Analysis,* 2nd ed., vol. 1, John Wiley and Sons, Chichester **1990**, p. 85–141.

2-5 M. P. Seah in J. M. Walls (ed.): *Methods of Surface Analysis,* Cambridge University Press, Cambridge **1989**.

2-6 B. P. Tonner, T. Droubay, J. Denlinger, W. Meyer-Ilse, T. Warwick, J. Rothe, E. Kneedler, K. Pecher, K. Nealson, T. Grundl, *Surf. Interface Anal.* **27 (1999)** 247–258.

2-7 J. H. Scofield, *J. Electron. Spectrosc. Relat. Phenom.* **8 (1976)** 129–137.

2-8 R. F. Reilman, A. Msezane, S. T. Manson, *J. Electron Spectrosc. Relat. Phenom.* **8 (1976)** 389–394.

2-9 M. P. Seah in: D. Briggs, M. P. Seah (eds.): *Practical Surface Analysis,* 2nd ed., vol. 1, John Wiley and Sons, Chichester **1990**, p. 201–255.

2-10 P. Brüsch, K. Müller, A. Atrens, H. Neff, *Appl. Phys. A* **38 (1985)** 1–18.

2-11 S. Hofmann in: D. Briggs, M. P. Seah (eds.): *Practical Surface Analysis,* 2nd ed., vol. 1, John Wiley and Sons, Chichester **1990**, p. 143–199.

2-12 S.W. Gaarenstroom, *J. Vac. Sci. Technol.* **16 (1979)** 600–604.

2-13 S. Hofmann, J. Steffen, *Surf. Interface Anal.* **14 (1989)** 59–65.

2-14 H. Bubert, H. Jenett, *Fresenius' Z. Anal. Chem.* **335 (1992)** 643–647.

2-15 C. Gatts, A. Zalar, S. Hofmann, M. Rühle, *Surf. Interface Anal.* **23 (1995)** 809–814.

2-16 J. Card, A. L. Testoni, L. A. Le Tarte, *Surf. Interface Anal.* **23 (1995)** 495–505.

2-17 T. Do, N. S. McIntyre, *Surf. Interface Anal.* **27 (1999)** 1037–1045.

2-18 R. Reiche, S. Oswald, K. Wetzig, M. Dobler, H. Reuther, M. Walterfang, *Nucl. Instrum. Methods Phys. Res. B* **160 (2000)** 397–407.

2-19 T. Morohashi, T. Hoshi, H. Nikaido, M. Kudo, *Appl. Surf. Sci.* **100/101 (1996)** 84–88.

2-20 M. J. G. Wenham, I. R. Barkshire, M. Prutton, R. H. Roberts, D. K. Wilkinson, *Surf. Interface Anal.* **23 (1995)** 858–872.

2-21 M. N. Souza, C. Gatts, M. A. Figueira, *Surf. Interface Anal.* **20 (1993)** 1047–1050.

2-22 H. Bubert, N. Niebuhr, *J. Microsc. Soc. Am.* **2 (1996)** 35–41.

2-23 T. Watanabe, S. Kishida, K. Ishihara, Y. Yamauchi, H. Tokutaka, *Hyomen Kagaku* **19 (1998)** 98–105.

2-24 P. Grunow, H. Paes, C. Gatts, W. Losch, *Thin Solid Films* **275 (1996)** 191–194.

2-25 H. Tokutaka, K. Yoshihara, K. Fujimura, K. Iwamoto, K. Obu-Cann, *Surf. Interface Anal.* **27 (1999)** 783–788.

2-26 B. Lesiak, J. Zemek, A. Jozwik, *Appl. Surf. Sci.* **135 (1998)** 318–330.

2-27 H. Bubert, H. Hillig, *Mikrochim. Acta* **133 (1999)** 95–103.

2-28 A. E. Munter, B. J. Heuser, K. M. Skulina, Physica B (Amsterdam) **221 (1996)** 500–506.

2-29 P. Coxon, J. Krizek, M. Humpherson, I. R. M. Wardell, *J. Electron Spectrosc. Relat. Phenom.* **52 (1990)** 821–836.

2-30 C. D. Wagner, *Faraday Discuss. Chem. Soc.* **60 (1975)** 291–300.

2-31 C. D. Wagner, L. H. Gale, R. H. Raymond, *Anal. Chem.* **51 (1979)** 466–482.

2-32 C. D. Wagner in: D. Briggs, M. P. Seah (eds.): *Practical Surface Analysis*, 2nd ed., vol. 1, John Wiley and Sons, Chichester **1990**, p. 595–634.

2-33 a I. Olefjord, L. Nyborg, D. Briggs (eds.): ECASIA 97, John Wiley and Sons, Chichester 1997.

2-33 b D. Briggs, J. P. Espinós, J. M. Sanz (eds.): ECASIA 99, *Surf. Interface Anal.* **30(1)**, Wiley Interscience **2000**.

2-34 J. M. Walls (ed.): *Methods of Surface Analysis*, Cambridge University Press, Cambridge **1989**.

2-35 D. P. Woodruff, T. A. Delchar: *Modern Techniques of Surface Analysis*, Cambridge University Press, Cambridge **1988**.

2-36 J. F. Watts: *An Introduction to Surface Analysis by Electron Spectroscopy*, Oxford University Press, Oxford **1990**.

2-37 G. Ertl, J. Küppers: *Low Energy Electrons and Surface Chemistry*, 2nd ed., VCH, Weinheim **1985**.

2-38 D. Briggs (ed.): *Handbook of X-Ray and Ultraviolet Photoelectron Spectroscopy*, Heyden, London **1977**.

2-39 J. Wolstenholme, *Mater. World.* **7 (1999)** 412–414.

2-40 J. M. López Nieto, V. Cortés Corberán, J. L. G. Fierro, *Surf. Interface Anal.* **17 (1991)** 940–946.

2-41 T. A. Patterson, J. C. Carver, D. E. Leyden, D. M. Hercules, *J. Phys. Chem.* **80 (1976)** 1700–1708.

2-42 A. Paul Pijpers, R. J. Meier, *Chem. Soc. Rev.* **28 (1999)** 233–238.

2-43 A. Boix, J. L. G. Fierro, *Surf. Interface Anal.* **27 (1999)** 1107–1113.

2-44 T. Fujitani, I. Nakamura, S. Ueno, T. Uchijiama, J. Nakamura, *Appl. Surf. Sci.* **121/122 (1997)** 583–586.

2-45 T. Miyazaki, T. Doi, M. Kato, T. Miyake, I. Matsuura, *Appl. Surf. Sci.* **121/122 (1997)** 492–495.

2-46 D. Robba, C. Casale, M. Notaro, D. M. Ori in: I. Olefjord, L. Nyborg, D. Briggs (eds.): ECASIA 97, John Wiley and Sons, Chichester **1997**, p. 169–72.

2-47 A. Galtayries, J. Grimblot, J.-P. Bonnelle, *Surf. Interface Anal.* **24 (1996)** 345–354.

2-48 G. Beamson, G. Briggs, *High Resolution XPS of Organic Polymers*, John Wiley and Sons, Chicester **(1992).**

2-49 D. Briggs, *Surface Analysis of Polymers by XPS and Static SIMS*, Cambridge University Press, Cambridge (UK) **(1998).**

2-50 A. A. Galuska, *Surf. Interface Anal.* **27 (1999)** 889–896.

2-51 D. Wolany, T. Fladung, L. Duda, J. W. Lee, T. Gantenfort, L. Wiedmann, A. Benninghoven, *Surf. Interface Anal.* **27 (1999)** 609–617.

2-52 J. Marchand-Brynaert, *Surfactant Sci. Ser.* **79 (1999)** 91–124.

2-53 L. Dai, A. W. H. Albert, *J. Phys. Chem. B* **104 (2000)** 1891–1915.

2-54 J. E. Fulghum, *J. Electron Spectrosc. Relat. Phenom.* **100 (1999)** 331–355.

2-55 K. Asami, K. Hashimoto, *Langmuir* **3 (1987)** 897–904.

2-56 R. Ramanauskas, *Appl. Surf. Sci.* **153 (1999)** 53–64.

2-57 G. Pajonk, H. Bubert, *Fresenius' J. Anal. Chem.* **365 (1999)** 236–243.

2-58 F. Bentiss, M. Traisnel, L. Gengembre, M. Lagrenee, *Appl. Surf. Sci.* **152 (1999)** 237–249.

2-59 J. H. Qiu, P. H. Chua, *Surf. Interface Anal.* **28 (1999)** 119–122.

2-60 J. F. Marco, A. C. Agudelo, J. R. Gancedo, D. Hanzel, *Surf. Interface Anal.* **26 (1998)** 667–673.

2-61 J. F. Marco, A. C. Agudelo, J. R. Gancedo, D. Hanzel, *Surf. Interface Anal.* **27 (1999)** 71–75.

2-62 L.-S. Johansson, T. Saastamoinen, *Appl. Surf. Sci.* **144/145 (1999)** 244–248.

2-63 G. Chen, C. R. Clayton, *Surf. Interface Anal.* **27 (1999)** 230–235.

2-64 D. Sprenger, H. Bach, W. Meisel, P. Guetlich, *Surf. Interface Anal.* **20 (1993)** 796–802.

2-65 K. Idla, A. Talo, H. E.-M. Niemi, O. Forsen, S. Ylasaari, *Surf. Interface Anal.* **25 (1997)** 837–854.

2-66 C. Ziegler, B. Reusch, J. Geis-Gerstorfer, *Fresenius' J. Anal. Chem.* **361 (1998)** 547–553.

2-67 E. McCafferty, *Proc. – Electrochem. Soc.* **98–17 (1999)** 42–55.

2-68 Chin-An Chang, Yong-Kil Kim, A. G. Schrott, *J. Vac. Sci. Technol. A* **8 (1990)** 3304–3309.

2-69 M. F. Fitzpatrick, J. F. Watts, *Surf. Interface Anal.* **27 (1999)** 705–715.

2-70 J. F. Watts, M. F. Fitzpatrick, T. J. Carney in: ECASIA97, John Wiley and Sons, Chichester **1997**, p. 113–116.

2-71 D. Wolany, T. Fladung, L. Duda, J. W. Lee, T. Gantenfort, L. Wiedmann, A. Benninghoven, *Surf. Interface Anal.* **27 (1999)** 609–617.

2-72 C. V. Bechtolsheim, V. Zaporojtchenko, F. Faupel, *Appl. Surf. Sci.* **151 (1999)** 119–128.

2-73 J. Zhang, C. Q. Cui, T. B. Lim, E.-T. Kang, K. G. Neoh, *Surf. Interface Anal.* **28 (1999)** 235–239.

2-74 H. Horn, S. Beil, D. A. Wesner, R. Weichenhain, E. W. Kreutz, *Nucl. Instrum. Methods Phys. Res., Sect. B* **151 (1999)** 279–284.

2-75 E. Capelli, S. Orlando, F. Pinzari, A. Napoli, S. Kaciulis, *Appl. Surf. Sci.* **138/139 (1999)** 376–382.

2-76 P. Steiner, V. Kinsinger, I. Sander, B. Siegwart, S. Huefner, C. Politis, R. Hoppe, H. P. Mueller, *Z. Phys. B* **67 (1987)** 497–502.

2-77 C. Ziegler, F. Baudenbacher, H. Karl, H. Kinder, W. Göpel, *Fresenius' J. Anal. Chem.* **341 (1991)** 308–313.

2-78 Y. Hwu, L. Lozzi, S. La Rosa, M. Onellion, H. Berger, F. Gozzo, F. Levy, G. Margaritondo, *Appl. Phys. Lett.* **59 (1991)** 979–981.

2-79 Y. Hwu, M. Marsi, C. Hwang, J. Seutjens, D. C. Larbalestier, M. Onellion, G. Margaritondo, *Appl. Phys. Lett.* **57 (1990)** 2139–2141.

2-80 Q. Y. Ma, M. T. Schmidt, L. S. Weinman, E. S. Yang, S. M. Sampere, S. W. Chan, *J. Vac. Sci. Technol. A* **9 (1991)** 390–393.

2-81 R. P. Vasquez, M. C. Foote, B. D. Hunt, L. Bajuk, *J. Vac. Sci. Technol. A* **9 (1991)** 570–573.

2-82 I. N. Shabanova, O. V. Popova, V. I. Kormilets, N. M. Nebogaticov, V. I. Kukuev in: I. Olefjord, L. Nyborg, D. Briggs (eds.): ECASIA 97, John Wiley and Sons, Chichester **1997**, p. 917–920.

2-83 H. Behner, G. Gieres, B. Sipos, *Fresenius' J. Anal. Chem.* **341 (1991)** 301–307.

2-84 W. A. M. Aarnink, J. Gao, H. Rogalla, A. van Silfhout, *J. Less Common Met.* **164/165 (1990)** 321–328.

2-85 C. C. Chang, M. S. Hedge, X. D. Wu, B. Dutta, A. Inam, T. Venkatesan, B. J. Wilkens, J. B. Wachtman Jr., *J. Appl. Phys.* **67 (1990)** 7483–7487.

2-86 G. Frank, C. Ziegler, W. Göpel, *Phys. Rev. B. Condens. Matter* **43 (1991)** 2828–2834.

2-87 S. Tanaka, T. Nakamura, M. Iiyama, N. Yoshida, S. Takano, *Jpn. J. Appl. Phys.* **30 (1991)** L 1458-L 1461.

2-88 S. Kohiki, S. Hatta, T. Kamada, A. Enokihara, T. Satoh, K. Setsune, K. Wasa, Y. Higashi, S. Fukushima, Y. Gohshi, *Appl. Phys. A* **50 (1990)** 509–514.

2-89 S. Kohiki, J. Kawai, S. Hayashi, H. Adachi, S. Hatta, K. Setsune, K. Wasa, *J. Appl. Phys.* **68 (1990)** 1229–1232.

2-90 M. A. Sobolewski, S. Semancik, E. S. Hellman, E. H. Hartford, *J. Vac. Sci. Technol. A* **9 (1991)** 2716–2720.

2-91 D. R. C. Hoad, W. R. Flavell, S. E. Male, I. W. Fletcher, G. Beamson, *Surf. Interface Anal.* **21 (1994)** 764–770.

2-92 G. Subramayam, F. Radpour, V. J. Kapoor, G. H. Lemon, *J. Appl. Phys.* **68 (1990)** 1157–1163.

2-93 K. Tanaka, H. Takaki, K. Koyama, S. Noguchi, *Jpn. J. Appl. Phys.* **29 (1990)** 1658–1663.

2-94 H. Yamanaka, H. Enomoto, J. S. Shin, T. Kishimoto, Y. Takano, N. Mori, H. Ozaki, *Jpn. J. Appl. Phys.* **30 (1991)** 645–649.

2-95 T. Suzuki, M. Nagoshi, Y. Fukuda, K. Ohishi, Y. Syono, M. Tachiki, *Phys. Rev. B Condens. Matter* **42 (1990)** 4263–4271.

2-96 A. Fujimori, Y. Tokura, H. Eisaki, H. Takagi, S. Uchida, E. Takayama-Muromachi, *Phys. Rev. B Condens. Matter* **42 (1990)** 325–328.

2-97 M. Klauda, J. P. Stroebel, J. Schloetterer, A. Grassmann, J. Markl, G. Saemann-Ischenko, *Physica C* **173 (1991)** 109–116.

2-98 K. Okada, Y. Seino, A. Kotani, *J. Phys. Soc. Jpn.* **60 (1991)** 1040–1050.

2-99 A. Hartmann, G. J. Russel, D. N. Matthews, J. W. Cochrane, *Surf. Interface Anal.* **24 (1996)** 657–661.

2-100 P. Adler, H. Buchkremer-Hermanns, A. Simon, *Z. Phys. B Condens. Matter* **81 (1990)** 355–363.

2-101 K. H. Young, E. J. Smith, M. M. Eddy, T. W. James, *Appl. Surf. Sci.* **52 (1991)** 85–89.

2-102 S. Gokhale, S. Mahamuni, K. Joshi, A. S. Nigavekar, S. K. Kulkarni, *Surf. Sci.* **257 (1991)** 157–166.

2-103 S.-I. Fujimori, Y. Saito, K.-I. Yamaki, T. Okane, N. Sato, T. Komatsubara, S. Suzuki, S. Sato, *Surf. Sci.* **444 (2000)** 180–186.

2-104 F. Esaka, K. Furuya, H. Shimada, M. Imamura, N. Matsubayashi, T. Kikuchi, H. Ichimura, A. Kawana, *Surf. Interface Anal.* **27 (1999)** 1098–1106.

2-105 L. Ren You, N. K. Huang, H. L. Zhan, B. Yang, D. Z. Wang, *Appl. Surf. Sci.* **150 (1999)** 39–42.

2-106 L. Ottaviano, L. Lozzi, A. Montefusco, S. Santucci, *Surf. Sci.* **443 (1999)** 227–237.

2-107 G. Iucci, G. Polzonetti, P. Altamura, G. Paolucci, A. Goldoni, R. D'Amato, M. V. Russo, *Appl. Surf. Sci.* **153 (1999)** 10–18.

2-108 L. L. Cao, Y. M. Sun, L. Q. Zheng, *Wear* **140 (1990)** 345–357.

2-109 S. Noël, L. Boyer, C. Bodin, *J. Vac. Sci. Technol. A* **9 (1991)** 32–38.

2-110 A. Osaka, Y.-H. Wang, Y. Miura, T. Tsugaru, *J. Mater. Sci.* **26 (1991)** 2778–2782.

2.111 S. M. Mukhopadhyay, S. H. Garofalini, *J. Non Cryst. Solids* **126 (1990)** 202–208.

2-112 E.-T. Kang, D. E. Day, *J. Non Cryst. Solids* **126 (1990)** 141–150.

2-113 Z. Hussain, E. E. Khawaja, *Phys. Scr.* **41 (1990)** 939–943.

2-114 R. K. Brow, Y. B. Peng, D. E. Day, *J. Non Cryst. Solids* **126 (1990)** 231–238.

2-115 B. A. van Hassel, A. J. Burggraaf, *Appl. Phys. A* **52 (1991)** 410–417.

2-116 P. Prieto, L. Galán, J. M. Sanz, *Surf. Sci.* **251/252 (1991)** 701–705.

2-117 K. Oyoshi, T. Tagami, S. Tanaka, *J. Appl. Phys.* **68 (1990)** 3653–3660.

2-118 Z.-H. Lu, A. Yelon, *Phys. Rev. B Condens. Matter* **41 (1990)** 3284–3286.

2-119 A. Carnera, P. Mazzoldi, A. Boscolo-Boscoletto, F. Caccavale, R. Bertoncello, G. Granozzi, I. Spagnol, G. Battaglin, *J. Non Cryst. Solids* **125 (1990)** 293–301.

2-120 R. W. Bernstein, J. K. Grepstad, *J. Appl. Phys.* **68 (1990)** 4811–4815.

2-121 A. Ermolieff, F. Martin, A. Amouroux, S. Marthon, J. F. M. Westendorp, *Semicond. Sci. Tech.* **6 (1991)** 98–102.

2-122 S. V. Hattangady, R. A. Rudder, M. J. Mantini, G. G. Fountain, J. B. Posthill, R. J. Markunas, *J. Appl. Phys.* **68 (1990)** 1233–1236.

2-123 Maria Faur, Mircea Faur, D. T. Jayne, M. Goradia, C. Goradia, *Surf. Interface Anal.* **15 (1990)** 641–650.

2-124 H. Hasegawa, M. Akazawa, H. Ishii, A. Uraie, H. Iwadate, E. Ohue, *J. Vac. Sci. Technol. B* **8 (1990)** 867–873.

2-125 L. A. Harris, *J. Appl. Phys.* **39 (1968)** 1419–1427, 1428–1431

2-126 R. E. Weber, W. T. Peria, *J. Appl. Phys.* **38 (1967)** 4355–4358.

2-127 P. W. Palmberg, G. K. Bohm, J. C. Tracy, *Appl. Phys. Lett.* **15 (1969)** 254–255.

2-128 E. Casnati, A. Tartari, C. Baraldi, *J. Phys. B* **15 (1982)** 155–167.

2-129 L. E. Davis, N. C. MacDonald, P. W. Palmberg, G. E. Riach, R. E. Weber: *Handbook of Auger Electron Spectroscopy*, Physical Electronic Ind., Minnesota **1976**.

2-130 (a) E. Kny, *J. Vac. Sci. Technol.* **17 (1980)** 658–660.
(b) J. Kleefeld, L. L. Levinson, *Thin Solid Films* **64 (1979)** 389–393.
(c) M. A. Smith, L. L. Levinson, *Phys. Rev. B Solid State* **16 (1977)** 1365–1369.

2-131 M. Gryzinski, *Phys. Rev.* **138 (1965)** A 305–321.

2-132 S. Ichimura, R. Shimizu. *Surf. Sci.* **112 (1981)** 386–408.

2-133 R. Payling, *J. Electron Spectros. Relat. Phenom.* **36 (1985)** 99–104.

2-134 S. Mroczkowski, D. Lichtman, *J. Vac. Sci. Technol. A* **3 (1985)** 1860–1865.

2-135 T. Sato, Y. Nagasawa, T. Sekine, Y. Sakai, *Surf. Interface Anal.* **14 (1989)** 787–793.

2-136 M. P. Seah, *Surf. Interface Anal.* **9 (1986)** 85–98.

2-137 H. Goretzki, *Fresenius' Z. Anal. Chem.* **329 (1987)** 180–189.

2-138 R. K. Wild, J. Hickey in: I. Olefjord, L. Nyborg, D. Briggs (eds.): ECASIA 97, John Wiley and Sons, Chichester **1997**, p. 643–646

2-139 D. Saidi, N. Souami, M. Oulladj, M. Negache, S. Lebaili in: I. Olefjord, L. Nyborg, D. Briggs (eds.): ECASIA 97, John Wiley and Sons, Chichester **1997**, p. 683–686

2-140 I. A. Vatter, J. M. Titchmarsh, *Surf. Interface Anal.* **25 (1997)** 760–776.

2-141 M. Norell, L. Nyborg in: I. Olefjord, L. Nyborg, D. Briggs (eds.): ECASIA 97, John Wiley and Sons, Chichester **1997**, p. 687–690.

2-142 B. M. Reichl, M. M. Eisl, T. Weis, H. Hutter, H. Stoeri, *Fresenius' Z. Anal. Chem.* **353 (1995)** 762–765.

2-143 T. H. Chuang, *Mat. Sci. Ens.* **A141 (1991)** 169–178.

2-144 D. J. Nettleship, R. K. Wild, *Surf. Interface Anal.* **16 (1990)** 552–558.

2-145 A. Bukaluk, *Appl. Surf. Sci.* **144/145 (1999)** 395–398.

2-146 E. M. Moser, M. Metzger, L. J. Gauckler, *Fresenius' Z. Anal. Chem.* **353 (1995)** 684–689.

2-147 A. Zalar, B. Baretzky, F. Dettenwanger, M. Rühle, P. Panjan, *Surf. Interface Anal.* **26 (1998)** 861–867.

2-148 H. Li, S. Jin, H. Bender, F. Lanckmans, I. Heyvaert, K. Maex, L. Froyen, *J. Vac. Sci. Technol. B* **18 (2000)** 242–251.

2-149 R. H. Roberts, M. Prutton, D. K. Wilkinson, I. R. Barkshire, C. J. Hill, P. J. Pearson, P. D. Augustus, D. K. Skinner, K. Stribley, *Surf. Interface Anal.* **26 (1998)** 461–470.

2-150 D. K. Fillmore, H. A. Krasinski, *Surf. Interface Anal.* **26 (1998)** 109–112.

2-151 C. Palacio, *Surf. Interface Anal.* **27 (1999)** 1092–1097.

2-152 L. Cao, F. Shi, W. Song, Y. Zhu, *Surf. Interface Anal.* **28 (1999)** 258–263.

2-153 Y. Zhu, W. Yao, B. Zheng, L. Cao, *Surf. Interface Anal.* **28 (1999)** 254–257.

2-154 A. D. Polli, T. Wagner, T. Gemming, M. Rühle, *Surf. Sci.* **448 (2000)** 279–289.

2-155 J. Marien, T. Wagner, G. Duscher, A. Koch, M. Rühle, *Surf. Sci.* **446 (2000)** 219–228.

2-156 R. Wagner, D. Schlatterbeck, K. Christmann, *Surf. Sci.* **440 (1999)** 231–251.

2-157 T. W. Kang, T. W. Kim, *Appl. Surf. Sci.* **150 (1991)** 190–194.

2-158 S. Bredendiek, H. Jenett, *Mikrochim. Acta* **107 (1992)** 219–225.

2-159 S. Bredendiek, H. Jenett, *Fresenius' J. Anal. Chem.* **346 (1993)** 315–317.

2-160 C. Uebing, *Surf. Sci.* **225 (1990)** 97–106.

2-161 C. Uebing in: I. Olefjord, L. Nyborg, D. Briggs (eds.): ECASIA 97, John Wiley and Sons, Chichester 1997, p. 58–67.

2-162 Z. Song, D. Tan, F. He, X. Bao, *Appl. Surf. Sci.* **137 (1999)** 142–149.

2-163 H. K. Lee, R. W. Hyland Jr., H. I. Aaronson, P. P. Wynblatt, *Surf. Sci.* **408 (1998)** 288–299.

2-164 G. Contini, V. Di Castro, N. Motta, A. Sgarlata, *Surf. Sci.* **405 (1998)** L 509-L 513.

2-165 D. T. L. van Agterveld, G. Palasantzas, J. T. M. De Hosson, *Appl. Surf. Sci.* **152 (1999)** 250–258.

2-166 E. Clauberg, C. Uebing, H. J. Grabke, *Appl. Surf. Sci.* **143 (1999)** 206–214.

2-167 R. Hamminger, G. Grathwohl, F. Thümmler, *J. Mater. Sci.* **18 (1983)** 353–364.

2-168 H. Bethe, *Ann. Phys.* **5 (1930)** 325.

2-169 M. Inokuti, *Rev. Mod. Phys.* **43 (1971)** 297.

2-170 P. Schattschneider: *Fundamentals of Inelastic Electron Scattering*, Springer-Verlag, Wien – New York **1986**.

2-171 R. F. Egerton: *Electron Energy-Loss Spectroscopy in the Electron Microscope*, 2nd edition, Plenum Press, New York and London **1996**.

2-172 J. J. Hren, J. I. Goldstein (eds.): *Introduction to Analytical Electron Microscopy*, Plenum Press, New York **1979**.

2-173 D. B. Williams, C. B. Carter: *Transmission Electron Microscopy – A Textbook for Materials Science*, Plenum Press, New York and London **1996**.

2-174 L. Reimer (ed.): *Energy-Filtering Transmission Electron Microscopy*, Springer-Verlag, Berlin–Heidelberg **1995**.

2-175 L. Reimer: *Transmission Electron Microscopy: Physics of Image Formation and Microanalysis*, Springer-Verlag, Berlin–Heidelberg–New York 1989.
2-176 H. Bethge, J. Heydenreich (eds.): *Electron Microscopy in Solid State Physics*, Elsevier **1987**.
2-177 S. Amelincks, D. van Dyck, J. van Landuyt, G. van Tendeloo (eds.): *Electron Microscopy: Principles and Fundamentals*, VCH Verlagsgesellschaft mbH, Weinheim **1997**.
2-178 R. M. Anderson, S. D. Walck (eds.): *Specimen Preparation for Transmission Electron Microscopy of Materials IV*, Materials Research Society, Pittsburgh **1997**.
2-179 A. J. F. Metherell, R. F. Cook, *Optik* **34 (1972)** 535.
2-180 J. Kramer, P. van Zuylen, D. F. Hardy in: *Proc. 8th Internat. Congr. El. Microsc.*, Canberra **1974**, Vol. 1, p. 372.
2-181 P. E. Batson in: *Proc. 42nd EMSA Meeting*, San Francisco **1984**, p. 558.
2-182 D. B. Wittry, *J. Phys. D (London)* **2 (1976)** 1757.
2-183 A. V. Crewe, M. Isaacson, D. Johnson, *Rev. Sci. Instrum.* **42 (1971)** 411.
2-184 H. T. Pearce-Percy, *J. Phys. D (London)* **9 (1976)** 135, 515.
2-185 R. F. Egerton, C. E. Lyman in: J. A. Venables (ed.): *Developments in Electron Microscopy and Analysis*, Academic Press, London and New York 1976, p. 35.
2-186 W. Engel in: *Proc. 9th Internat. Congr. El. Microsc.*, Toronto **1978**, Vol. 1, p. 48.
2-187 J. W. Andrew, F. P. Ottensmeyer, E. Mortell in: *Proc. 9th Internat. Congr. El. Microsc.*, Toronto **1978**, Vol. 1, p. 40.
2-188 R. Castaing, L. Henry, *J. Microsc. (Paris)* **3 (1964)** 133.
2-189 B. Jouffrey in: *Proc. 5th Eur. Conf. El. Microsc.*, Manchester **1972**, p. 190.
2-190 R. M. Henkelman, F. P. Ottensmeyer, *J. Microsc. (London)* **102 (1974)** 79.
2-191 T. Ichinokawa, Jap. *J. Appl. Phys.* **7 (1968)** 799.
2-192 H. Rose, E. Plies, *Optik* **40 (1974)** 336.
2-193 G. Zanchi, J. P. Perez, J. Sevely, *Optik* **43 (1975)** 495.
2-194 H. T. Pearce-Percy, D. Krahl, J. Jäger in: *Proc. Eur. Conf, El. Microsc.*, Jerusalem **1976**, p. 348.
2-195 D. Krahl, K.-H. Herrmann, W. Kunath in: *Proc. 9th Internat. Cong. El. Microsc.*, Toronto **1978**, Vol. 1, p. 42.
2-196 G. Zanchi, J. Sevely, B. Jouffrey, *J. Microsc. Spectrosc. Electron.* **2 (1977)** 95.
2-197 H. Rose, *Ultramicroscopy* **78 (1999)** 13.
2-198 P. W. Hawkes, E. Kasper: *Principles of Electron Optics*, Vol. 2, Academic Press, London–San Diego–New York 1996.
2-199 W. Rechner in: *Veröff. 10. Tagung Elektronenmikroskopie*, Leipzig **1981**, S. 254.
2-200 R. Schneider, W. Rechner, *Mikrochim. Acta* **12** [Suppl.] **(1992)** 197.
2-201 W. Probst, G. Benner, J. Mayer, *Inst. Phys. Conf. Ser.* **130 (1993)** 295.
2-202 K. Tsuno, T. Kaneyama, T. Honda, Y. Ishida, K. Tsuda, M. Terauchi, M. Tanaka in: *Proc. Internat. Cong. El. Microsc.*, Vol. I, Cancun **1998**, p. 253.
2-203 H. Raether in: *Excitations of Plasmons and Interband Transitions by Electrons*, Springer-Verlag, Berlin–Heidelberg–New York, **1980**.
2-204 R. H. Ritchie, *Phys. Rev.* **106 (1957)** 874.
2-205 D. R. Spalding, A. J. F. Metherell, *Phil. Mag.* **18 (1968)** 41.
2-206 R. F. Cook, S. L. Cundy, *Phil. Mag.* **20 (1969)** 665.
2-207 P. Schattschneider, B. Jouffrey in: L. Reimer (ed.): *Energy-Filtering Transmission Electron Microscopy*, Springer-Verlag, Berlin – Heidelberg **1995**, p. 151.
2-208 G. Brockt, H. Lakner in: *Proc. Internat. Cong. El. Microsc.*, Vol. III, Cancun **1998**, p. 625.
2-209 C. C. Ahn, O. L. Krivanek: *EELS Atlas – A Reference Guide of Electron Energy Loss Spectra Covering All Stable Elements*, GATAN Inc.
2-210 P. Rez in: M. M. Disko, C. C. Ahn, B. Fultz (eds.): *Transmission Electron Energy Loss Spectrometry in Materials Science*, The Minerals, Metals and Materials Society, Warrendale **1992**, p. 107.
2-211 F. Hofer in: L. Reimer (ed.): *Energy-Filtering Transmission Electron Microscopy*, Springer-Verlag, Berlin–Heidelberg–New York **1995**, p. 225.
2-212 R. Brydson, H. Sauer, W. Engel in: M. M. Disko, C. C. Ahn, B. Fultz (eds.): *Transmission Electron Energy Loss Spectrometry in Materials Science*, The Minerals, Metals and Materials Society, Warrendale **1992**, p. 131.
2-213 L. Reimer, U. Zepke, St. Schulze-Hillert, M. Ross-Messemer, W. Probst, E. Wei-

MER: *EELSpectroscopy – A Reference Handbook of Standard Data for Identification and Interpretation of Electron Energy Loss Spectra and for Generation of Electron Spectroscopic Images*, Carl Zeiss, Electron Optics Division, Oberkochen **1992**.

2-214 K. H. JOHNSON IN: P. O. LOEWDIN (ed.): *Advances in Quantum Chemistry*, Vol. 7, Academic Press, New York **1973**, p. 143.

2-215 D. DILL, J. L. DEHMER, *J. Chem. Phys.* **61 (1974)** 692.

2-216 W. WURTH, J. STÖHR, *Vacuum* **41 (1990)** 237.

2-217 P. J. DURHAM, *Computer Phys. Commun.* **25 (1982)** 193.

2-218 D. D. VVEDENSKY, D. K. SALDIN, J. B. PENDRY, *Computer Phys. Commun.* **40 (1986)** 421.

2-219 R. W. JANSEN, O. F. SANKEY, *Phys. Rev. B* **36 (1987)** 6520.

2-220 X. WENG, P. REZ, O. F. SANKEY, *Phys. Rev. B* **40 (1989)** 5694.

2-221 R. D. LEAPMAN, J. SILCOX, *Phys. Rev. Lett.* **42 (1979)** 1361.

2-222 R. SCHNEIDER, F. SYROWATKA, A. RÖDER, H. P. ABICHT, D. VÖLTZKE, J. WOLTERSDORF in: Committee of Eur. Soc. Microsc. (eds.): *Proc. 11th Eur. Cong. Electr. Microsc.*, Brussels **1998**, I-287.

2-223 R. BRYDSON, H. SAUER, W. ENGEL, F. HOFER, *J. Phys.: Conden. Matter* **4 (1992)** 3429.

2-224 C. COLLIEX in: R. BARER, V. E. COSSLETT (eds.): *Advances in Optical and Electron Microscopy*, Academic Press, London and New York **1984**.

2-225 B. K. TEO, D. C. JOY (eds.): *EXAFS Spectroscopy, Techniques and Applications*, Plenum Press, New York **1981**.

2-226 E. A. STERN in: D. C. KONINGSBERGER, R. PRINS (eds.): *X-ray Absorption*, John Wiley and Sons, New York **1988**, p. 3.

2-227 R. F. EGERTON, *Phil. Mag.* **31 (1975)** 199.

2-228 R. F. EGERTON, *Ultramicroscopy* **4 (1979)** 169.

2-229 R. D. LEAPMAN, C. E. FIORI, C. R. SWYT, *J. Microsc. (London)* **133 (1984)** 239.

2-230 D. C. JOY IN: D. C. JOY, A. D. ROMIG JR., J. I. GOLDSTEIN (eds.): *Principles of Analytical Electron Microscopy*, Plenum Press, New York and London **1986**, p. 249.

2-231 R. F. EGERTON, R. D. LEAPMAN in: L. REIMER (ed.): *Energy-Filtering Transmission Electron Microscopy*, Springer-Verlag, Berlin–Heidelberg–New York 1995, p. 269.

2-232 R. D. LEAPMAN, S. B. ANDREWS, *J. Microsc.* **161 (1991)** 3.

2-233 S. J. PENNYCOOK, D. E. JESSON, N. D. BROWNING, M. F. CHISHOLM, *Mikrochim. Acta* **114/115 (1994)** 195.

2-234 C. JEANGUILLAUME, C. COLLIEX, *Ultramicroscopy* **28 (1989)** 252.

2-235 R. SCHNEIDER, J. WOLTERSDORF, A. RÖDER, *Mikrochim. Acta* **114/115 (1994)** 545.

2-236 H. ROSE, *Ultramicroscopy* **78 (1999)** 13.

2-237 W. SIGLE in: L. FRANK, F. CIAMPOR (eds.): *Proc. 12th Eur. Cong. Electr. Microsc.*, Vol. II, Czechoslovak Soc. for Electron Microscopy, Brno, **2000**, P 101.

2-238 C. DAVISSON, L. H. GERMER, *Nature* (London) **119 (1927)** 558.

2-239 C. DAVISSON, L. H. GERMER, *Phys. Rev.* **29 (1927)** 908.

2-240 C. DAVISSON, L. H. GERMER, *Phys. Rev.* **30 (1927)** 705.

2-241 J. B. PENDRY, *Low Energy Electron Diffraction*, Academic Press, London, **1974**.

2-242 M. A. VAN HOVE AND S. Y. TONG, *Surface Crystallography by LEED*, Springer-Verlag, Berlin, **1979**.

2-243 M. A. VAN HOVE, W. H. WEINBERG, C.-M. CHAN, *Low-Energy Electron Diffraction*, Springer-Verlag, Berlin, **1986**.

2-244 L. J. CLARKE, *Surface Crystallography – An Introduction to Low Energy Electron Diffraction*, John Wiley and Sons, Chichester, **1985**.

2-245 U. SCHEITHAUER, G. MEYER, M. HENZLER, *Surf. Sci.* **178 (1986)** 441.

2-246 R. L. PARK, J. E. HOUSTON, D. G. SCHREINER, *Rev. Sci. Instrum.* **42 (1971)** 60.

2-247 E. A. WOOD, *J. Appl. Phys.* **35 (1964)** 1306.

2-248 M. HENZLER in: H. IBACH (ed.): *Electron Spectroscopy for Surface Analysis*, Springer-Verlag, Berlin, **1977**, p. 117.

2-249 M. HENZLER, W. GÖPEL, *Oberflächenphysik des Festkörpers*, Teubner Verlag, Stuttgart, **1991**.

2-250 P. R. WATSON, M. A. VAN HOVE, K. HERMANN, *NIST Surface Structure Database (SSD)*, Version 3.0, NIST, Gaithersburg, **1999**.

2-251 H. PFNÜR, P. PIERCY, *Phys. Rev.* **B40 (1989)** 2515.

2-252 H. PFNÜR, P. PIERCY, *Phys. Rev.* **B41 (1990)** 582.

2-253 G. Hoogers, D. A. King, *Surf. Sci.* **286 (1993)** 306.
2-254 E. Bauer, H. Poppa, Y. Visvanath, *Surf. Sci.* **58 (1976)** 517.
2-255 C. Zhang, M. A. Van Hove, G. A. Somorjai, *Surf. Sci.* **149 (1985)** 326.
2-256 S. Fölsch, U. Barjenbruch, M. Henzler, *Thin Solid Films* **172 (1989)** 123.
2-257 H. Neureiter, S. Spranger, M. Schneider, U. Winkler, M. Sokolowski, E. Umbach, *Surf. Sci.* **388 (1997)** 186.
2-258 M. Sokolowski, H. Neureiter, M. Schneider, E. Umbach, S. Tatarenko, *Appl. Surf. Sci.* **123–124 (1998)** 71.
2-259 H. Neureiter, S. Schinzer, M. Sokolowski, E. Umbach, *Journal of Crystal Growth* **201–202 (1999)** 93.
2-260 J. B. Pendry, *J. Phys.* **C 13 (1980)** 937.
2-261 W. H. Press, B. P. Flannery, S. A. Teukolsky, W. T. Vetterling, *Numerical Recipes in C*, Cambridge University Press, Cambridge, **1988**.
2-262 D. E. Goldberg, *Genetic Algorithms in Search, Optimization and Machine Learning*, Addison-Wesley, Reading (Mass.), **1989**.
2-263 W. Braun, G. Held, H.-P. Steinrück, C. Stellwag, D. Menzel, *Surf. Sci.* **475 (2001)** 18.
2-264 K. Heinz, U. Starke, F. Bothe, *Surf. Sci. Lett.* **243 (1991)** L70.
2-265 K. Heinz, U. Starke, M. A. Van Hove, G. A. Somorjai, *Surf. Sci.* **261 (1992)** 57.
2-266 E. Lang, P. Heilmann, G. Hanke, K. Heinz, K. Müller, *Appl. Phys.* **19 (1979)** 287.
2-267 G. Held, S. Uremovic, C. Stellwag, D. Menzel, *Rev. Sci. Instrum.* **67 (1996)** 378.
2-268 P. J. Rous, J. B. Pendry, *Surf. Sci.* **219 (1989)** 355.
2-269 P. J. Rous, J. B. Pendry, *Surf. Sci.* **219 (1989)** 373.
2-270 P. J. Rous, *Prog. Surf. Sci.* **39 (1992)** 3.
2-271 LEED I-V program packages are distributed by:
G. Held: *gh10009@cam.ac.uk*
V. Blum, K. Heinz: *Comp. Phys. Comm.* 134 **(2001)** 392.
M. A. Van Hove: *vanhove@lbl.gov*
2-272 K. Heinz, *Surf. Sci.* **299/300 (1994)** 433.
2-273 K. Heinz, S. Müller, L. Hammer, *J. Phys. Condens. Matter* 11 **(1999)** 9437.
2-274 H. Over, *Prog. Surf. Sci.* **58 (1998)** 249.
2-275 A. Kahn, *Surf. Sci.* **299/300 (1994)** 469.

References to Chapter 3

3-1 A. Benninghoven, *Z. Phys.* **220 (1969)** 159–180.
3-2 P. Sigmund in: N. H. Tolk, J. C. Tully, W. Heiland, C. W White (eds.): *Inelastic Ion-Surface Collisions*, Academic Press. New York 1977.
3-3 J. C. Rivière: *Surface Analytical Techniques*, Oxford University Press, Oxford **1990**.
3-4 D. Stapel, O. Brox, A. Benninghoven, *Appl. Surf. Sci.* **140 (1999)156–167.**
3-5 K. Kötter, A. Benninghoven, *Appl. Surf. Sci.* **133 (1998)** 47–57.
3-6 A. Benninghoven, F. G. Rüdennauer, H. W. Werner: *Secondary Ion Mass Spectrometry*, John Wiley and Sons, New York **1987**.
3-7 P. D. Prewett, D. K. Jefferies, *J. Phys. D* **13 (1980)** 1747–1755.
3-8 A. R. Krauss, D. M. Gruen, *Appl. Phys.* **14 (1977)** 89–97.
3-9 P. Steffens, E. Niehuis, T. Friese, D. Greifendorf, A. Benninghoven, *J. Vac. Sci. Technol. A* **3 (1985)** 1322–1325.
3-10 E. Niehuis, T. Heller, H. Feld, A. Benninghoven, *J. Vac. Sci. Technol. A* **5 (1987)** 1243.
3-11 V. I. Karataev, B. A. Mamyrin, D. V. Shmikk, *Sov. Phys. Techn. Phys.* **16 (1972)** 1177.
3-12 B. W. Schueler, *Microsc. Microanal. Microstruct.* **3 (1992)** 119.
3-13 T. Sakurai, T. Matsuo, H. Matsuda, *Int. J. Mass. Spectrom. Ion Phys.* **63 (1985)** 273.
3-14 K. Iltgen, C. Bendel, E. Niehuis, A. Benninghoven, *J. Vac. Sci. Technol. A* **15 (1997)** 460.
3-15 I. V. Bletsos, D. M. Hercules, A. Benninghoven, D. Greifendorf in: A. Benninghoven, R. J. Colton, D. S. Simons, H. W Werner (eds.): *SIMS V*, Springer-Verlag, Berlin 1986.
3-16 P. H. McBreen, S. Moore, A. Adnot, D. Roy, *Surf. Sci.* **194 (1988)** L112-L118.
3-17 D. van Leyen, B. Hagenhoff, E. Niehuis, A. Benninghoven, I. V. Bletsos, D. M. Hercules, *J. Vac. Sci. Technol. A* **7 (1989)** 1790.
3-18 J. A. Treverton, J. Ball, D. Johnson, J. C. Vickerman, R. H. West, *Surf. Interface Anal.* **15 (1990)** 369–376.
3-19 B. Hagenhoff, D. Rading in: J. C. Rivière, S. Myhra (eds): *Handbook of Sur-*

face and Interface Analysis, Marcel Dekker, New York **1998**, p. 209–253.

3-20 M. Morra, E. Occhiello, F. Garbassi, *Surf. Interface Anal.* **16 (1990)** 412–417.

3-21 D. E. Fowler, R. D. Johnson, D. van Leyen, A. Benninghoven, *Surf. Interface Anal.* **17 (1991)** 125–136.

3-22 G. J. Leggett, D. Briggs, J. C. Vickermnan, *Surf. Interface Anal.* **17 (1991)** 737–744.

3-23 W. D. Ramsden, *Surf. Interface Anal.* **17 (1991)** 793–802.

3-24 D. J. Pawson, A. P. Ameen, R. D. Short, *Surf. Interface Anal.* **18 (1992)** 13–22.

3-25 H. F. Arlinghaus, M. N. Kwoka, K. B. Jacobson, *Anal. Chem.* **69 (1997)** 3747–3753.

3-26 H. F. Arlinghaus, C. Höppener, J. Drexler in: A. Benninghoven, P. Bertrand, H. N. Migeon, H. W. Werner (eds.): *SIMS XII*, Elsevier Science, Amsterdam **2000**, p. 951.

3-27 C. Höppener, J. Drexler, M. Ostrop, H. F. Arlinghaus in: A. Benninghoven, P. Bertrand, H. N. Migeon, H. W. Werner (eds.): *SIMS XII*, Elsevier Science, Amsterdam **2000**, p. 915.

3-28 R. A. Cocco, B. J. Tatarchuk, *Surf. Sci.* **218 (1989)** 127–146.

3-29 B. H. Sakakini, I. A. Ransley, C. F. Oduoza, J. C. Vickerman, M. A. Chesters, *Surf. Sci.* **271 (1992)** 227–236.

3-30 X.-L. Zhou, C. R. Flores, J. M. White, *Surf. Sci. Lett.* **268 (1992)** L267-L273.

3-31 M. E. Castro, J. M. White, *Surf. Sci.* **257 (1991)** 22–32.

3-32 D. Briggs, M. J. Hearn, *Surf. Interface Anal.* **13 (1988)** 181–185.

3-33 A. Benninghoven, *Angew. Chem.*, (Int. Ed. Engl.) **33 (1994)** 1023–1043.

3-34 A. Benninghoven, B. Hagenhoff, E. Niehuis, *Anal. Chem.* **65 (1993)** 630A-640A.

3-35 N. Bourdos, F. Kolmer, A. Benninghoven, M. Sieber, H.-J. Galla, *Langmuir* **16 (2000)** 1481–1484.

3-36 M. Deimel, H. Rulle, V. Liebing, A. Benninghoven, *Appl. Surf. Sci.* **134 (1998)** 271–274.

3-37 D. Briggs, M. J. Hearn, I. W. Fletcher, A. R. Waugh, B. J. McIntosh, *Surf. Interface Anal.* **15 (1990)** 62–65.

3-38 M. J. Hearn, D. Briggs, *Surf. Interface Anal.* **17 (1991)** 421–429.

3-39 A. Licciardello, C. Puglisi, D. Lipinsky, E. Niehuis, A. Benninghoven in: A. Benninghoven, B. Hagenhoff, H. W. Werner (eds.): *SIMS X*, John Wiley and Sons, New York **1997**.

3-40 F. Kollmer, R. Kamischke, R. Ostendorf, A. Schnieders, C. Y. Kim, J. W. Lee, A. Benninghoven in: A. Benninghoven, P. Bertrand, H. N. Migeon, H. W. Werner (eds.): *SIMS XII*, Elsevier Science, Amsterdam **2000**, p. 329.

3-41 K. Iltgen, A. Benninghoven, E. Niehuis in: G. Gillen, R. Lareau, J. Bennett, F. Stevie (eds.): *SIMS XI*, John Wiley and Sons, New York **1998**, p. 367–370.

3-42 R. F. K. Herzog, F. P. Viehböck, *Phys. Rev.* **76 (1949)** 855L.

3-43 R. F. K. Herzog, H. Liebl, *J. Phys.* **34 (1963)** 2893.

3-44 R. Castaing, G. Slodzian, *J. de Microscopie* **1 (1962)** 395.

3-45 A. Benninghoven, *Z. Phys.* **239 (1970)** 403.

3-46 J. M. Schroeer, T. Rodin, R. Bradley, *Surf. Sci.* **34 (1973)** 511.

3-47 Z. Sroubek, *Phys. Rev. B* **25 (1983)** 604.

3-48 C. A. Anderson, J. R. Hinthorne, *Anal. Chem.* **45 (1973)** 1421.

3-49 J. Mattauch, R. Herzog, *Z. Phys.* **89 (1934)** 786.

3-50 H. Hutter, M. Grasserbauer, *Mikrochim. Acta* **107 (1992)** 137.

3-51 J. M. Schroeer, H. Gnaser, H. Oechsner in: A. Benninghoven, Y. Nihei, R. Shimizu, H. W. Werner (eds.): *SIMS IX*, John Wiley and Sons, New York **1994**, p. 394.

3-52 Ch. Bruner, H. Hutter, K. Piplits, M. Gritsch, G. Pöckl, M. Grasserbauer, *Fresenius' J. Anal. Chem.* **361 (1998)** 667–671.

3-53 T. Stubbings, M. Wolkenstein, H. Hutter, *J. Trace and Microprobe Techniques* **17 (1999)** 1–16.

3-54 M. Gritsch, K. Piplits, R. Barbist, P. Wilhartitz, H. Hutter, *Microchim. Acta* **133 (2000)** 89–93.

3-55 M. Grasserbauer, H. Hutter in: A. Benninghoven, Y. Nihei, R. Shimizu, H. W. Werner (eds.): *SIMS IX*, John Wiley and Sons, New York 1994, p. 545.

3-56 F. G. Ruedenauer, W. Steiger in: A. Benninghoven, A. M. Huber, H. W. Werner (eds.): *SIMS VI*, John Wiley and Sons, New York **1987**, p. 361.

3-57 C. Jarms, H.-R. Stock, P. Mayr, *Surf. Coat. Technol.* **108–109 (1998)** 206–210.

3-58 C. Pollak, B. Kriszt, H. Hutter, *Mikrochim. Acta* **133 (2000)** 261–266.

3-59 Th. Albers, M. Neumann, D. Lipinsky, L. Wiedmann, A. Benninghoven, *Surf. Interface Anal.* **22 (1994)** 9.

3-60 D. Lipinsky, R. Jede, O. Ganschow, A. Benninghoven, *J. Vac. Sci. Technol. A* **3 (1985)** 2007–2017.

3-61 A. R. Bayly, J. Wolstenholme, C. R. Petts, *Surf. Interface Anal.* **21 (1993)** 414–417.

3-62 W. Lotz, *Z. Phys.* **232 (1970)** 101.

3-63 R. Jede, O. Ganschow, U. Kaiser in: D. Briggs, M. P. Seah (eds.): *Practical Surface Analysis*, 2nd ed., vol. 2, John Wiley and Sons, Chichester **1992**.

3-64 H. Oechsner, L. Reichert, *Phys. Lett.* **23 (1966)** 90–92.

3-65 H. Oechsner, *Plasma Phys.* **835 (1974)** 16.

3-66 E. Stumpe, H. Oechsner, H. Schoof, *Appl. Phys.* **20 (1979)** 55.

3-67 H. Oechsner, M. Müller, *J. Vac. Sci. Technol. A* **17 (1999)** 3401–3405.

3-68 H. Gnaser, H.-J. Schneider, H. Oechsner, *Nucl. Instrum. Methods Phys. Res. B* **136–138 (1998)** 1023–1027.

3-69 A. Wucher, *J. Vac. Sci. Technol. A* **6 (1988)** 2287.

3-70 H. Jenett in: H. Günzler, A. M. Bahadir, R. Borsdorf, K. Danzer, W. Fresenius, R. Galensa, W. Huber, M. Linscheid, I. Lüderwald, G. Schwedt, G. Tölg, H. Wisser (eds.): *Analytiker-Taschenbuch* **16**, Springer-Verlag, Berlin, Heidelberg, New York **1997**, p. 43–117.

3-71 M. Fichtner, J. Goschnick, U. C. Schmidt, A. Schweiker, H. J. Ache, *J. Vac. Sci. Technol. A* **10 (1992)** 362.

3-72 R. Jede, H. Peters, G. Dünnebier, O. Ganschow, U. Kaiser, K. Seifert, *J. Vac. Sci. Technol. A* **6 (1988)** 2271.

3-73 J. Bartella, R. Fuchs, J. Goschnick, D. Grunenberg, *Fresenius' J. Anal. Chem.* **346 (1993)** 131.

3-74 A. Wucher, F. Novak, W. Reuter, *J. Vac. Sci. Technol. A* **6 (1988)** 2265.

3-75 H. Oechsner in: M. Grasserbauer, H. W. Werner (eds.): *Analysis of Microelectronic Materials an Devices*, John Wiley and Sons, Chichester **1991**.

3-76 H. Jenett, J. D. Sunderkötter, M. F. Stroosnijder, *Fresenius' J. Anal. Chem.* **358 (1997)** 225–229.

3-77 J. Goschnick, C. Natzeck, M. Sommer, *Appl. Surf. Sci.* **144–145 (1999)** 201–207.

3-78 S. Hopfe, N. Kallis, H. Mai, W. Pompe, R. Scholz, S. Völlmar, B. Wehner, P. Weissbrot, *Fres. J. Anal. Chem.* **346 (1993)** 14–22.

3-79 R. W. Martin, Teak Kim, D. Burns, I. M. Watson, M. D. Dawson, T. F. Krauss, J. H. Marsh, R. M. De la Rue, S. Romani, H. Kheyrandish, *Phys. Status Solidi A* **176 (1999)** 67–71.

3-80 S. Dreer, *Fresenius' J. Anal. Chem.* **365 (1999)** 85–95.

3-81 I. Galesic, B. O. Kolbesen, *Fresenius' J. Anal. Chem.* **365 (1999)** 199–202.

3-82 T. Tharigen, G. Lippold, V. Riede, M. Lorenz, K. J. Koivusaari, D. Lorenz, S. Mosch, P. Grau, R. Hesse, P. Streubel, R. Szargan, *Thin Solid Films* **348 (1999)** 103–113.

3-83 M. Witthaut, R. Cremer, A. von Richthofen, D. Neuschütz, *Fresenius' J. Anal. Chem.* **361 (1998)** 639–641.

3-84 H. Boerner, H. Jenett, V.-D. Hodoroaba, *Fresenius' J. Anal. Chem.* **361 (1998)** 590–591.

3-85 B. Hueber, G. H. Frischat, A. Maldener, O. Dersch, F. Rauch, *J. Non-Cryst. Solids* **256 & 257 (1999)** 130–134.

3-86 R. Getto, J. Freytag, M. Kopnarski, H. Oechsner, *Mater. Sci. Eng. B* **61–62 (1999)** 270–274.

3-87 S. Kruijer, O. Nikolov, W. Keune, H. Reuther, S. Weber, S. Scherrer, *J. Appl. Phys.* **84 (1998)** 6570–6581.

3-88 G. A. Seryogin, S. A. Nikishin, H. Temkin, R. Schlaf, L. I. Sharp, Y. C. Wen, B. Parkinson, V. A. Elyukhin, Yu. A. Kudriavtsev, A. M. Mintairov, N. N. Faleev, M. V. Baidakova, *J. Vac. Sci. Technol. B* **16 (1998)** 1456–1458.

3-89 M. C. Simmonds, A. Savan, H. van Swygenhoven, E. Pfluger, *Thin Solid Films* **354 (1999)** 59–65.

3-90 A. Vogt, A. Simon, H. L. Hartnagel, J. Schikora, V. Buschmann, M. Rodewald, H. Fuess, S. Fascko, C. Koerdt, H. Kurz, *J. Appl. Phys.* **83 (1998), 7715–7719.**

3-91 J. Goschnick, U. Maeder, M. Sommer, *Fresenius' J. Anal. Chem.* **361 (1998)** 707–709.

3-92 J. A. Rebane, O. Yu. Gorbenko, S. G. Suslov, N. V. Yakovlev, I. E. Korsakov, V. A. Amelichev, Yu. D. Tretyakov, *Thin Solid Films* **302 (1997)** 140–146.

3-93 C. Degueldre, D. Buckley, J. C. Dran, E. Schenker, *J. Nucl. Mater.* **252 (1998)** 22–27.

3-94 D. Sommer, A. Essing, *Rep. Eur. Comm.* **EUR 18864 (1999)** 1–126.

3-95 T. A. Dang, T. A. Frisk, M. W. Grossman, C. H. Peters, *J. Electrochem. Soc.* **146 (1999)** 3896–3902.

3-96 I. Baumann, F. Cusso, B. Herreros, H. Holzbrecher, H. Paulus, K. Schaefer, W. Sohler, *Appl. Phys. A* **68 (1999)** 321–324.

3-97 M. Miloshova, E. Bychkov, V. Tsegelnik, V. Strykanov, H. Klewe-Nebenius, M. Bruns, W. Hoffmann, P. Papet, J. Sarradin, A. Pradel, M. Ribes, *Sens. Actuators* **B7 (1999)** 171–178.

3-98 R. B. Vasiliev, M. N. Rumyantseva, S. E. Podguzova, A. S. Ryzhikov, L. I. Ryabova, A. M. Gaskov, *Mater. Sci. Eng. B* **57 (1999)** 241–246.

3-99 J. Goschnick, C. Natzeck, M. Sommer, F. Zudock, *Thin Solid Films* **332 (1998)** 215–219.

3-100 J. Goschnick, M. Sommer, *Fresenius' J. Anal. Chem.* **361 (1998)** 704–707.

3-101 H. Nickel, D. Clemens, W. J. Quadakkers, L. Singheiser, *PVP* (Am. Soc. Mech. Eng.) **374 (1998)** 357–361.

3-102 C. H. Becker, K. T. Gillen, *Anal. Chem.* **56 (1984)** 1671.

3-103 C. H. Becker, *J. Vac. Sci. Technol. A* **5 (1987)** 1181.

3-104 J. B. Pallix, C. H. Becker, N. Newman, *J. Vac. Sci. Technol. A* **6 (1988)** 1049.

3-105 M. Terhorst, R. Möllers, A. Schnieders, E. Niehuis, A. Benninghoven in: A. Benninghoven, Y. Nihei, R. Shimizu, H. W. Werner (eds.): *SIMS IX*, John Wiley and Sons, Chichester **1994**, p. 434–437.

3-106 A. Schnieders, R. Möllers, M. Terhorst, H.-G. Cramer, E. Niehuis, A. Benninghoven, *J. Vac. Sci. Technol. B* **14 (1996)** 2712.

3-107 N. Thonnard, J. E. Parks, R. D. Willis, L. J. Moore, H. F. Arlinghaus, *Surf. Interface Anal.* **14 (1989)** 751.

3-108 H. F. Arlinghaus, M. T. Spaar, N. Thonnard, *J. Vac. Sci. Technol. A* **8 (1990)** 2318.

3-109 H. F. Arlinghaus, M. T. Spaar, N. Thonnard, A. W. McMahon, T. Tanigaki, H. Shichi, P. H. Holloway, *J. Vac. Sci. Technol. A* **11 (1993)** 2317.

3-110 D. L. Pappas, D. M. Hrubowchak, M. H. Ervin, N. Winograd, *Science* **243 (1989)** 64.

3-111 M. J. Pellin, C. E. Young, W. F. Calaway, J. E. Whitten, D. M. Gruen, J. B. Blum, I. D. Hutcheon, G. J. Wasserburg, *Phil. Trans. R. Soc. Lond. A* **333 (1990)** 133.

3-112 S. W. Downey, A. B. Emerson, R. F. Kopf, *Inst. Phys. Conf. Ser.* **114** (Resonance Ionization Spectroscopy) **(1991)** p. 401.

3-113 N. Thonnard, R. D. Willis, M. C. Wright, W. A. Davis, B. E. Lehman, *Nucl. Instr. and Meth. Phys. Res. B* **29 (1987)** 398–406.

3-114 J. Zehnpfenning, E. Niehuis, H.-G. Cramer, T. Heller, R. Möllers, M. Terhorst, A. Schnieders, Zhiyuan Zhang, A. Benninghoven in: A. Benninghoven, Y. Nihei, R. Shimizu, H. W. Werner (eds.): *SIMS IX*, John Wiley and Sons, Chichester **1994**, p. 561–564.

3-115 H. F. Arlinghaus, C. F. Joyner, *J. Vac. Sci. Technol. B* **14 (1996)** 294–300.

3-116 H. F. Arlinghaus, C. F. Joyner, J. Tower, S. Sen in: A. Benninghoven, B. Hagenhoff, H. W. Werner (eds): *SIMS X*, John Wiley and Sons, New York **1997**, p. 463–466.

3-117 H. F. Arlinghaus, M. T. Spaar, R. C. Switzer, G. W. Kablaka, *Anal. Chem.* **69 (1997)** 3169–3176.

3-118 H. F. Arlinghaus, M. P. Kwoka, X. Q. Guo, K. B. Jacobson, *Anal. Chem.* **69 (1997)** 1510–1517.

3-119 H. F. Arlinghaus, M. P. Kwoka, K. B. Jacobson, *Anal. Chem.* **69 (1997)** 3747–3753.

3-120 L. C. Feldman, J. W. Mayer: *Fundamentals of Surface and Thin Film Analysis*, North Holland, New York 1986.

3-121 J. A. Leavitt, L. C. McIntyre, M. R. Weller in: J. R. Tesmer, M. Nastasi (eds.): *Handbook of Modern Ion Beam Materials Analysis*, Materials Research Society, Pittsburgh **1995**, p. 37–81.

3-122 M. L. Swanson in: J. R. Tesmer, M. Nastasi (eds.): *Handbook of Modern Ion Beam Materials Analysis*, Materials Research Society, Pittsburgh **1995**, p. 231–300.

3-123 S. Humphries: *Principles of Charged Particle Acceleration*, John Wiley and Sons, New York **1986**.

3-124 E. Steinbauer, P. Bauer, M. Geretschläger, G. Bortels, J. P. Biersack, P. Burger, *Nucl. Instr. Meth. B* **85 (1994)** 642–649.

3-125 J. Vrijmoeth, P. M. Zagwin, J. W. M. Frenken, J. F. van der Veen, *Phys. Rev. Lett.* **67 (1991)** 1134–1137.

3-126 J. C. Banks, B. L. Doyle, J. A. Knapp, D. Werho, R. B. Gregory, M. Anthony, T. Q. Hurd, A. C. Diebold, *Nucl. Instr. Meth. B* **136–138 (1998)** 1223–1228.

3-127 R. S. Bhattacharya, P. P. Pronko, *Appl. Phys. Lett.* **40 (1982)** 890–892.

3-128 L. R. Doolittle, *Nucl. Instr. Meth. B* **9 (1985)** 344–351; *Nucl. Instr. Meth. B* **15 (1986)** 227–231.

3-129 J. A. Davies, W. N. Lennard, I. V. Mitchell in: J. R. Tesmer, M. Nastasi (eds.): *Handbook of Modern Ion Beam Materials Analysis,* Materials Research Society, Pittsburgh **1995**, p. 343–363.

3-130 S. Mantl, *Nucl. Instr. Meth. B* **80/81 (1993)** 895–900.

3-131 N. R. Parikh, G. S. Sandhu, N. Yu, W. K. Chu, T. E. Jackman, J. M. Baribeau, D. C. Houghton, *Thin Solid Films* **163 (1988)** 455–460.

3-132 K. Nakajima, A. Konishi, K. Kimura, *Phys. Rev. Lett.* **83 (1999)** 1802–1805.

3-133 J. W. M. Frenken, J. F. van der Veen, *Phys. Rev. Lett.* **54 (1985)** 134–137.

3-134 E. Taglauer in: J. C. Vickerman (ed.): *Surface Analysis – The principle techniques,* John Wiley and Sons, New York **1997**, p. 215.

3-135 H. Niehus, W. Heiland, E. Taglauer, *Surf. Sci. Rep.* **17 (1993)** 213–303.

3-136 E. S. Parilis, L. M. Kishinevsky, N. Yu. Turnaev, B. E. Baklitzky, F. F. Umarov in: S. I. Nizhnaya, I. S. Bitensky (eds.): *Atomic Collicions on Atomic Surfaces,* North-Holland, Amsterdam 1993.

3-137 E. C. Goldberg, R. Monreal, F. Flores, H. H. Brongersma, P. Bauer, *Surface Science* **440 (1999)** L875 –L880.

3-138 S. Tsuneyuki, M. Tsukada, *Phys. Rev. B* **34 (1986)** 5758–5768.

3-139 L. C. A. van den Oetelaar, S. N. Mikhailov, H. H. Brongersma, *Nucl. Instr. Meth. B* **93 (1994)** 210–214.

3-140 H. H. Brongersma, P. A. C. Groenen, J.-P. Jacobs, *Mater. Sci. Monogr.* **81 (1994)** 113–182.

3-141 A. W. Denier van der Gon, R. Cortenraad, W. P. A. Jansen, M. A. Reijme, H. H. Brongersma, *Nucl. Instr. Meth. B* **161–163 (2000)** 56–64.

3-142 G. C. van Leerdam: *PhD Thesis,* University of Technology, Eindhoven, The Netherlands **1991**.

3-143 J.-P. Jacobs: *PhD Thesis,* University of Technology, Eindhoven, The Netherlands **1995**.

3-144 L. C. A. van den Oetelaar, J.-P. Jacobs, M. J. Mietus, H. H. Brongersma, V. N. Semenov, V. G. Glebovsky, *Appl. Surf. Sci.* **70/71 (1993)** 79–84.

3-145 H. C. Yao, M. Shelef, *J. Phys. Chem.* **78 (1974)** 2490.

3-146 H. H. Brongersma, J.-P. Jacobs, *Appl. Surf. Sci.* **75 (1994)** 133–138.

3-147 M. M. Viitanen, W. P. A. Jansen, R. van Welzenis, H. H. Brongersma, *J. Phys. Chem. B* **103 (1999)** 6025–6029.

3-148 H. Niehus, R. Spitzl, *Surf. Interface Anal.* **17 (1991)** 287–307.

3-149 J. W. Rabalais, *Surf. Interface Anal.* **27 (1999)** 171–178.

3-150 A. Krauss, O. Auciello, J. A. Schultz, *MRS Bull.* **20 (1995)** 18–23.

3-151 J. L'Ecuyer, C. Brassard, C. Cardinal, J. Chabbat, L. Deschenes, J. Labrie, *J. Appl. Phys.* 47 **(1976)** 381.

3-152 J. Barbour, B. Doyle in: J. Tesmer, M. Nastasi (eds.): *Handbook of Modern Ion Beam Analysis,* Materials Research Society, Pittsburgh **1995**, p. 83.

3-153 W. Assmann, H. Huber, C. Steinhausen, M. Dobler, H. Glückler, A. Weidinger, *Nucl. Instr. Meth. B* **89 (1994)** 131.

3-154 F. Habraken, *Nucl. Instr. Meth. B* **68 (1992)** 181.

3-155 M. Bozoian, K. Hubbard, M. Nastasi, *Nucl. Instr. Meth. B* **51 (1990)** 311.

3-156 J. Baglian, A. Kellok, M. Crockett, A. Shih, *Nucl. Instr. Meth. B* **64 (1992)** 469.

3-157 Y. Kido, S. Miyauchi, O. Takeda, Y. Nagayama, M. Sato, K. Kusao, *Nucl. Instr. Meth. B* **82 (1993)** 474.

3-158 V. Quillet, F. Abel, M. Schott, *Nucl. Instr. Meth. B* **83 (1993)** 47.

3-159 C. Kim, S. Kim, H. Choi, *Nucl. Instr. Meth. B* **155 (1999)** 229.

3-160 H. Andersen, F. Besenbacher, P. Loftager, W. Moller, *Phys. Rev. A* **21 (1980)** 1891.

3-161 S. Nagata, S. Yamaguchi, Y. Fujino, Y. Hori, N. Sugiyama, K. Kamada, *Nucl. Instr. Meth. B* **6 (1985)** 533.

3-162 J. Genzer, J. Rothman, R. Composto, *Nucl. Instr. Meth. B* **86 (1994)** 345.

3-163 H. Ermer, O. Pfaff, W. Straub, M. Geoghean, R. Brenn, *Nucl. Instr. Meth. B* **134 (1998)** 237.
3-164 E. Steinbauer, O. Benka, M. Steinbatz, *Nucl. Instr. Meth. B* **136 (1998)** 695.
3-165 A. Behrooz, R. Headrick, L. Seiberling, R. Zurmühle, *Nucl. Instr. Meth. B* **28 (1987)** 108.
3-166 R. Elliman, H. Timmers, T. Ophel, T. Weijers, L. Wielunski, G. Harding, *Nucl. Instr. Meth. B* **161–163 (2000)** 231.
3-167 J. Stoquert, G. Guillaume, M. Hagi-Ali, J. Grob, C. Ganter, P. Siffert, *Nucl. Instr. Meth. B* **44 (1989)** 184.
3-168 W. Anold Bik, C. deLaat, F. Habraken, *Nucl. Instr. Meth. B* **64 (1992)** 832.
3-169 M. Wielunski, M. Mayer, R. Behrisch, J. Roth, A. Scherzer, *Nucl. Instr. Meth. B* **122 (1997)** 133.
3-170 J. Thomas, M. Fallavier, A. Ziani, *Nucl. Instr. Meth. B* **15 (1986)** 443.
3-171 W. Bohne, J. Röhrich, G. Röschert, *Nucl. Instr. Meth. B* **136–138 (1998)** 633.
3-172 Y. Zhang, H. Whitlow, T. Winzell, I. Bubb, T. Sajavaara, K. Arstila, J. Keinonen, *Nucl. Instr. Meth. B* **149 (1999)** 477.
3-173 P. Johnston, M. El Bouanani, W. Stannard, I. Bubb, D. Cohen, N. Dytlewski, R. Siegele, *Nucl. Instr. Meth. B* **136–138 (1998)** 669.
3-174 H. Timmers, R. Elliman, G. Palmer, T. Ophel, D. O'Connor, *Nucl. Instr. Meth. B* **136–138 (1998)** 611.
3-175 M. Mayer, *Technical Report* IPP9/113, Max-Planck-Institut für Plasmaphysik, Garching (Germany), **1997**.
3-176 E. Szilagyi, F. Paszti and G. Amsel, *Nucl. Instr. Meth. B* **100 (1995)** 103.
3-177 I. Bogdanovic-Radovic, E. Steinbauer, O. Benka, *Nucl. Instr. Meth B* **170 (2000)** 163.
3-178 R. Elliman, H. Timmers, G. Palmer, T. Ophel, *Nucl. Instr. Meth. B* **136–138, (1998)** 649.
3-179 Y. Kim, J. Kim, H. Choi, G. Kim, H. Woo, *Nucl. Instr. Meth. B* **136–138 (1998)** 724.
3-180 S. Grigull, U. Kreissig, H. Huber, W. Assmann, *Nucl. Instr. Meth. B* **132 (1997)** 709.
3-181 G. Dollinger, C. Frey, A. Bergmaier, T. Faestermann, *Nucl. Instr. Meth. B* **136–138 (1998)** 603.
3-182 J. Tesmer, M. Nastasi (eds.) in: *Handbook of Modern Ion Beam Materials Analysis*, Materials Research Society, Pittsburgh **1985**, p. 139–204.
3-183 B. Maurel, G. Amsel, J. Nadai, *Nucl. Instr. Meth.* **197 (1982)** 1.
3-184 W. Lanford, *Nucl. Instr. Meth. B* **66 (1992)** 65.
3-185 I. Vickridge, G. Amsel, *Nucl. Instr. Meth. B* **45 (1990)** 6.
3-186 M. Mayer, *Technical Report* IPP9/113, Max-Planck-Institut für Plasmaphysik, Garching (Germany), **1997**.
3-187 G. Amsel, E. d'Artemare, G. Battistig, E. Girard, L. Gosset, P. Reveresz, *Nucl. Instr. Meth. B* **136–138 (1998)** 545.
3-188 J. Meier, F. Richter, *Nucl. Instr. Meth. B* **47 (1990)** 303.
3-189 A. Sagura, K. Kamada, S. Yamaguchi, *Nucl. Instr. Meth. B* **34 (1988)** 465.
3-190 D. Kuhn, F. Rauch, H. Baumann, *Nucl. Instr. Meth. B* **45 (1990)** 252.
3-191 Z. Elkes, A. Kiss, G. Gyürky, E. Somorjai, I. Uzonyi, *Nucl. Instr. Meth. B* **158 (1999)** 209.
3-192 B. Hartmann, S. Kalbitzer, M. Behar, *Nucl. Instr. Meth. B* **103 (1995)** 494.
3-193 I. Vickridge, *Nucl. Instr. Meth. B* **34 (1988)** 470.
3-194 E. Ioannidou, D. Bourgarit, T. Calligaro, J. Dran, M. Dubus, J. Salomon, P. Walter, *Nucl. Instr. Meth. B* **161–163 (2000)** 730.
3-195 K. Ecker, J. Krauser, A. Weidinger, H. Weise, K. Maser, *Nucl. Instr. Meth. B* **161–163 (2000)** 682.
3-196 R. K. Marcus (ed.): *Glow Discharge Spectroscopies*, Plenum Press, New York 1993.
3-197 C. M. Barshick, D. C. Duckworth, D. H. Smith (eds.): *Practical Spectroscopy Series Vol. 23*, Marcel Dekker, New York, **2000**.
3-198 J. S. Becker, H.-J. Dietze, *Int. J. Mass. Spectrom.* **197 (2000)** 1.
3-199 N. Jakubowski, D. Stuewer, *J. Anal. Atom. Spektrom.* **7 (1992)** 951.
3-200 C. Venzago, M. Weigert, *Fresenius' J. Anal. Chem.* **350 (1994)** 303.
3-201 A. Raith, R. C. Hutton, J. C. Huneke, *J. Anal. Atom. Spektrom.* **8 (1993)** 867.
3-202 N. Jakubowski, I. Feldmann, D. Stuewer, *J. Anal. Atom. Spektrom.* **12 (1997)** 151.
3-203 C. Yang, M. Mohill, W. W. Harrison, *J. Anal. Atom. Spektrom.* **15 (2000)** 1255.

References to Chapter 4

4-1 M. I. Pupin, *Science* **3** (1896) 538–544.

4-2 H. G. J. Moseley, *Philos. Mag.* **26 (1913)** 1024; **27 (1914)** 703.

4-3 Y. Yoneda, T. Horiuchi, *Rev. Sci. Instrum.* **42 (1971)** 1069.

4-4 P. Wobrauschek, P. Kregsamer, C. Streli, H. Aiginger, *X-Ray Spectrom.* **20 (1991)** 23–28.

4-5 P. Kregsamer, *Spectrochim. Acta B* **46 (1991)** 1333–1340.

4-6 H. Aiginger, *Spectrochim. Acta B* **46 (1991)** 1313–1321.

4-7 H. Aiginger, P. Wobrauschek, *Nucl. Instrum. Methods* **114 (1974)** 157.

4-8 J. Knoth, H. Schwenke, *Fresenius' Z. Anal. Chem.* **291 (1978)** 200.

4-9 H. Schwenke, J. Knoth, *Nucl. Instrum. Methods* **193 (1982)** 239.

4-10 Semiconductor Equipment and Materials International, SEMI Standards E 45.

4-11 Semiconductor Equipment and Materials International, SEMI Standards M 33.

4-12 R. S. Becker, J. A. Golovchenko, J. R. Patel, *Phys. Rev. Lett.* **50 (1983)** 153.

4-13 P. Wobrauschek, R. Görgl, P. Kregsamer, C. Streli, S. Pahlke, L. Fabry, M. Haller, A. Knöchel, M. Radtke, *Spectrochim. Acta Part B* **52 (1997)** 901–906.

4-14 L. Fabry, S. Pahlke, L. Kotz, P. Wobrauschek, Ch. Streli, P. Eichinger, *Fresenius' J. Anal. Chem.* **363 (1999)** 98–102.

4-15 R. Klockenkämper, A. von Bohlen, *Spectrochim. Acta B* **54 (1999)** 1385–1392.

4-16 Ch. Streli, P. Kregsamer, P. Wobrauschek, H. Gatterbauer, P. Pianetta, S. Pahlke, L. Fabry, L. Palmetshofer, M. Schmeling, *Spectrochim. Acta B* **54 (1999)** 1433–1441.

4-17 X. Xiao, S. Hayakawa, Y. Goshi, M. Oshima, *Anal. Sci. (Jpn)* **14 (1998)** 1139–1144.

4-18 P. K. de Bokx, S. J. Kidd, G. Wiener, H. P. Urbach, S. de Gendt, P. W. Mertens, M. M. Heyns, *Proc. – Electrochem. Soc.* **(1998)**, PV 98–1 (Silicon Materials Science and Technology, Vol. 2) p. 1511–1523.

4-19 H. Schwenke, *GKSS*, Geesthacht, *priv. comm.*, 1999; Proc. of GMM Workshop, Erlangen Feb. 23–24, **2000**, Gesellschaft für Mikroelektronik VDE/VDI.

4-20 P. Wobrauschek, C. Streli: *Advanced Light Sources, ADLIS,* TU Wien **1999** pp. 284–300.

4-21 R. Klockenkämper: *Total Reflection X-ray Analysis,* Wiley Interscience **1997**.

4-22 P. J. Potts, A. T. Ellis, P. Kregsamer, C. Streli, M. West, P. Wobrauschek, *J. Anal. At. Spectrom.* **14 (1999)** 1773–1799.

4-23 P. Wobrauschek, Ch. Streli in: *Encyclopedia of Anal. Chem.* in press.

4-24 K. N. Stoev, K. Sakurai, *Spectrochim. Acta B* **54 (1999)** 41–82.

4-25 K. Yakushiji, S. Ohkawa, A. Yoshinaga, J. Harada, *Jpn. J. Appl. Phys.* **32 (1993)** 1191–1196.

4-26 K. Yakushiji, S. Ohkawa, A. Yoshinaga, J. Harada, *Jpn. J. Appl. Phys.* **33 (1994)** 1130–1135.

4-27 B. W. Liou, C. L. Lee, *Chin. J. Phys.* **37 (1999)** 623–630.

4-28 K. Yakushiji, S. Ohkawa, A. Yoshinaga, J. Harada, *Jpn. J. Appl. Phys.* **32 (1993)** 4750–4751.

4-29 K. L. Williams: *An Introduction to X-Ray Spectrometry,* Allen and Unwin, London **1987**.

4-30 L. Strüder, P. Lechner, P. Leutenegger, *Naturwissenschaften* **85 (1998)** 539–543.

4-31 K. Yakushiji, S. Ohkawa, A. Yoshinaga, J. Harada, *Anal. Sci. (Jpn.)* **11 (1995)** 505–511.

4-32 J. Knoth, H. Schwenke, P. Eichinger in: M. Heyns, M. Heuris, P. Mertens (eds.): *Ultra-Clean Processing (UCPSS '94, IMEC),* ACCO Leuven **1994**, p. 107.

4-33 G. Y. Tao, S. J. Zhuo, A. Ji, K. Norrish, P. Fazey, U. E. Senff, *X-Ray Spectrom.* **27 (1998)** 357–366.

4-34 K. Tsuji, T. Yamada, T. Utaka, K. Hirokawa, *J. Appl. Phys.* **78 (1995)** 969–973.

4-35 Y. Mori, K. Shimanoe, *Anal. Sci. (Jpn)* **12 (1996)** 277–279.

4-36 L. Fabry, S. Pahlke, L. Kotz, *Fresenius' J. Anal. Chem.* **354 (1996)** 266–270.

4-37 P. Mertens, S. DeGendt, K. Kenis: *Calibration accuracy of different Atomika TXRF instruments,* IMEC, Leuven **1996**.

4-38 I. Rink, H. Thewissen, *Spectrochim. Acta B* **54 (1999)** 1427–1431.

4-39 Y. Mori, K. Shimanoe, T. Sakon, *Anal. Sci. (Jpn)* **11 (1995)** 499.

4-40 R. Hölzl, K. J. Range, L. Fabry, D. Huber, *J. Electrochem. Soc.* **146 (1999)** 2245.

4-41 A. C. Diebold, *J. Vac. Sci. Technol. A* **14** **(1996)** 1919.
4-42 R. S. Hockett, *Anal. Sci. (Jpn)* **11** **(1995)** 511.
4-43 H. Schwenke, J. Knoth, U. Weisbrod, *X-Ray Spectrom.* **20** **(1991)** 277.
4-44 U. Weisbrod, R. Gutschke, J. Knoth, H. Schwenke, *Fresenius' J. Anal. Chem.* **341** **(1991)** 83.
4-45 W. Berneike, *Spectrochim. Acta B* **48** **(1993)** 269–275.
4-46 H. Schwenke, R. Gutschke, J. Knoth, M. Kock, *Appl. Phys. A* **54** **(1992)** 460.
4-47 L. Fabry, Y. Matsushita, *Proc. – Electrochem. Soc.* **(1998)**, PV 98–1 (Silicon Materials Science and Technology, Vol. 2) p. 1459.
4-48 P. Eichinger, H.-J. Rath, H. Schwenke in: D. C. Gupta (ed.): *Semiconductor Fabrication: Technology and Metrology, ASTM STP 990,* ASTM **1989** p. 305.
4-49 R. S. Hockett, *Adv. in X-Ray Anal.* **37** **(1994)** 565–565.
4-50 V. Penka, W. Hub, *Fresenius' Z. Anal. Chem.* **333** **(1989)** 586–589.
4-51 J. A. Sees, L. H. Hall, *J. Electrochem. Soc.* 142 **(1995)** 1238–1241.
4-52 L. Fabry, *Accred. Qual. Assur.* **1** **(1996)** 99.
4-53 L. Fabry, L. Köster, S. Pahlke, L. Kotz, J. Hage, *IEEE Trans. Semicond. Manuf.* **9** **(1996)** 428.
4-54 C. Streli, P. Wobrauschek, V. Bauer, P. Kregsamer, R. Görgl, P. Pianetta, R. Ryon, S. Pahlke, L. Fabry, *Spectrochim. Acta B* **52** **(1997)** 861–872.
4-55 P. Wobrauschek, P. Kregsamer, W. Ladisich, C. Streli, S. Pahlke, L. Fabry, S. Garbe, M. Haller, A. Knöchel, M. Radtke, *Nucl. Instr. Meth. A* **363** **(1995)** 619–620.
4-56 F. Comin, M. Navizet, P. Mangiagalli, G. Aposolo, *Nucl. Instr. Meth. B* **150** **(1999)** 538–542.
4-57 H. Binder, *M+W Zander,* Stuttgart, *priv. comm.* **2000**.
4-58 S. A. McHugo, A. C. Thompson, C. Flink, E. R. Weber, G. Lambie, B. Gunion, A. MacDowell, R. Celestre, H. A. Padmore, Z. Hussain, *J. Cryst. Growth* **210** **(2000)** 395–400.
4-59 H. Schwenke, W. Berneike, J. Knoth, U. Weisbrod, *Adv. X-Ray Anal.* **32** **(1989)** 105.
4-60 H. Schwenke, J. Knoth **(1993)** in: R. van Grieken, A. Markowicz (eds.): *Handbook on X-Ray Spectrometry,* Practical Spectroscopy Ser. Vol. 14, Dekker, New York, p. 453.
4-61 A. D. Dane, H. A. van Sprang, L. M. C. Buydens, *Anal. Chem.* **71** **(1999)** 4580–4586.
4-62 H. Schwenke, J. Knoth, L. Fabry, S. Pahlke, R. Scholz, L. Frey, *J. Electrochem. Soc.* **144** **(1997)** 3979–3983.
4-63 R. Klockenkämper, A. von Bohlen, H. W. Becker, L. Palmetshofer, *Surf. Interface Anal.* **27** **(1999)** 1003–1008.
4-64 R. Klockenkämper, A. von Bohlen, *J. Anal. At. Spectrom.* **14** **(1999)** 571–576.
4-65 R. Klockenkämper, A. von Bohlen, *Anal. Commun.* **36** **(1999)** 27–29.
4-66 T. Shiraiwa, N. Fujino, S. Sumita, Y. Tanizoe in: D. C. Gupta (ed.): *Semiconductor Fabrication: Technology and Metrology, ASTM STP 990,* ASTM 1989, p. 314.
4-67 A. Huber, H. J. Rath, P. Eichinger, T. Bauer, L. Kotz, R. Staudigl, *Proc. – Electrochem. Soc.* **(1988)**, PV 88–20 (Diagnostic Techniques for Semicond. Materials and Devices) p. 109–112.
4-68 C. Neumann, P. Eichinger, *Spectrochim. Acta B* **46** **(1991)** 1369–1377.
4-69 G. Buhrer, *Spectrochim. Acta B* **54** **(1999)** 1399.
4-70 A. C. Diebold, P. Maillot, M. Gordon, J. Baylis, J. Chacon, R. Witowski, H. F. Arlinghaus, J. A. Knapp, B. L. Doyle, *J. Vac. Sci. Technol. A* **10** **(1992)** 2945–2952.
4-71 S. Metz, G. Kilian, G. Mainka, C. Angelkort, A. Fester, B. O. Kolbesen, *Proc. – Electrochem. Soc.* **(1997)** PV 97–22 (Crystalline Defects and Contamination: Their Impact and Control in Device Manufacturing II) p. 458–467.
4-72 L. Fabry, S. Pahlke, L. Kotz, E. Schemmel, W. Berneike, *Proc. – Electrochem. Soc.* **(1993)**, PV 93–15(Crystalline Defects and Contamination) p. 232–239.
4-73 U. Reus, K. Freitag, A. Haase, J. F. Alexandre, *Spectr. 2000* **143** **(1989)** 42.
4-74 A. Prange, H. Schwenke, *Adv. X-Ray Anal.* **35B** **(1992)** 899.
4-75 L. Fabry, S. Pahlke, L. Kotz, *Adv. in X-Ray Chem. Anal. Jpn.* **27** **(1995)** 345.
4-76 H. Kondo, J. Ryuta, E. Morita, T. Yoshimi, Y. Shimanuki, *Jpn. J. Appl. Phys. Part 2 No. 1A/B,* **31** **(1992)** L11.
4-77 R. Klockenkämper, A. von Bohlen, *Spectrochim. Acta B* **48** **(1989)** 461.
4-78 I. Rink, H. Thewissen, *Spectrochim. Acta B* **54** **(1999)** 1427–1431.

4-79 M. Matsumura, T. Sakoda, Y. Nishioka, *Jpn. J. Appl. Phys.* **37 (1998)** 5963–5964.

4-80 H. Yamaguchi, S. Itoh, S. Igarashi, K. Naitoh, R. Hasegawa, *Anal. Sci. (Jpn)* **14 (1998)** 909.

4-81 H. Yamaguchi, S. Itoh, S. Igarashi, K. Naitoh, R. Hasegawa, *Bunseki Kagaku* **48 (1999)** 855L.

4-82 M. Yamagami, M. Nonoguchi, T. Yamada, T. Shoji, T. Utaka, S. Nomura, K. Taniguchi, H. Wakita, S. Ikeda, *X-Ray Spectrom.* **26 (1999)** 451.

4-83 M. Yamagami, M. Nonoguchi, T. Yamada, T. Shoji, T. Utaka, S. Nomura, K. Taniguchi, H. Wakita, S. Ikeda, *Bunseki Kagaku* **48 (1999)** 1005.

4-84 G. Vereecke, M. Schaekers, K. Verstraete, S. Arnauts, M. M. Heyns, W. Plante, *J. Electrochem. Soc.* **147 (2000)** 1499.

4-85 L. Reimer in: *Transmission Electron Microscopy: Physics of Image Formation and Microanalysis*, Springer-Verlag, Berlin–Heidelberg–New York, **1989**.

4-86 L. Reimer in: *Scanning Electron Microscopy: Physics of Image Formation and Microanalysis*, Springer-Verlag, Berlin–Heidelberg–New York, **1998**.

4-87 H. Bethge, J. Heydenreich (eds.): *Electron Microscopy in Solid State Physics*, Elsevier, **1987**.

4-88 S. Amelincks, D. van Dyck, J. van Landuyt, G. van Tendeloo (eds.): *Electron Microscopy: Principles and Fundamentals*, VCH Verlagsgesellschaft mbH, Weinheim **1997**.

4-89 J. J. Hren, J. I. Goldstein (eds.): *Introduction to Analytical Electron Microscopy*, Plenum Press, New York, **1979**.

4-90 D. B. Williams in: *Practical Analytical Electron Microscopy in Materials Science*, Verlag Chemie International, Weinheim, **1984**.

4-91 D. B. Williams, C. B. Carter in: *Transmission Electron Microscopy – A Textbook for Materials Science*, Plenum Press, New York and London, **1996**.

4-92 B. K. Agarwal in: *X-Ray Spectroscopy*, Springer-Verlag, Berlin–Heidelberg–New York, **1991**.

4-93 L. V. Azaroff in: *X-Ray Spectroscopy*, McGraw-Hill Book Company, New York, 1974.

4-94 J. A. Bearden, *Rev. Mod. Phys.* **39 (1967)** 78

4-95 K. F. J. Heinrich, D. E. Newbury, H. Yakowitz in: *NBS Special Publication 460*, U.S. Dept. of Commerce, Washington, D.C., **1975**.

4-96 D. C. Joy in: *Monte Carlo Modeling for Electron Microscopy and Microanalysis*, Oxford University Press, New York, **1995**.

4-97 U. Werner, H. Johansen in: H. Bethge, J. Heydenreich (eds.): *Electron Microscopy in Solid State Physics*, Elsevier, **1987**, p. 170.

4-98 T. Johansson, *Zeit. Phys.* **82 (1933)** 507.

4-99 E. P. Bertin in: *Principles and Practice of X-Ray Spectrometric Analysis*, Plenum Press, New York, **1975**.

4-100 K. F. J. Heinrich, *Electron Beam X-Ray Microanalysis*, Van Nostrand, New York, **1981**.

4-101 D. C. Joy, A. D. Romig, J. I. Goldstein (eds.): *Principles of Analytical Electron Microscopy*, Plenum Press, New York and London, **1986**.

4-102 S. J. B. Reed in: *Electron Microprobe Analysis*, Cambridge University Press, London, **1975**.

4-103 E. Gatti, P. Rehak, *NIMA* **225 (1984)** 608.

4-104 L. Strüder, N. Meidinger, D. Stötter, J. Kemmer, P. Lechner, P. Leutenegger, H. Soltau, F. Eggert, M. Rohde, T. Schülein, *Microscopy and Microanalysis* **4 (1999)** 622.

4-105 RÖNTEC GmbH, X-Flash Detector, Product Information 1.1(4), March **1997**.

4-106 E. Lifshin, M. F. Ciccarelli, R. B. Bolon in: J. I. Goldstein, H. Yakowitz: *Practical Scanning Electron Microscopy*, Plenum Press, New York, **1975**, p. 363.

4-107 T. J. White, D. R. Cousens, G. J. Auchterlonie, *J. Microsc.* **162 (1991)** 379.

4-108 H. A. Kramers, *Phil. Mag.* **46 (1923)** 836.

4-109 R. Castaing, PhD thesis, University of Paris, **1951**.

4-110 P. Duncamb, S. J. B. Reed in: K. F. J. Heinrich (ed.): *The Calculation of Stopping Power and Backscatter Effects in Electron Probe Microanalysis*, NBS Special Publ. 298, Washington, **1968**.

4-111 J. Philibert in: *Proc. Symp. X-Ray Optics and Microanalysis*, Stanford, **1962**, p. 379.

4-112 S. B. J. Reed, *Brit. J. Appl. Phys.* **16 (1965)** 913.

4-113 T. V. Ziebold, R. E. Ogilvie, *Anal. Chem.* **36 (1964)** 323.

4-114 J. Philibert, R. Tixier, *NBS Tech. Publ.* **298 (1968)** 13.

4-115 D. R. Beamon, J. A. Isasi, *Anal. Chem.* **42 (1970)** 1540.

4-116 H. H. Weinke, H. Malissa jun., F. Kluger, W. Kiesel, *Microchim. Acta Suppl.* **5 (1974)** 233.

4-117 A. R. Büchner, J. P. M. Stienen, *Mikrochim. Acta Suppl.* **6 (1975)** 227.

4-118 G. Cliff, G. W. Lorimer, *J. Microsc.* **103 (1975)** 203.

4-119 Z. Horita, T. Sano, M. Nemoto, *Ultramicroscopy* **21 (1987)** 271.

4-120 E. van Cappellen, *Microsc. Microanal. Microstruct.* **1 (1990)** 1.

4-121 E. van Cappellen, J. C. Doukhan, *Ultramicroscopy* **53 (1994)** 343.

4-122 C. Jeanguillaume, C. Colliex, *Ultramicroscopy* **28 (1989)** 252.

4-123 R. Schneider, J. Wolterdorf in: J. Heydenreich, W. Neumann (eds.): *Proc. Analytical Transmission electron Microscopy in Materials Science – Fundamentals and Techniques*, Elbe Druckerei, Wittenberg, **1993**.

4-124 R. Schneider, H. Kirmse, I. Hähnert, W. Neumann, *Fresenius' J. Anal. Chem.* **365 (1999)** 217.

4-125 A. H. Compton, *Philos. Mag.* **45 (1923)** 1121–1131.

4-126 H. Kiessig, *Ann. Physik* **10 (1931)** 715–768.

4-127 H. Kiessig, *Ann. Physik* **10 (1931)** 769–788.

4-128 L. G. Parratt, *Phys. Rev.* **95 (1954)** 359–369.

4-129 G. Lim, W. Parrish, C. Ortiz, M. Bellotto, M. Hart, *J. Mater. Res.* **2 (1987)** 471–477.

4-130 W. C. Marra, P. Eisenberger, A. Y. Cho, *J. Appl. Phys.* **50 (1979)** 6927–6933.

4-131 R. Feder, B. S. Berry, *J. Appl. Cryst.* **3 (1970)** 372–379.

4-132 P. A. Lee, J. B. Pendry, *Phys. Rev. B* 11(8) **(1975)** 2795–2811.

4-133 P. Eisenberger, B. M. Kincaid, *Science* **200 (1978)** 1441–1447.

4-134 P. Eisenberger, P. Citrin, R. Hewitt, B. Kincaid, *CRC Critical Reviews in Solid State and Materials Science* **10**(2) **(1981)** 191–207.

4-135 R. Barchewitz, M. Cremonese-Visicato, G. Onori, *J. Phys. C* **11 (1978)** 4439–4445.

4-136 R. Fox, S. J. Gurman, *J. Phys. C* **13 (1980)** L249-L253.

4-137 S. M. Heald, W. Keller, E. A. Stern, *Phys. Lett.* **103A**(3) **(1984)** 155–158.

4-138 R. A. Cowley, T. W. Ryan, *J. Phys. D* **20 (1987)** 61–68.

4-139 P. Croce, L. Névot, *Revue Phys. Appl.* **11 (1976)** 113–125.

4-140 L. Névot, P. Croce, *Revue Phys. Appl.* **15 (1980)** 761–779.

4-141 T. A. Crabb, PhD thesis **(1993)** Univ. Strathclyde UK.

4-142 Th. H. De Keijser, E. J. Mittemeijer, H. C. F. Rozendaal, *J. Appl. Cryst.* **16 (1983)** 309–316.

4-143 S. Enzo, S. Polizzi, A. Benedetti, *Zeitschrift für Kristallographie* **170 (1985)** 275–287.

4-144 N. Wainfan, N. J. Scott, L. G. Parratt, *J. Appl. Phys.* **30 (1959)** 1604–1609.

4-145 N. Wainfan, L. G. Parratt, *J. Appl. Phys.* **31 (1960)** 1331–1337.

4-146 A. Wagendristel, H. Bangert, W. Tonsern, *Surface Science* **86 (1979)** 68–74.

4-147 M. G. Le Boité, A. Traverse, L. Névot, B. Pardo, J. Corno, *J. Mat. Res.* **3**(6) **(1988)** 1089–1096.

4-148 M. G. Le Boité, A. Traverse, L. Névot, B. Pardo, J. Corno, *Nucl. Instrum. Methods B* **29 (1988)** 653–660.

4-149 M. E. Raggio, Degree thesis **(1994)** Univ. degli Studi di Milano, Italy.

4-150 L. Névot, B. Pardo, J. Corno, *Revue Phys. Appl.* **23 (1988)** 1675–1686.

4-151 M. Pomerantz, A. Segmüller, *Thin Solid Films* **68 (1980)** 33–45.

4-152 L. Bosio, J. J. Benattar, F. Rieutord, *Revue Phys. Appl.* **22 (1987)** 775–778.

4-153 K. Kjaer, J. Als-Nielsen, C. A. Helm, P. Tippman-Krayer, H. Möhwald, *J. Phys. Chem.* **93 (1989)** 3200–3206.

4-154 A. Schalchli, J. J. Benattar, P. Tchoreloff, P. Zhang, A. W. Coleman, *Langmuir* **9 (1993)** 1968–1970.

4-155 T. C Huang, B. R. York, *Appl. Phys. Lett.* **50**(7) **(1987)** 389–391.

4-156 T. A. Crabb, P. N. Gibson, E. McAlpine, *Corrosion Science* **34**(9) **(1993)** 1541–1550.

4-157 D. G. Rickerby, P. N. Gibson, W. Gissler, J. Haupt, *Thin Solid Films* **209 (1992)** 155–160.

4-158 W. Gissler, P. N. Gibson, *Ceramics International* **22 (1996)** 335–340.

4-159 W. Gissler, M. A. Baker, J. Haupt, P. N. Gibson, R. Gilmore, T. P. Mollart, *Dia-*

mond Films and Technology 7(3) **(1997)** 165–180.
4-160 P. N. Gibson, M. E. Özsan, D. Lincot, P. Cowache, D. Summa, *Thin Solid Films* **361–362 (2000)** 34–40.
4-161 M. G. Samant, M. F. Toney, G. L. Borges, L. Blum, O. R. Melroy, *J. Phys. Chem.* **92 (1988)** 220–225.
4-162 P. Eisenberger, W. C. Marra, *Phys. Rev. Lett.* **46**(16) **(1981)** 1081–1084.
4-163 I. K. Robinson, *Phys. Rev. Lett.* **50**(15) **(1983)** 1145–1148.
4-164 W. C. Marra, P. H. Fuoss, P. E. Eisenberger, *Phys. Rev. Lett.* **49**(16) **(1982)** 1169–1172.
4-165 P. H. Fuoss, L. J. Norton, S. Brennan, A. Fischer-Colbrie, *Phys. Rev. Lett.* **60**(7) **(1988)** 600–603.
4-166 A. Segmüller, *Thin Solid Films* **154 (1987)** 33–42.
4-167 P. N. Gibson, T. A. Crabb, E. McAlpine, R. Falcone, *Mat. Sci. Forum* **126–128 (1993)** 595–598.
4-168 G. Martens, P. Rabe, *Phys. Stat. Sol. (a)* **58 (1980)** 415–424.
4-169 G. Martens, P. Rabe, *J. Phys. C* **14 (1981)** 1523–1534.
4-170 J. Goulon, C. Goulon-Ginet, R. Cortes, J. M. Dubois, *J. Physique* **43 (1982)** 539–548.
4-171 L. Bosio, R. Cortes, A. Defrain, M. Froment, *J. Electroanal. Chem.* **180 (1984)** 265–271.
4-172 H. Chen, S. M. Heald, *Solid State Ionics* **32/33 (1989)** 924–929.
4-173 F. R. Thornley, N. T. Barrett, G. N. Greaves, G. M. Antonini, *J. Phys. C* **19 (1986)** L563-L569.
4-174 N. T. Barrett, P. N. Gibson, G. N. Greaves, P. Mackle, K. J. Roberts, M. Sacchi, *J. Phys. D* **22 (1989)** 542–546.
4-175 P. N. Gibson, T. A. Crabb, *Nucl. Instr. And Meth. In Phys. Res. B* **97 (1995)** 495–498.
4-176 M. J. Cristóbal, P. N. Gibson, M. F. Stroosnijder, *Corrosion Science* **38**(6) **(1996)** 805–822.
4-177 S. Pizzini, K. J. Roberts, I. S. Dring, P. J. Moreland, R. J. Oldman, J. Robinson, *J. Mater. Chem.* **2**(1) **(1992)** 49–55.
4-178 B. Chapman: *Glow Discharge Processes*, John Wiley and Sons, New York 1980, p. 77–138.
4-179 W. Grimm, *Spectrochim. Acta B* **23 (1968)** 443.
4-180 M. R. Winchester, C. Lazik, R. K. Marcus, *Spectrochim. Acta B* **46 (1991)** 483.
4-181 V. Hoffmann, H.-J. Uhlemann, F. Prässler, K. Wetzig, D. Birus, *Fresenius' J. Anal.Chem.* **355 (1996)** 826.
4-182 H. Becker-Ross, S. Florek, H. Franken, B. Radziuk, M. Zeiher, *J. Anal. At. Spectrom.* **15 (2000)** 851.
4-183 F. M. Pennebaker, D. A. Jones, C. A. Gresham, R. H. Williams, R. E. Simon, M. F. Schappert, M. B. Denton, *J. Anal. At. Spectrom.* **13 (1998)** 821.
4-184 A. Bengtson, A. Eklund, M. Lundholm, A. Saric, *J. Anal. At. Spectrom.* **5 (1990)** 563.
4-185 A. Bengtson, *Spectrochim. Acta B* **49 (1994)** 411.
4-186 A. Bengtson, S. Hänström, *J. Anal. At. Spectrom.* **13 (1998)** 437.
4-187 A. Bengtson, S. Hänström, E. Lo Piccolo, N. Zacchetti, R. Meilland, H. Hocquaux, *Surf. Interface Anal.* **27 (1999)** 743.
4-188 Z. Weiss, P. Šmid, *J. Anal. At. Spectrom.* **15 (2000)** 1485.
4-189 M. R. Winchester, U. Beck, *Surf. Interface Anal.* **27 (1999)** 930.
4-190 M. Ives, D. B. Lewis, C. Lehmberg, *Surf. Interface Anal.* **25 (1999)** 191.
4-191 K. Shimuzu, H. Habazaki, P. Skeldon, G. E. Thompson, G. C. Wood, *Surf. Interface Anal.* **27 (1999)** 1046.
4-192 A. Quentmeier, *J. Anal. At. Spectrom.* **9 (1994)** 355.
4-193 K. Shimuzu, H. Habazaki, P. Skeldon, G. E. Thompson, G. C. Wood, *Surf. Interface Anal.* **27 (1999)** 950.
4-194 V.-D. Hodoroaba and T. Wirth, *J. Anal. At. Spectrom.* **14 (1999)** 1533
4-195 F. Prässler, V. Hoffmann, J. Schumann, K. Wetzig, *Fresenius' J. Anal. Chem.* **355 (1996)** 840.
4-196 T. R. Harville, R. K. Marcus, *Anal. Chem.* **67 (1995)** 1271.
4-197 M. R. Winchester, C. Maul, *J. Anal. At. Spectrom.* **12 (1997)** 1297.
4-198 M. R. Winchester, *J. Anal. At. Spectrom.* **13 (1998)** 235.
4-199 K. A. Marshall, *J. Anal. At. Spectrom.* **14 (1999)** 923.
4-200 W. W. Harrison, K. R. Hess, R. K. Marcus, F. L. King, *Anal. Chem.* **58 (1986)** 341A.
4-201 N. Jakubowski, D. Stüwer, G. Tölg, *Int. J. Mass Spectrom. Ion Process.* **71 (1986)** 183.

4-202 R. Payling, D. Jones, A. Bengtson (eds.): *Glow Discharge Optical Emission Spectrometry*, John Wiley and Sons, Chichester **1997**.

4-203 E. Jansen, *Materials Sci. Eng.* **42 (1980)** 309.

4-204 R. Berneron, J. C. Charbonnier, *Surf. Interface Anal.* **3 (1981)** 134.

4-205 R. Payling, D. G. Jones, *Surf. Interface Anal.* **20 (1993)** 787.

4-206 E. Rose, P. Mayr, *Mikrochim. Acta* **I (1989)** 197.

4-207 H. Böhm in: R. Payling, D. G. Jones, A. Bengtson (eds.): *Glow Discharge Optical Emission Spectrometry*, John Wiley and Sons, Chichester **1997**, p. 676.

4-208 X. Pan, B. Hu, Y. Ye and R. K. Marcus, *J. Anal. Atom. Spectrom.* **13 (1998)** 1159.

4-209 R. Payling, D. G. Jones, S. A. Gower, *Surf. Interface Anal.* **20 (1993)** 959.

4-210 C. Pérez, R. Pereiro, N. Bordel, A. Sans-Medel, *J. Anal. At. Spectrom.* **15 (2000)** 1247.

4-211 K. Shimizu, H. Habazaki, P. Skeldon, G. E. Thompson, G. C. Wood, *Surf. Interface Anal.* **29 (2000)** 155.

4-212 R. Payling, M. Aeberhard, D. Delfosse, *J. Anal. At. Spectrom.* **16** (2001) 50.

4-213 M. L. Hartenstein, S. J. Christopher, R. K. Marcus, *J. Anal. At. Spectrom.* **14 (1999)** 1039.

4-214 R. K. Marcus, *J. Anal. At. Spectrom.* **11 (1996)** 821.

4-215 M. Fernández, N. Bordel, R. Pereiro, A. Sans-Medel, *J. Anal. At. Spectrom.* **12 (1997)** 1209.

4-216 R. K. Marcus, *J. Anal. At. Spectrom.* **15 (2000)** 1271.

4-217 A. Bengtson, C. Yang, W. W. Harrison, *J. Anal. At. Spectrom.* **15 (2000)** 1279.

4-218 H. Nickel, W. Fischer, D. Guntur, A. Naoumidis, *J. Anal. At. Spectrom.* **7 (1992)** 239.

4-219 C. Yang, K. Ingeneri, M. Mohill, W. W. Harrison, *J. Anal. At. Spectrom.* **15 (2000)** 73.

4-220 A. Bengtson, *J. Anal. At. Spectrom.* **11 (1996)** 829.

4-221 H. Böhm, *Oberflächen Werkstoffe* **4 (1994)** 8.

4-222 K. Shimizu, H. Habazaki, P. Skeldon, G. E. Thompson, G. C. Wood, *Surf. Interface Anal.* **29 (2000)** 151.

4-223 L. Moenke-Blankenburg, *Spectrochim. Acta Rev.* **15 (1993)** 1–37.

4-224 S. A. Darke, J. F. Tyson, *J. Anal. At. Spectrom.*, **8 (1993)** 145–209.

4-225 R. E. Russo, *Appl. Spec.*, **49**(9) **(1995)** 14A-28A.

4-226 D. Günter, R. Frischknecht, C. Heinrich, H. Kahlert, *J. Anal. At. Spectrom.*, **12, (1997)** 939.

4-227 GeoLas, The high resolution UV laser Ablation system, Prospect of the Company MicroLas Lasersystem GmbH, Goettingen, Germany.

4-228 Company INO: www.ino.qc.ca/en/syst_et_compo/grm.asp .

4-229 S. Amoruso, R. Bruzzese, N. Spinelli, R. Velotta, *J. Phys. B*, **32 (1999)** R131.

4-230 E. Cromwell, P. Arrowsmith, *Appl. Spectrosc.* **49 (1995)** 1652.

4-231 B. N. Chichkov, S. Momma, F. von Alvensleben, A. Tünnermann, *Appl. Phys. A*, **63 (1996)** 109–115.

4-232 H. K. Tönshoff, F. von Alvensleben, A. Ostendorf, G. Kamlage, S. Nolte, *International Journal of Electrical Machining IJEM* **4 (1999)** 1–6.

4-233 CPA-10, 1000, 2000, Company Clark-MXR Inc., MI, USA.

4-234 Company Spectra Physics.

4-235 H. Bauer, F. Leis, K. Niemax, *Spectrochim. Acta Part B*, **53 (1998)** 1815.

4-236 A. Bengtson, M. Lundholm, *J. Anal. At. Spectrom.* **3 (1988)** 879.

4-237 N. Jakubowski, D. Stuewer, *J. Anal. At. Spectrom.*, **7 (1992)** 951.

4-238 S. Oswald, V. Hoffmann, G. Ehrlich, *Spectrochim. Acta Part B*, **49** (1994) 1123.

4-239 D. Anderson, C. McLeod, T. English, A. Trevor, *Appl. Spectrosc.* **49 (1995)** 691.

4-240 V. Kanicky, I. Novotny, J. Musil, J. Mermet, *Appl. Spectrosc.*, **51 (1997)** 1037.

4-241 V. Kanicky, J. Musil, I. Novotny, J. Mermet, *Appl. Spectrosc.*, **51 (1997)** 1042.

4-242 J. Vadillo, C. Garcia, S. Palanco, J. Lazerna, *J. Anal. At. Spectrom.*, **13 (1998)** 793.

4-243 C. Garcia, M. Corral, J. Vadillo, J. Lazerna, *Appl. Spectrosc.*, **54 (2000)** 1027.

4-244 D. Bleiner, A. Plotnikov, C. Vogt, K. Wetzig, D. Günter, *Fresenius' J. Anal. Chem.*, **368 (2000)** 221.

4-245 V. Margetic, M. Bolshov, A. Stockhaus, K. Niemax, R. Hergenröder, *J. Anal. At. Spectr.* **16 (2001)** 616–621.

4-246 R. Goutte, C. Guillaud, R. Javelas, J. P. Meriaux, *Optik* **26 (1967)** 574.

4-247 C. W. White, D. L. Simms, N. H. Tolk, *Science* **177 (1972)** 481.

4-248 G. Blaise, *Surf. Sci.* **60 (1976)** 65.

4-249 G. E. Thomas, *Surf. Sci.* **90 (1979)** 381.

4-250 I. S. T. Tsong in: E. Taglauer, W. Heiland (eds.): *Inelastic Particle-Surface Collisions*, Springer-Verlag, Heidelberg **1981**.

4-251 W. F. v.d.Weg, P. K. Rol, *Nucl. Instr. Meth.* **38 (1965)** 274.

4-252 H. Bach, *Glass Technology* **30 (1989)** 75.

4-253 V. Rupertus, U. Rothhaar, P. Köpfer, A. Lorenz, H. Oechsner, *Vakuum in der Praxis* **3 (1993)** 183.

4-254 R. Behrisch: *Sputtering by particle bombardment II*, Springer-Veralg, Heidelberg **1983**.

4-255 A. Benninghoven, F. G. Rüdenauer, H. W. Werner: *Secondary Ion Mass Spectrometry*, John Wiley and Sons, New York **1987**.

4-256 W. F. v.d.Weg, P. K. Rol, *Nucl. Instr. Meth.* **38 (1965)** 274.

4-257 G. E. Thomas, *Surf. Sci.* **90 (1979)** 381.

4-258 P. J. Martin, R. J. MacDonald, *Surf. Sci.* **62 (1977)** 551.

4-259 C. A. Andersen, J. R. Hinthorne, *Anal. Chem.* **45 (1973)** 1421.

4-260 H. E. Bauer, H. Seiler, *Scan. Elec. Micros.* **III (1984)** 1081.

4-261 R. Kollath in: *Handuch der Physik XXI*, Springer-Verlag, Berlin **1956**.

4-262 A. N. Zaidel, V. K. Prokofev, S. M. Raiskii, V. A. Slavnyi, E. Ya. Shreider, *Tables of spectral lines*, IFI, Plenum New York 1970.

4-263 H. Oechsner, *Thin film and depth profile analysis*, Springer-Verlag, Heidelberg **1984**.

4-264 O. Dersch, M. Laube, F. Rauch, O. Becker, *Glastechn. Ber. Glass Sci. Technol.* **72**(10) **(1999)** 329.

4-265 B. E. Hayden in: J. T. Yates, T. E. Madey (eds.): *Vibrational Spectroscopy of Molecules on Surfaces*, Plenum Press, New York , London **1987**, p. 267–344.

4-266 M. D. Porter, *Anal. Chem.* **60 (1988)** 1143A–1155A.

4-267 H. Brunner, U. Mayer, H. Hoffmann, *Appl. Spectr.* **51 (1997)** 209–217.

4-268 P. Tengvall, I. Lundström, B. Lindberg, *Biomaterials* **19 (1998)** 407–422.

4-269 C. A. Melendres, G. A. Bowmaker, J. M. Leger, B. Beden, *J. of Electroanlytical Chem.* **499 (1998)** 215–218.

4-270 W. Richter, D. R. T. Zahn in: G. Bauer, W. Richter (eds.): *Optical Characterization of Epitaxial Semiconductor Layers*, Springer-Verlag Berlin **1996**, p. 60–62.

4-271 G. A. Beitel, A. Laskov, H. Oosterbeek, E. W. Kuipers, *J. Phys. Chem.* **100 (1996)** 12494–12502.

4-272 R. G. Greenler, *J. Chem. Phys.* **44 (1966)** 310–314.

4-273 C. R. Flach, A. Gericke, R. Mendelsohn, *J. Phys. Chem. B* **101 (1997)** 58–65.

4-274 A. N. Parikh, D. L. Allara, *J. Chem. Phys.* **96 (1992)** 927–945.

4-275 T. Buffeteau, D. Blaudez, E. Pere, B. Desbat, *J. Phys. Chem. B* **103 (1999)** 5020–5027.

4-276 K. Yamamoto, H. Ishida, *Appl. Spectr.* **48 (1994)** 775–787.

4-277 P. Christensen, A. Hamnett, *Electrochim. Acta* **45 (2000)** 2443–2459.

4-278 C. Sammon, J. Yarwood, N. Everall, *Polymer Degradation and Stability* **67 (2000)** 149–158.

4-279 H. Lavoie, J. Gallant, M. Grandbois, D. Blaudez, B. Desbat, F. Boucher, C. Salesse, *Mat. Sci. Eng. C* **10 (1999)** 147–154.

4-280 Y. Urai, C. Ohe, K. Itoh, M. Yoshida, Ken-ichi Iimura, T. Kato, *Langmuir* **16 (2000)** 3920–3926.

4-281 T. Buffeteau, E. Le Calvez, S. Castano, B. Desbat, D. Blaudez, J. Dufourcq, *J. Phys. Chem. B* **104 (2000)** 4537–4544.

4-282 S. C. Street, A. J. Gellman, *J. Phys. Chem.* **100 (1996)** 8338–8348.

4-283 J. A. Mielczarski, *J. Phys. Chem.* **97 (1993)** 2649–2663.

4-284 P. Dumas, Y. J. Chabal, P. Jakob, *Surf. Sci.* **269/270 (1992)** 867–878.

4-285 R. F. Hicks, H. Qi, Q. Fu, B.-K. Han, L. Li, *J. Chem. Phys.* **110 (1999)** 10498–10508.

4-286 Q. Fu, L. Li, M. J. Begarney, B.-K. Han, D. C. Law, R. F. Hicks, *J. Phys. IV* **9** (Pr8, Proc. of the 12th European Conf. on Chemical Vapor Deposition, Vol. 1) **(1999)** 3–14.

4-287 C. V. Raman, K. S. Krishnan, *Nature* **121 (1928)** 501.

4-288 M. Fleischman, J. P. Hendra, A. J. McQuillan, *Chem. Phys. Lett.* **26 (1974)** 163.

4-289 D. J. Jeanmaire, R. P. Van Duyne, *J. Electroanal. Chem.* **84 (1977)** 1.

4-290 S. Nie, S. R. Emroy, *Science* **275 (1997)** 1102.

4-291 M. J. Weaver, S. Zou, H. Y. H. Chan, *Anal. Chem.* **72 (2000)** 38A.
4-292 Y.-C. Liu, R. L. McCreery, *J. Am. Chem. Soc.* **117 (1995)** 11254.
4-293 K. R. Ray, R. L. McCreery, *J. Electroanal. Chem.* **469 (1999)** 150.
4-294 L. Xia, R. L. McCreery, *J. Electrochem. Soc.* **146 (1999)** 3696.
4-295 I. E. Wachs, *Catalysis Today* **27 (1996)** 437.
4-296 H. Knozinger, *Catalysis Today* **32 (1996)** 70.
4-297 C. Bremard, D. Bougeard, *Adv. Mater.* **7 (1995)** 10.
4-298 W. Hill, F. Marlow, J. Kornatowski, *Appl. Spectrosc.* **48 (1994)** 224.
4-299 P. A. Mosier-Boss, R. Newbery, S. Szpak, S. H. Lieberman, J. W. Rovang, *Anal. Chem.* **68 (1996)** 3277.
4-300 J. Souto, R. Aroca, J. A. Desaja, *J. Raman Spectrosc.* **25 (1994)** 435.
4-301 W. Hill, B. Wehling, C. G. Gibbs, C. D. Gutsche, D. Klockow, *Anal. Chem.* **67 (1995)** 3187.
4-302 W. Hill, V. Fallourd, D. Klockow, *J. Phys. Chem. B* **103 (1999)** 4707.
4-303 W. Hill, B. Wehling, D. Klockow, *Appl. Spectrosc.* **53 (1999)** 547.
4-304 S. R. Emory, S. Nie, *Anal. Chem.* **69 (1997)** 2631.
4-305 R. M. Stöckle, Y. D. Suh, V. Deckert, R. Zenobi, *Chem. Phys. Lett.* **318 (2000)** 131.
4-306 B. Pettinger, G. Picardi, R. Schuster, G. Ertl, *Electrochem.* **68 (2000)** 942.
4-307 N. Bloembergen, P. S. Pershan, *Phys. Rev.* **128 (1962)** 606.
4-308 G. J. Simpson, *Appl. Spectrosc.* **55 (2001)** 16A.
4-309 Q. Du, E. Freysz, Y. R. Shen, *Science* **264 (1994)** 826.
4-310 G. L. Richmond, *Anal. Chem.* **69 (1997)** 536A.
4-311 R. M. A. Azzam, N. M. Bashara: *Ellipsometry and Polarized Light*, North Holland, Amsterdam **1977**, p. 287.
4-312 J. Leng, J. Opsal, H. Chu, M. Senko, D. E. Aspnes, *Thin Solid Films* **313–314 (1998)** 132.
4-313 G. E. Jellison Jr., F. A. Modine, *Appl. Phys. Lett.* **69 (1996)** 371.
4-314 B. Johs, J. A. Woollam, C. M. Herzinger, J. Hilfiker, R. Synowicki, C. L. Bungay, *Crit. Rev. Opt. Sci. Technol.*, **CR72 (1999)** 29–58.
4-315 A. Röseler, *Thin Solid Films* **234 (1993)** 307–313.
4-316 A. Canillas, E. Pascual, B. Drevillon, *Rev. Sci. Instr.* **64 (1993)** 2153–2159
4-317 M. Born, E. Wolf: *Principles of Optics*, Pergamon Press, Oxford 1964.
4-318 H. Angermann, W. Henrion, A. Röseler, M. Rebien, *Mater. Sci. Eng. B* **73 (2000)** 178–183
4-319 A. R. Reinberg, *Appl. Opt.* **11 (1972)** 1273.

References to Chapter 5

5-1 G. Binnig, H. Rohrer, Ch. Gerber, E. Weibel, *Phys. Rev. Lett.* **49 (1982)** 57.
5-2 D. A. Bonnell (ed.): *Scanning tunneling microscopy and spectroscopy. Theory, Techniques and Applications*, VCH Publishers Inc., New York **1993**.
5-3 R. Wiesendanger: *Scanning probe microscopy and spectroscopy. Methods and applications*, Cambridge University Press, Cambridge **1994**.
5-4 G. Binnig, C. F. Quate, Ch. Gerber, *Phys. Rev. Lett.* **56 (1986)** 930.
5-5 D. Sarid: *Scanning Force Microscopy*, Oxford University Press, New York **1991**.
5-6 Q. Zhong, D. Inniss, K. Kjoller, V. B. Elings, *Surf. Sci. Lett.* **290 (1993)** L688.
5-7 P. K. Hansma, J. P. Cleveland, M. Radmacher, D. A. Walters, P. E. Hillner, M. Bezanilla, M. Fritz, D. Vie, H. G. Hansma, C. B. Prater, J. Massie, L. Fukunaga, J. Gurley, V. Elings, *Appl. Phys. Lett.* **64 (1994)** 1738.
5-8 T. R. Albrecht, S. Akamine, T. E. Carver, C. F. Quate, *J. Vac. Sci. Technol. A* **8 (1990)** 3386.
5-9 O. Wolter, Th. Bayer, J. Greschner, *J. Vac. Sci. Technol. B* **9 (1991)** 1353.
5-10 G. Meyer, N. M. Amer, *Appl. Phys. Lett.* **53 (1988)** 2400.
5-11 S. Manne, P. K. Hansma, J. Massie, V. B. Elings, A. A. Gewirth, *Science* **251 (1991)** 183.
5-12 A. A. Gewirth, *AIP Conference Proceedings* **241 (1992)** 253.
5-13 T. Prohaska, G. Friedbacher, M. Grasserbauer, H. Nickel, R. Lösch, W. Schlapp, *Anal. Chem.* **67 (1995)** 1530.

5-14 G. Friedbacher, T. Prohaska, M. Grasserbauer, *Mikrochim. Acta* **113 (1994)** 179.
5-15 J. Frommer, *Angew. Chem.* **104 (1992)** 1325.
5-16 H. Fuchs, *J. Mol. Struct.* **292 (1993)** 29.
5-17 H. G. Hansma, *J. Vac. Sci. Technol. B* **14 (1996)** 1930.
5-18 S. Kasas, N. H. Thomson, B. L. Smith, P. K. Hansma, J. Miklossy, H. G. Hansma, *Int. J. Imaging Systems and Technology* **8 (1997)** 151.
5-19 S. N. Magonov, *Appl. Spectroc. Rev.* **28 (1993)** 1.
5-20 P. E. Hillner, A. J. Gratz, S. Manne, P. K. Hansma, *Geology* **20 (1992)** 359.
5-21 I. Schmitz, T. Prohaska, G. Friedbacher, M. Schreiner, M. Grasserbauer, *Fresenius' J. Anal. Chem.* **353 (1995)** 666.
5-22 T. Leitner, G. Friedbacher, T. Vallant, H. Brunner, U. Mayer, H. Hoffmann, *Mikrochim. Acta* **133 (2000)** 331.
5-23 C. M. Mate, G. M. McClelland, R. Erlandsson, S. Chiang, *Phys. Rev. Lett.* **59 (1987)** 1942.
5-24 R. M. Overney, T. Bonner, E. Meyer, M. Rüetschi, R. Lüthi, L. Howald, J. Frommer, H.-J. Güntherodt, M. Fujihira, H. Takano, *J. Vac. Sci. Technol. B* **12 (1994)** 1973.
5-25 C. D. Frisbie, L. F. Rozsnyai, A. Noy, M. S. Wrighton, C. S. Lieber, *Science* **265 (1994)** 2071.
5-26 P. Maivald, H.-J. Butt, S. A. C. Gould, C. B. Prater, B. Drake, J. A. Gurley, V. B. Elings, P. K. Hansma, *Nanotechnology* **2 (1991)** 103.
5-27 H.-J. Butt, *Biophys. J.* **60 (1991)** 777.
5-28 A. L. Weisenhorn, P. Maivald, H.-J. Butt, P. K. Hansma, *Phys. Rev. B* **45 (1992)** 11226.
5-29 J. H. Hoh, J. P. Cleveland, C. B. Prater, J.-P. Revel, P. K. Hansma, *J. Am. Chem. Soc.* **114 (1992)** 4917.
5-30 J. E. Stern, B. D. Terris, H. J. Manin, D. Rugar, *Appl. Phys. Lett.* **53 (1988)** 2717.
5-31 B. D. Terris, J. E. Stern, D. Rugar, H. J. Manin, *J. Vac. Sci. Technol. A* **8 (1990)** 374.
5-32 Y. Martin, H. K. Wickramasinghe, *Appl. Phys. Lett.* **50 (1987)** 1455.
5-33 G. Persch, H. Strecker, *Ultramicroscopy* **42–44 (1992)** 1269.
5-34 G. F. A. Van der Walle, E. J. Van Loenen in: M. Grasserbauer, H. W. Werner (eds.): *Analysis of Microelectronic Materials and Devices*, John Wiley and Sons, Chichester **1991**.
5-35 P. K. Hansma, J. Tersoff, *J. Appl. Phys.* **61 (1987)** R1.
5-36 J. A. Golovchenko, *Science* **232 (1986)** 48.
5-37 O. M. Magnussen, J. Hotlos, G. Beitel, D. M. Kolb, R. J. Behm, *J. Vac. Sci. Technol. B* **9 (1991)** 969.
5-38 H. Neddermeyer, *Trends Analyt. Chem.* **8 (1989)** 230.
5-39 R. M. Feenstra, J. A. Stroscio, J. Tersoff, A. P. Fein, *Phys. Rev. Lett.* **58 (1987)** 1192.
5-40 Ph. Avouris, *J. Phys. Chem.* **94 (1990)** 2246.
5-41 Ph. Avouris, I. W. Lyo, F. Bozso, *J. Vac. Sci. Technol. B* **9 (1991)** 424.
5-42 Ph. Avouris, I.-W. Lyo, *AIP Conference Proceedings* **241 (1992)** 283.
5-43 S. M. Lindsay, L. A. Nagahara, T. Thundat, U. Knipping, R. L. Rill, B. Drake, C. B. Prater, A. L. Weisenhorn, S. A. C. Gould, P. K. Hansma, *J. Biomol. Struct. Dyn.* **7 (1989)** 279.
5-44 R. J. Driscoll, M. G. Youngquist, J. D. Baldeschwieler, *Nature* **346 (1990)** 294.
5-45 P. G. Arscott, V. A. Bloomfield, *STM SFM Biol.* **(1993)** 259–272.
5-46 M. Amrein, H. Gross, R. Guckenberger, *STM SFM Biol.* **(1993)** 127–175.
5-47 S. F. Alvarado, *J. Appl. Phys.* **64 (1988)** 5931.
5-48 R. Wiesendanger, H.-J. Güntherodt, G. Güntherodt, R. J. Gambino, R. Ruf, *Phys. Rev. Lett.* **65 (1990)** 247.
5-49 J. K. Gimzewski, B. Reihl, J. H. Coombs, R. R. Schlittler, *Z. Phys. B* **72 (1988)** 497.
5-50 R. Berndt, R. R. Schlittler, J. K. Gimzewski, *J. Vac. Sci. Technol. B* **9 (1991)** 573.
5-51 M. Völcker, W. Krieger, T. Suzuki, H. Walther, *J. Vac. Sci. Technol. B* **9 (1991)** 541.
5-52 Y. Kuk, R. S. Becker, P. J. Silverman, G. P. Kochanski, *J. Vac. Sci. Technol. B* **9 (1991)** 545.
5-53 S. Akari, M. Ch. Lux-Steiner, M. Vögt, M. Stachel, K. Dransfeld, *J. Vac. Sci. Technol. B* **9 (1991)** 561.
5-54 D. G. Cahill, R. J. Hamers, *J. Vac. Sci. Technol. B* **9 (1991)** 564.

Index